Knowledge Engineering and Data Mining

Knowledge Engineering and Data Mining

Editors

Agnieszka Konys
Agnieszka Nowak-Brzezińska

MDPI • Basel • Beijing • Wuhan • Barcelona • Belgrade • Manchester • Tokyo • Cluj • Tianjin

Editors
Agnieszka Konys
West Pomeranian University
of Technology Szczecin
Szczecin
Poland

Agnieszka
Nowak-Brzezińska
University of Silesia
Sosnowiec
Poland

Editorial Office
MDPI
St. Alban-Anlage 66
4052 Basel, Switzerland

This is a reprint of articles from the Special Issue published online in the open access journal *Electronics* (ISSN 2079-9292) (available at: https://www.mdpi.com/journal/electronics/special_issues/Knowledge_Data_Mining).

For citation purposes, cite each article independently as indicated on the article page online and as indicated below:

LastName, A.A.; LastName, B.B.; LastName, C.C. Article Title. *Journal Name* **Year**, *Volume Number*, Page Range.

ISBN 978-3-0365-6788-4 (Hbk)
ISBN 978-3-0365-6789-1 (PDF)

Contents

About the Editors

Agnieszka Konys

Agnieszka Konys is an Assistant Professor in the Faculty of Computer Science and Information Technology at the West Pomeranian University of Technology in Szczecin, Poland (since 2013). Her research studies the topics of ontology, knowledge representation methods, semantic web technologies, knowledge management and reasoning, and sustainability assessment.

Agnieszka Nowak-Brzezińska

Agnieszka Nowak-Brzezińska is an Associate Professor in the Faculty of Science and Technology at the University of Silesia in Katowice, Poland (since 2002). In 2019, she was awarded the academic degree of Habilitated Doctor of Philosophy in Engineering and Technology Science in the field of information and communication technology by the Polish Academy of Sciences. Her research studies the topics of outlier detection algorithms, clustering algorithms for complex data structures, knowledge engineering, and information retrieval systems.

electronics

Editorial

Knowledge Engineering and Data Mining

Agnieszka Konys [1] and Agnieszka Nowak-Brzezińska [2,*]

[1] Faculty of Computer Science and Information Technology, West Pomeranian University of Technology Szczecin, Zolnierska 49, 71-210 Szczecin, Poland
[2] Institute of Computer Science, Faculty of Science and Technology, University of Silesia, ul. Będzińska 39, 41-200 Sosnowiec, Poland
* Correspondence: agnieszka.nowak-brzezinska@us.edu.pl

Knowledge engineering and data mining are the two biggest pillars of modern intelligent systems. Knowledge induction from data is often based on using a wide range of machine learning algorithms and feature selection or extraction algorithms. When we collect various data types, we need solutions that will allow us to supervise these data correctly. Recently, machine-learning-based methods are increasingly employed to solve such problems; however, the selection of an appropriate feature selection technique, sampling mechanism, and/or classifiers for building decision support systems is very challenging. To address this challenging task, article [1] examines the effectiveness of various data science techniques concerning the issue of credit decision support. In particular, a processing pipeline was designed that consists of methods for data resampling, feature discretization, feature selection, and binary classification.

The capability of machine learning to discover hidden patterns in large datasets encourages researchers to invent data with high-dimensional features. In contrast, not all features are needed by machine learning, and, in many cases, high-dimensional features decrease the performance of machine learning. The research presented in paper [2] investigates and proposes methods to determine the best feature selection method in the domain of psychosocial education.

Recommendation systems are powerful tools that are integral parts of a great many websites. Most often, recommendations are presented in the form of a list that is generated by using various recommendation methods. Typically, however, these methods do not generate identical recommendations, and their effectiveness varies between users. In order to solve this problem, the application of aggregation techniques was suggested in article [3], the aim of which is to combine several lists into one, which, in theory, should improve the overall quality of generated recommendations.

Ontologies, and especially formal ones, have traditionally been investigated as a means with which to formalize an application domain, so as to carry out automated reasoning on it. The union of the terminological part of an ontology and the corresponding assertional part is known as a knowledge graph. On the other hand, database technology has often focused on the optimal organization of data, so as to boost efficiency in their storage, management, and retrieval. Graph databases are a recent technology that specifically focus on element-driven data browsing rather than on batch processing.

Paper [4] proposes an intermediate format that can be easily mapped onto a formal ontology on the one hand, so as to allow complex reasoning, and onto a graph database on the other, so as to benefit from efficient data handling. Selecting the right supplier is a critical decision in sustainable supply chain management. Paper [5] proposes and implements an ontology-based approach for knowledge acquisition from the text for a sustainable supplier selection domain. This approach is dedicated to acquiring complex relationships from texts and coding these in the form of rules.

Whenever we need to analyze big data we need to do it effectively, with the shortest possible time and the highest possible accuracy. If we deal with multidimensional data that

Citation: Konys, A.; Nowak-Brzezińska, A. Knowledge Engineering and Data Mining. *Electronics* **2023**, *12*, 927. https://doi.org/10.3390/electronics12040927

Received: 6 February 2023
Accepted: 8 February 2023
Published: 13 February 2023

are computing-intensive, applications should be parallelized and run on modern multicore machines to reduce the execution time. In paper [6] the authors demonstrate how to apply an affine transformation framework and generate parallel 2D tiled code computing GLREs (general linear recurrence equations).

The most popular classification techniques are decision trees, k-nearest neighbor classifiers, naive Bayes classifiers, or neural networks. A very interesting approach is presented in paper [7], where the study developed an autocorrect system for UAV smoke tracing. An AI model was used to calculate smoke tube angle corrections, such that smoke tube angles could be immediately corrected when smoke is sprayed.

Another interesting approach was presented in [8]. The exploration of oil and gas in offshore regions is increasing due to global energy demand. The weather in offshore areas is truly unpredictable due to the sparsity and unreliability of metocean data. Using metocean data, offshore wave height and period are predicted from the wind speed by three state-of-the-art machine learning algorithms (an artificial neural network, a support vector machine, and random forest).

Another interesting research is presented in [9], where the authors present an original concept of the classification of types of project tasks, which will allow for the more beneficial use of collected data in management support systems in the IT industry. The classification algorithms presented in the article are based on the manual recognition of task types. Rules based on keywords are created, which allow for the automatic recognition of task types at subsequent occurrences, which will allow for the fully automated operation of a task classification as well as subtask classification algorithm on a real-time basis and, finally, for the comprehensive support of the management of the development process.

A knowledge-mining- and graph-convolutional-network-based method is described in paper [10], where the authors propose a novel graph-convolutional-network-based method for the knowledge mining of interactions between drugs from the extensive literature. Thus, identifying possible drug–drug interactions (DDIs) has always been a crucial research topic in the field of clinical pharmacology.

A convolutional neural network is also used by the authors of paper [11], in which a neural network helps to discern the morphological information hidden in Chinese characters and a pretrained model obtains vectors with medical features. The different vectors are stitched together to form a multi-feature vector. Deep learning requires a large amount of annotated data to train the model, as does the proposed model, but large-scale annotated data in the Chinese electronic medical record domain require medical experts for annotation annotate, which can be time-consuming.

The healthcare sector is one of the most sensitive sectors in our society, and it is believed that the application of specific and detailed database creation and design techniques can improve the quality of patient care. In this sense, the better management of emergency resources should be achieved. Paper [12] presents an optimized database designed for emergency care. The general objective of the project was to create a database that was as complete as possible and with a great diversity of information, which would represent, in detail, all possible aspects of emergency health activity. A multi-model database allowed for the exploitation of information with predictive models.

Knowledge delivery is the topic which has recently been explored in an enormous way. The reason for this is the post-COVID-19 era in university education, where instructors around the world were at the forefront of implementing hybrid learning spaces for knowledge delivery. The purpose of the study presented in paper [13] is not only to divert the primary use of a YouTube channel into a tool to support asynchronous teaching, it also aims to provide feedback to instructors and suggest steps as well as actions to implement in their teaching modules to ensure students' access to new knowledge while promoting their engagement and satisfaction, regardless of the learning environment, i.e., face-to-face, distance, and hybrid. By analyzing and interpreting data directly from YouTube channel reports, six variables were identified and tested to quantify the lack of statistically significant changes in learners' viewing habits.

In facial aesthetics, soft-tissue landmark recognition and linear as well as angular measurements play critical roles in treatment planning. Visual identification and judgment by hand are time-consuming and prone to errors. As a result, user-friendly software solutions are required to assist healthcare practitioners in improving treatment planning. Paper [14] presents "A Computational Tool for Detection of Soft Tissue Landmarks and Cephalometric Analysis". The goal of the authors is to create a computational tool that may be used to identify and save critical landmarks from patient X-ray pictures. The second goal is to create automated software that can assess the soft-tissue facial profiles of patients in both linear and angular directions by using the landmarks that have been identified.

A variety of different techniques with which to support decisions requires deep knowledge about the advantages and disadvantages of these techniques, especially when we need to deal with multicriteria tasks. Multicriteria methods have gained traction in academia and industry practices for effective decision making. Paper [15] provides a complete overview of multicriteria methods through a bibliometric study, enabling scholars to comprehend the current state and future development patterns of multicriteria decision-making methods research.

We believe that this Special Issue covers the entire knowledge engineering pipeline: from data acquisition and data mining to knowledge extraction and exploitation. For this reason, we tried to gather the many researchers operating in the field to contribute to a collective effort in understanding the trends and future questions in the fields of knowledge engineering and data mining.

Conflicts of Interest: The authors declare no conflict of interest.

References

1. Ziemba, P.; Becker, J.; Becker, A.; Radomska-Zalas, A.; Pawluk, M.; Wierzba, D. Credit Decision Support Based on Real Set of Cash Loans Using Integrated Machine Learning Algorithms. *Electronics* **2021**, *10*, 2099. [CrossRef]
2. Muttakin, F.; Wang, J.-T.; Mulyanto, M.; Leu, J.-S. Evaluation of Feature Selection Methods on Psychosocial Education Data Using Additive Ratio Assessment. *Electronics* **2022**, *11*, 114. [CrossRef]
3. Bałchanowski, M.; Boryczka, U. Aggregation of Rankings Using Metaheuristics in Recommendation Systems. *Electronics* **2022**, *11*, 369. [CrossRef]
4. Ferilli, S. Integration Strategy and Tool between Formal Ontology and Graph Database Technology. *Electronics* **2021**, *10*, 2616. [CrossRef]
5. Konys, A. An Ontology-Based Approach for Knowledge Acquisition: An Example of Sustainable Supplier Selection Domain Corpus. *Electronics* **2022**, *11*, 4012. [CrossRef]
6. Bielecki, W.; Błaszyński, P. Parallel Tiled Code for Computing General Linear Recurrence Equations. *Electronics* **2021**, *10*, 2050. [CrossRef]
7. Chao, P.-Y.; Hsu, W.-C.; Chen, W.-Y. Design of Automatic Correction System for UAV's Smoke Trajectory Angle Based on KNN Algorithm. *Electronics* **2022**, *11*, 3587. [CrossRef]
8. Azad, M.; Uddin, M.A. Prediction of Offshore Wave at East Coast of Malaysia—A Comparative Study. *Electronics* **2022**, *11*, 2527. [CrossRef]
9. Wysocki, W.; Miciuła, I.; Mastalerz, M. Classification of Task Types in Software Development Projects. *Electronics* **2022**, *11*, 3827. [CrossRef]
10. Xu, X.; Meng, F.; Sun, L. Knowledge Mining of Interactions between Drugs from the Extensive Literature with a Novel Graph-Convolutional-Network-Based Method. *Electronics* **2023**, *12*, 311. [CrossRef]
11. Li, J.; Liu, R.; Chen, C.; Zhou, S.; Shang, X.; Wang, Y. An RG-FLAT-CRF Model for Named Entity Recognition of Chinese Electronic Clinical Records. *Electronics* **2022**, *11*, 1282. [CrossRef]
12. Arias, J.C.; Cubillas, J.J.; Ramos, M.I. Optimising Health Emergency Resource Management from Multi-Model Databases. *Electronics* **2022**, *11*, 3602. [CrossRef]
13. Kanetaki, Z.; Stergiou, C.; Bekas, G.; Jacques, S.; Troussas, C.; Sgouropoulou, C.; Ouahabi, A. Acquiring, Analyzing and Interpreting Knowledge Data for Sustainable Engineering Education: An Experimental Study Using YouTube. *Electronics* **2022**, *11*, 2210. [CrossRef]
14. Azad, M.; Elaiwat, S.; Alam, M.K. A Computational Tool for Detection of Soft Tissue Landmarks and Cephalometric Analysis. *Electronics* **2022**, *11*, 2408. [CrossRef]
15. Basílio, M.P.; Pereira, V.; Costa, H.G.; Santos, M.; Ghosh, A. A Systematic Review of the Applications of Multi-Criteria Decision Aid Methods (1977–2022). *Electronics* **2022**, *11*, 1720. [CrossRef]

electronics

Article

Credit Decision Support Based on Real Set of Cash Loans Using Integrated Machine Learning Algorithms

Paweł Ziemba [1,*], Jarosław Becker [2,*], Aneta Becker [3], Aleksandra Radomska-Zalas [2], Mateusz Pawluk [4] and Dariusz Wierzba [5]

1 Institute of Management, University of Szczecin, Aleja Papieża Jana Pawła II 22A, 70-453 Szczecin, Poland
2 Faculty of Technology, The Jacob of Paradies University, Chopina 52, 66-400 Gorzów Wielkopolski, Poland; aradomska-zalas@ajp.edu.pl
3 Faculty of Economics, West Pomeranian University of Technology, Janickiego 31, 71-210 Szczecin, Poland; abecker@zut.edu.pl
4 Faculty of Mathematics and Information Science, Informatics, Warsaw University of Technology, Koszykowa 75, 00-662 Warsaw, Poland; m.pawluk@mini.pw.edu.pl
5 Faculty of Economic Sciences, University of Warsaw, Długa 44/50, 00-241 Warsaw, Poland; dariusz.wierzba@fulbrightmail.org
* Correspondence: pawel.ziemba@usz.edu.pl (P.Z.); jbecker@ajp.edu.pl (J.B.)

Abstract: One of the important research problems in the context of financial institutions is the assessment of credit risk and the decision to whether grant or refuse a loan. Recently, machine learning based methods are increasingly employed to solve such problems. However, the selection of appropriate feature selection technique, sampling mechanism, and/or classifiers for credit decision support is very challenging, and can affect the quality of the loan recommendations. To address this challenging task, this article examines the effectiveness of various data science techniques in issue of credit decision support. In particular, processing pipeline was designed, which consists of methods for data resampling, feature discretization, feature selection, and binary classification. We suggest building appropriate decision models leveraging pertinent methods for binary classification, feature selection, as well as data resampling and feature discretization. The selected models' feasibility analysis was performed through rigorous experiments on real data describing the client's ability for loan repayment. During experiments, we analyzed the impact of feature selection on the results of binary classification, and the impact of data resampling with feature discretization on the results of feature selection and binary classification. After experimental evaluation, we found that correlation-based feature selection technique and random forest classifier yield the superior performance in solving underlying problem.

Keywords: credit scoring; cash loans; machine learning; decision model; classification; feature selection; resampling; discretization

Citation: Ziemba, P.; Becker, J.; Becker, A.; Radomska-Zalas, A.; Pawluk, M.; Wierzba, D. Credit Decision Support Based on Real Set of Cash Loans Using Integrated Machine Learning Algorithms. *Electronics* **2021**, *10*, 2099. https://doi.org/10.3390/electronics10172099

Academic Editor: Jaime Lloret

Received: 5 August 2021
Accepted: 26 August 2021
Published: 30 August 2021

Publisher's Note: MDPI stays neutral with regard to jurisdictional claims in published maps and institutional affiliations.

1. Introduction

Nowadays, banks and financial institutions carefully analyze the credit risk of their clients [1]. The current world situation, i.e., COVID-19 pandemic, affects not only people's lives, but also has a negative impact on economic factor, especially related to paying liabilities by potential borrowers [2]. According to that issue, credit scoring systems [1] are needed by such organizations in order to select the most promising clients to work with and offer well-tailored services for them. These models are particularly suited for financial institutions, due to the ability of assessing the numerical score of individual customers, which determines their loan repayment probability [3]. Under the hood the final decision is made—whether loan granting is justified or not. Most often, credit risk is assessed on the basis of historical data, using mainly statistical or machine learning methods [4], among them, e.g., rough sets [5], usually combined with: probability theory [6], fuzzy sets [7], decision trees [8], Neural Networks and Support Vector Machines [9], or genetic algorithms [10].

Of particular importance in the problems of credit scoring are classification models that play role of decision models [11], usually, supported by feature selection, data resampling and feature discretization methods [12]. There exist many applications of above techniques in numerous publications [1–4,13–18]. Reduction of computational burden and significant improvement of model efficiency and understandability can be achieved when relevant feature subset is selected [19]. Moreover, credit scoring models may be sensitive due to the dataset imbalance, i.e., the number of positive and negative cases is not equally distributed—in that situation, their overall performance may be improved by data resampling [20]. The use of discretization may also have a positive impact on credit scoring models by increasing the efficiency of certain classification algorithms [21]. Unfortunately, when analyzing the literature on credit scoring, there is a shortage of research in which all the indicated techniques (feature selection, resampling, discretization, classification) would be used in one process of processing a dataset and building a classification model. In connection with the identified research gap, the question arises whether the combined use of the indicted methods and techniques in the process of dataset processing will increase the effectiveness of classification models.

The aim of this article is to analyze the effectiveness of various classification models in supporting credit decisions. Contribution includes:

- creation of decision models using different binary classifiers, feature selection methods, as well as data resampling and feature discretization methods;
- evaluation of models on dataset containing real data of cash loans.

It is important to note that the presented research is a significant extension of the earlier works in which we examined only selected classifiers and feature selection methods [22] as well as rough set approach [23].

Section 2 discusses the problem of credit risk assessment and reviews the literature on the subject. Section 3 presents a review of useful methods for classification task, feature selection, data resampling, and feature discretization incorporated in the study, as well as proven measures for assessment of classification models. Section 4 contains a description and explanation of the adopted test procedure. The general results of the research carried out are included in Section 5, while the more detailed results are included in the Appendices A–G. The paper is summarized with conclusions and proposals for further research presented in Section 6.

2. Literature Review

The subject of interest of authors dealing with financial issues is often credit risk, generally defined as the risk of a business partner who does not fully meet its obligations on time and avoids such activities altogether [24]. Credit risk can also be understood as the risk of changes in the value of the company's equity as a result of changes in the creditworthiness of its debtors. It is noted that in recent years a lot of attention has been paid to the methods and algorithms for assessing financial credit risk. This was due, among others, to the occurrence of global financial crises, but also to the need for a thorough assessment of such threats and forecasting business failures. It should be added that the above-mentioned factors have an impact on the functioning of the economy and financial decisions made by societies [25].

Due to the fact that financial credit risk indicates a risk related to financing, its assessment is aimed at solving the following two categories of problems: credit rating or scoring and predicting bankruptcy of forecasting a financial crisis of enterprises. Historically, research on financial credit risk assessment was initiated in the 1930s [26] and continued over the years with considerable success in the 1960s [27]. Nowadays, apart from taking into account the achievements obtained with the use of traditional statistical methods, the research focuses primarily on the use of advanced machine learning methods. This approach, without the need to follow strict assumptions, results in an improvement in the accuracy of the results obtained in a conventional manner. At the same time, it is impossible to indicate the only effective method that is superior to others. On the other hand, the most recently used intelligence techniques include: artificial neural networks (ANNs), fuzzy set theory (FST), decision trees (DTRs), case-based reasoning (CBR), support vector machines (SVMs), rough set theory (RST), genetic programming (GP), hybrid learning, and ensemble computing [25].

The traditional approach to credit risk assessment focuses on obtaining the optimal linear combination of the input explanatory variables. It is expected that thanks to these variables it will be possible to: model, analyze and predict the risk of corporate insolvency. Their use is determined by popularity, but attention is paid, for example, to the fact that they do not take into account complex relationships between variables. To assess credit risk using statistical models, among others, linear discrimination analysis (LDA), logistic regression (LR), multivariate discriminant analysis (MDA), quadratic discriminant analysis (QDA), factor analysis (FA), risk index models, and conditional probability model are used [25]. Among the works pointing to the domination of statistical methods over other approaches, there are [28,29].

The group of methods that combine the traditional and intelligent approaches are semi-parametric method, which are characterized by greater flexibility of the model structure, clearly interpret the modelled process and show greater accuracy. More information on this can be found in [30,31]. In the literature on the subject, there are many interesting combinations of parametric, non-parametric and semi-parametric models, for example, the Klein and Spady model [32], Logit model and the CART model [33]. Another proposal is the integration of a parametric binary logistic regression model (BLRM) and non-parametric models (e.g., SVM, DTR) [34].

Many publications report good results obtained with the use of artificial neural networks [35–37]. The feature of networks that makes them useful for the assessment of credit risk is the ability to process non-linear data and approximate most of the functions. In this way, internal patterns can be found from complex financial data [38]. There are also some limitations to their use, such as difficulty in explaining the black box algorithm, time-consuming learning, not providing optimal solutions, and too much adjustment to the training data.

Another proposal for credit risk assessment are SVMs, which transform non-linear input vectors into a multidimensional feature space. It is possible with the use of kernel functions, which means that the data can be separated by linear models. The interest in SVMs is due to their good performance, the possibility of generalizing a small set of high-value data [39]. Their effectiveness is noticeable when the input data are non-linear and non-stationary, which results in obtaining models supporting credit decisions [40].

The classical classification approach is represented by decision trees. In the case of credit risk, their usefulness results from: easy interpretation of the obtained results, non-linear estimation, non-parametric form, accuracy, possibility of application in the case of continuous and categorical variables, as well as the indication of significant variables. In the discussed field, for example, ID3, C4.5, CART, CHAID, MARS, ADTree [33] can be used.

In the literature on the subject [25], it is possible to note the use of CBR in the subject of credit risk. This approach makes it possible to propose problem-solving by recalling similar experiences. All activities are based on the principle of k-nearest neighbors (kNN), which in the case of classification includes the identified object in the class to which most of its k-nearest neighbors belong. It is suggested to use CBR in the case of small data sets, although it is less precise in relation to other methods used in this type of problem and its improvement is proposed [41].

There have been many interesting publications on credit risk assessment recently. In their work, Wang et al. (2020) [42] presented the results of a study on the assessment of credit risk in the supply chain of commercial banks online. The authors used the literature induction method, the non-linear LS-SVM model and compared the obtained results with the results of the logistic regression model. They found that the LS-SVM evaluation model had a higher classification accuracy than the logistic regression model. In addition, they found that it has a strong generalization capacity and can comprehensively identify credit risk and provide sound, scientific analysis, and is an effective tool supporting the credit risk assessment of small and medium-sized enterprises.

The article by Arora and Kaur (2020) [43], which confirmed the usefulness of modern data mining and machine learning techniques, is also worth mentioning. According to the authors, these methods show precision in predicting credit risk and support taking appropriate decisions. Bolasso (Bootstrap-Lasso) was used in the research. In order to test the predictive accuracy, the functions obtained by Bolasso were applied to the following classification algorithms: Random Forest (RF), Support Vector Machine (SVM), Naïve Bayes (NB), and kNN. The authors concluded that the Random Forest algorithm (BS-RF) with Bolasso enabled provides the best credit risk assessment results.

Other conclusions were reached by Froelich and Hajek (2019) [44], who proposed in their previous studies to automate credit risk assessment by using systems based on machine learning methods. The authors concluded that the obtained results are difficult to interpret and do not fully take into account the expert knowledge. In the next step, they applied multi-criteria group decision making methods (MCGDM) to simulate the assessment process performed by a team of credit risk experts. According to the authors, standard MCGDM methods do not take into account high uncertainty and are not effective in the case of a significant impact of the assessed credit risk criteria. Therefore, they proposed an MCGDM model that combines fuzzy sets and fuzzy cognitive maps with the traditional TOPSIS approach. In turn, Heidary Dahooie et al. (2021) [45] proposed a combination of Data Envelopment Analysis (DEA) with the dynamic multi-attribute decision-making method (DMADM), considering it an innovative dynamic decision-making method for assessing loan applications. The credit performance criteria were distinguished on the basis of a literature review and expert opinion. In contrast, the criteria weights were calculated using the dynamic approach to the common set of DEA weights. Then, candidates were prioritized using five Gray MADM methods (including SAW-G, VIKOR-G, TOPSIS-G, ARAS-G and COPRAS-G). In the final study, a new method called the correlation coefficient and standard deviation (CCSD) was used to determine the aggregate rank.

In the summary of the review of credit risk assessment methods, it should be added that in recent years, in line with the observations of Bellacos (2018) [46], efforts to improve the traditional approach to credit scoring have not always been successful. Compared to traditional credit models, the data used in the new credit models is much more precise, comprehensive and holistic. These data, combined with modern machine learning (ML) algorithms and artificial intelligence (AI), provide much better calibrated risk assessment models. On the other hand, when comparing ML and AI methods with expert credit risk assessment, it should be noted that modern methods take into account many more decision-making factors than a human can do. The expert has knowledge based on his previous experience, but classification models have much more knowledge. The knowledge of classifiers is also based on previous experiences, in this case written as a set of training cases, but their ability to process information is much greater than that of an expert who has limited perception. Moreover, ML methods, unlike humans, do not get tired, do not get sick, etc. Additionally, in the literature, the advantage of machine learning and data mining methods over expert assessment in complex problems requiring the processing of many data is noticed [47]. On the other hand, there are still areas where the expert outweighs ML and AI methods [48].

The banking sector already has some characteristics such as: advanced computerization (available computing power, modern analytical tools), large amounts of transaction data, financial history of customers, which make it the preferred field for implementing credit risk assessment models based on machine learning and artificial intelligence. The content of the Digital Banking report (2021) [49] presenting current trends and priorities in retail banking shows that most banking institutions know what is needed, and many of them even know how to face the current challenges. The problem, however, is that current banking standards keep organizations from doing this. In the area of credit decisions, this applies to solutions with a very complicated, difficult or even impossible explanation mechanism. An example is neural networks seen as black boxes. What is happening inside such a network cannot be fully explained. Banks in Poland refuse to use such tools, as it is difficult to justify a specific credit decision made on their basis before the Polish Financial Supervision Authority (PFSA). PFSA is sympathetic to traditional scoring and other methods whose results are intuitive, easily interpreted, and easy to argue and explain.

3. Materials and Methods

3.1. Classification Methods

Machine learning can be used for various tasks, among others, in classification problems, consisting in predicting the belonging of an object to a certain class on the basis of well-defined characteristics of this object. Usually, discrimination of selected object is based on the earlier training of the classifier, during which the classification algorithm attempts to "learn", what are the real classes of training objects and what features determine whether the objects belong to specific classes [47,50]. Methods for classification task are, e.g., C4.5 decision tree (C4.5), random forest (RF), decision table (DT), naive Bayes (NB) classifier, logistic regression (LR), or k-nearest neighbors (kNN) algorithm. The characteristics of selected classification methods are presented in Table 1.

Table 1. Characteristics of selected classification methods.

Method	Essence of the Method	Advantages	Disadvantages	Ref.
C4.5	The C4.5 algorithm is based on dataset splits according to individual variables, works in a recursive manner when visiting each decision node and proposing optimal division according to selected criterion.	• C4.5 is not built by binary splits only, therefore, varied shape of model is obtained. • When categorical variable is analyzed, branching based on each level of attribute is made. This results that tree, when all possible divisions are made, has greater depth.	• Assigning one value to dependent variable. • Significant change of predicted value when value of one of the features changes slightly.	[51,52]
RF	RF is a complex classifier, consisting of multiple instances of decision trees, which is trained without supervision. One tree can be grown by obtaining a randomly drawn subset of data with replacement from the training dataset. Then the decision tree is created for the selected subset. Training finishes when the number of trees has reached its maximum or error in testing set has stopped decreasing.	• Possibility of enabling parallel computation for each tree, due to independence of trees. • This approach has more stability than simple decision tree model, providing improved classification accuracy. • Some of frequent issues are addressed by random forest: incomplete data, irrelevant and redundant explanatory variables, sophisticated and large dependency structure of features.	• The main disadvantage can be loss of interpretability for trained classifier model. • High computational complexity.	[53–56]
DT	DT is an accurate method for numeric prediction from decision trees and it is an ordered set of *If-Then* rules that have the potential to be more compact and therefore more understandable than the decision trees. The entire problem of learning DT consists of selecting the right attributes to be included. Usually this is done by measuring the tables cross validation performance for different subsets of attributes and choosing the best performing subset.	• DT is one of the simplest hypothesis spaces possible and usually they are easy to understand. • It is a simpler, less compute intensive algorithm than the decision-tree-based approach. • Leave-one-out cross-validation is very cheap for this kind of classifier.	• The TD algorithm very rarely achieves above-average classification accuracy. • There are always the same number of evaluation conditions and actions to be performed in the decision table. • DT does not depict the flow of logic for the solution to a given problem.	[57,58]

Table 1. *Cont.*

Method	Essence of the Method	Advantages	Disadvantages	Ref.
NB	It is a family of algorithms based on a common principle, that the value of a given feature is independent of the value of any other feature, taking into account the class variable. The purpose of NB algorithm is to assess conditional probability of occurring events.	• The NB classifier is considered to be relatively simple, effective algorithm. • NB is able to analyze any number of independent, continuous and categorical variables. • It can be used for tasks with two or more classes for output variable, assuming complete independence of individual variables. • It only requires a small number of training data to estimate the parameters necessary for classification. • It is not sensitive to insignificant features.	• NB assumes that all features are independent, what rarely happening in real (it limits the applicability of this algorithm). • There is a problem of 'zero frequency' in the NB, where it assigns zero probability to a categorical variable whose category in the test data set wasn't available in the training dataset.	[59–61]
LR	LR is one of the classification methods used when each sample is assigned to one of two classes (binary classification). This model assesses the probability of an event that dependent variable is equal to 1.	• LR takes into account all significant variables and excludes all irrelevant features from model. • The resulting model is easy to interpret, because each feature has one weight assigned.	• The LR model does not explain interactions between independent variables and data cannot be collinear. • In case of outliers LR model efficiency deteriorates much, so that they should be removed before starting the analysis.	[62,63]
kNN	kNN is a nonparametric method. The algorithm assumes that similar objects are in the same class and the prediction of belonging to the class of a new object is based on a comparison with a set of prototype objects.	• kNN can be used both for regression and classification tasks. • It does not require learning as it uses the idea of prototypes. • No need for parameter optimization. • Possibly huge number of classes. • Very fast evaluation of new samples. • Ease of implementation.	• kNN treats all the attributes of the feature space equally important, which increases risk of domination irrelevant or redundant features over significant ones, leading to inferior classification. To avoid such situation, an appropriate set of features should be selected [39].	[64,65]

3.2. Feature Selection Methods

One of the basic issues in classification task is the multidimensionality of the object to be assigned to a specific class. This is a serious obstacle decreasing accuracy of classification algorithms, known as the "dimensional curse" [66]. Dimensionality reduction of feature space allows lowering the computational and data collection costs, which eventually improves predictions [67]. Tools, which can be used for that task are called feature selection methods.

The feature selection process focuses on identifying relevant features in dataset as significant and rejecting redundant features [68]. For this purpose, various algorithms are used to assess the importance of particular features in the classification task. The feature selection methods are divided into three categories: filters, wrappers, and embedded methods [69]. Filters and wrappers are usually composed of four elements (steps), such as: generation of feature subset, evaluation of the subset, stopping criterion, result validation [70]. By describing individual elements of the feature selection methods, it is possible to point out significant differences between these groups of methods.

Filters are based on independent evaluation of features using general data characteristics. For example, Pearson correlation coefficients between each input and selected output can be used. Feature subset is determined by defining threshold for minimum value of correlation or particular number of features to be selected before training the machine learning algorithm [71].

Wrappers evaluate individual feature subsets using machine learning algorithms, which algorithms will eventually be used in the classification or regression task. In this case, training algorithm is included in the feature selection procedure, therefore, cross-validation based on set of training cases is usually used to estimate the accuracy of the classifier using a specific feature subset [72].

Embedded methods are similar to wrappers in that they use classification to perform the task of feature selection. The main difference between wrappers and embedded methods is "embedding" of selection procedure into the selected classifier. In other words, the dimensions of training objects subject to classification are reduced while building classifier model [73]. For instance, in decision trees unnecessary features are eliminated by trimming and defining the minimum number of objects in the node.

Wrappers differ only in the applied machine learning algorithms, so, as in the case of embedded methods, the results obtained using them depend solely on the quality of the machine learning algorithm and the algorithm fit to a specific classification task. Wrappers and embedded methods analyze the features of the objects contained in the training set only in terms of obtaining the maximum number of correct classifications, omitting other characteristics of the features. Meanwhile, the general characteristics of the features seem so important that they should affect the selection of individual features that determine the training and test cases. Therefore, filtration procedures that determine the significance of individual attributes using measures other than classifier's accuracy seem to be more interesting. Filter methods are using various measures to assess relevance of each feature, e.g., distance function and different correlation measures.

Popular filter technique that uses the distance function is ReliefF [74]. On the other hand, the most numerous groups of filters are correlation procedures, among them the most promising are: Symmetrical Uncertainty (SU) [75], Correlation-based Feature Selection (CFS) [76], Fast Correlation-Based Filter (FCBF) [77], and Significance Attribute (SA) [78]. The basis characteristics of each method are presented in Table 2.

Table 2. Characteristics of selected feature selection methods.

Method	Group of Methods	Methodological Basics	Applied Heuristics	Essence of the Method	Ref.
ReliefF	distance based	k-nearest neighbors	good attributes should discriminate objects belonging to other classes and should have the same value for objects being similar and belonging to the same class	introduces hits and misses concepts, which improves or deteriorates classifier's accuracy	[79] [80] [81] [74]
SU	correlation based	entropy, information gain	'1' means that we are fully informed based on the attribute, allowing us to predict the class of the object; '0' means there is no information after analyzing the attribute and prediction is not possible	compensates a deviation of information gain towards multi-valued attributes and normalizes final score to range [0, 1]	[75] [82]
CFS	correlation based	SU, Pearson linear correlation	good subset of features contains attributes that are strongly correlated with a specific class of objects and not correlated with other classes and attributes	matrix of mutual correlation between attributes and correlation between attributes and classes of objects are initially computed, forward search is performed using the "Best First" algorithm	[76] [83]

Table 2. *Cont.*

Method	Group of Methods	Methodological Basics	Applied Heuristics	Essence of the Method	Ref.
FCBF	correlation based	SU	only the attributes whose SU values are above defined threshold are selected for further consideration	The procedure employs sets of redundant features separately for each feature; selected attributes are sorted based on descending order of SU score and feature set is examined, whether redundancy of features exists	[77]
SA	correlation based	probability theory	if selected attribute is significant, then there is a high chance for that elements with complementary sets of values for this attribute will belong to complementary sets of classes; if class decisions for two sets of elements are different, then significant attribute values for these two sets of elements should also be different	significance of each attribute is calculated as the average value of general associations: given attribute with classes and classes with given attribute; the attribute is relevant when both values of associations are high	[78]

3.3. Resampling Methods

In binary classification, when number of classes in training set is unbalanced, i.e., class distribution is strongly skewed, conventional classifiers maximizing their accuracy usually build models that tend to classify all objects as belonging to the majority class. This results in low accuracy for the minority class, whose objects are underrepresented in training set, whereas such class is often of uttermost importance [84]. To overcome this issue, **resampling** methods are commonly used for training set. The two most popular in machine learning, yet very simple, are techniques of random undersampling and random oversampling [20]. In addition to the resampling methods already aforementioned, another interesting approach is Synthetic Minority Over-sampling Technique (**SMOTE**) [85]. Table 3 lists the main advantages and disadvantages of each of these approaches.

Table 3. Characteristics of selected resampling methods.

Method	Essence of the Method	Advantages	Disadvantages	Ref.
Random undersampling	assumes that multiple objects of the majority class are redundant and random deletion of them will not significantly change data distribution	reduces representation of the majority class by removing random objects of such class until number of classes is balanced	there is possible risk of certain objects removal, which have positive impact on accuracy of classifier	[84]
Random oversampling	increases size of the minority class by replicating objects belonging to such class	-	puts at risk of overfitting the classifier model by shifting the model towards the minority class; not add any new valuable objects, of the minority class; classifier training is significantly extended by increasing the size of the training set	[20] [84]
SMOTE oversampling	the minority class is oversampled by generating synthetic objects in neighborhood of the real objects; among k nearest neighbors, $n \leq k$ neighbors are randomly selected and one synthetic object is generated similar to each of them	using interpolation instead of replication, as opposed to random oversampling, SMOTE avoids problem of overfitting	shifts the decision boundaries of the minority class towards space of the majority class	[20] [84] [85]

3.4. Discretization Methods

Some classification algorithms improve their performance by using feature discretization. Moreover, certain classifiers cannot work without data discretization. Such methods bin continuous features, dividing them into ranges or intervals, resulting in conversion of numerical data to nominal data. Here, main issue with feature discretization is appropriate choice of cutpoints, because continuous data can be discretized in an infinite number of ways. Perfect discretization method should find a relatively small number of cutpoints, dividing data into relevant bins. Among discretization techniques, there are supervised and unsupervised methods. First group results are superior to second group, because it uses class distribution to which each object belongs as additional information. Great number of methods perform discretization based on class entropy, which is a measure of uncertainty in finite range of classes. Entropy is calculated for different splits and compared to entropy of dataset without splits. It is run recursively until the search stop criterion is meet [86]. For instance, heuristic method of Minimal Description Length Principle (MDLP) can be used, here. This technique determines whether or not to accept current cut-off point candidate, thus, stopping recursion if specified condition is not met [87]. The entropy-based discretization with MDLP stop criterion is considered to be one of the best supervised discretization methods [71]. It measures information gain score of possible cutpoint by comparing entropy value. For each considered cutpoint, entropy of input interval is compared to the weighted sum of entropies for two output intervals. There are several different criteria for MDLP stopping condition, including Fayyad criterion [88] and Kononenko criterion [89].

3.5. Classification Evaluation Metrics

The quality of the classification can be evaluated by, e.g., Receiver Operating Characteristic curve (ROC), Area Under Receiver Operating Characteristic curve (AUROC) and Gini coefficient (GC). Another interesting measure is Precision-Recall Curve (PRC).

ROC is the graphic representation of the predictive model effectiveness made by sketching the quantitative characteristics of binary classifiers derived from such model using variety of cut-off points. This shows the relationship between True Positive Rate (TPR) and False Positive Rate (FPR). *TPR* can be calculated as follows by Equation (1) [85]:

$$TPR = \frac{TP}{TP + FN} \tag{1}$$

where *TP* indicates number of true positives, i.e., model predicts positive class correctly and *FN* indicates number of false negatives, i.e., model predicts negative class incorrectly. In turn, *FPR* is defined as Equation (2) [85]:

$$FPR = \frac{FP}{FP + TN} \tag{2}$$

where *FP* indicates number of false positives, i.e., model predicts positive class incorrectly and *TN* indicates number of true negatives, i.e., model predicts negative class correctly.

AUROC measures the classifier's accuracy. It is calculated as probability thresholds for following event—considered object belongs to negative or positive class. Geometrically, this is area below ROC. The higher value of AUROC, the better classification results of model are, where AUROC < 0.5 means invalid classifier, i.e., worse than random, AUROC = 0.5 means random classifier, and AUROC = 1 means ideal classifier [85].

GC is a measure of model's quality, interpreted as degree of ideality for classifier. GC is calculated based on the following Equation (3):

$$GC = 2 * AUROC - 1 \tag{3}$$

The higher value of GC, the better classifier is, where GC = 0 means random classifier, and GC = 1 means ideal classifier [90].

PRC shows dependence between precision (Positive Predictive Value—PPV) and recall (TPR) for the classifier, where former is calculated as follows Equation (4) [91]:

$$PPV = \frac{TP}{TP + FP} \quad (4)$$

Big area under PRC (AUPRC) represents both high precision and high recall, where high precision corresponds to low false positive frequency and high recall corresponds to low false negative frequency. High scores for precision and recall indicate that classifier predicts accurate results and also most of them are positive [91]. PRCs are often zigzag curves with oscillations. Due to that fact, they tend to cross over much more than ROCs, therefore, leaving researcher difficult comparison. It is recommended to use PRCs in addition to ROCs for obtaining complete overview while evaluation and comparison of classifier models [92].

4. Research Procedure

The dataset on which the experiment was conducted describes anonymized data about loan repayment and borrowers. This set consists of 91,759 records described by 272 conditional attributes (features) and the decision attribute. It was divided in proportion 70/30% into training set (64,230 records) and testing set (27,529 records) [93].

Final research was preceded by a series of preliminary tests, during which following were selected:

- the most promising and various filter methods for feature selection;
- different classifiers, bearing in mind their core algorithm, way of knowledge representation and ability to explain classification of cases.

During preliminary tests, it was noticed that one of the models with outstanding classification results can be random forest, therefore, its more detailed examination allowed to select optimal parameters, i.e., number of iterations = 239 and maximum tree depth = 13 [22].

In this research study it was assumed that various combinations will be tested, consisting in filter methods (SU, FCBF, CFS, SA, ReliefF), classifiers models (C4.5, DT, kNN, LR, NB, RF, optimized random forest (ORF)), resampling methods (without resampling, random undersampling, SMOTE) and feature discretization (without discretization, Fayyad criterion, Kononenko criterion). Taking into account the number of methodological approaches considered in each group, this gives 315 different scenarios and the same number of classification models supporting credit decisions. In practice, this number was smaller due to the fact that the number of conducted scenarios was limited, because of omitting selected resampling and discretization algorithms. Here, following heuristics was used, according to which, if specific preprocessing method, i.e., resampling or discretization, does not give satisfactory results, then there is no reason for its inclusion in subsequent scenario. Moreover, due to the high computational complexity, some scenarios did not use ReliefF. It should be noted that in case of large training dataset, this method performed in general time-consuming calculations, not yielding acceptable results. Therefore, all scenarios included at least 4 filter methods (SU, FCBF, CFS, SA) and all seven classifiers. Additionally, it should be clarified that for case of random undersampling, each scenario was repeated three times, building three different classification models and averaging results, eventually. The above approach was followed in order to minimize the impact of training cases random selection on classification results. The research study was divided into four general scenarios in which following combinations of methods were applied:

1. without resampling, without discretization, feature selection, classification method;
2. resampling, without discretization, feature selection, classification method;
3. without resampling, discretization, feature selection, classification method;
4. resampling, discretization, feature selection, classification method.

Furthermore, at the beginning, classification was performed without using filter methods, i.e., scenario 0. Results of this study were reference to subsequent scenarios in which filter methods were used. According to such approach all research scenarios allowed to define:

- the effect of feature selection on classification;
- the effect of data resampling on classification with feature selection;
- the effect of feature discretization on classification with feature selection;
- the effect of data resampling with feature discretization on classification with feature selection.

Figure 1 depicts the research study, which was carried out. Figure 1 shows that processing techniques including feature discretization and feature selection were applied to training set and results were used in testing set. This was necessary step to allow full consistency between training set and testing set. For instance, binning of training data was achieved and then the same bins were adopted to testing data. Likewise, selection of relevant features was done based on training set and redundant features were removed from testing set. Only one processing method used on training cases without testing cases was data resampling.

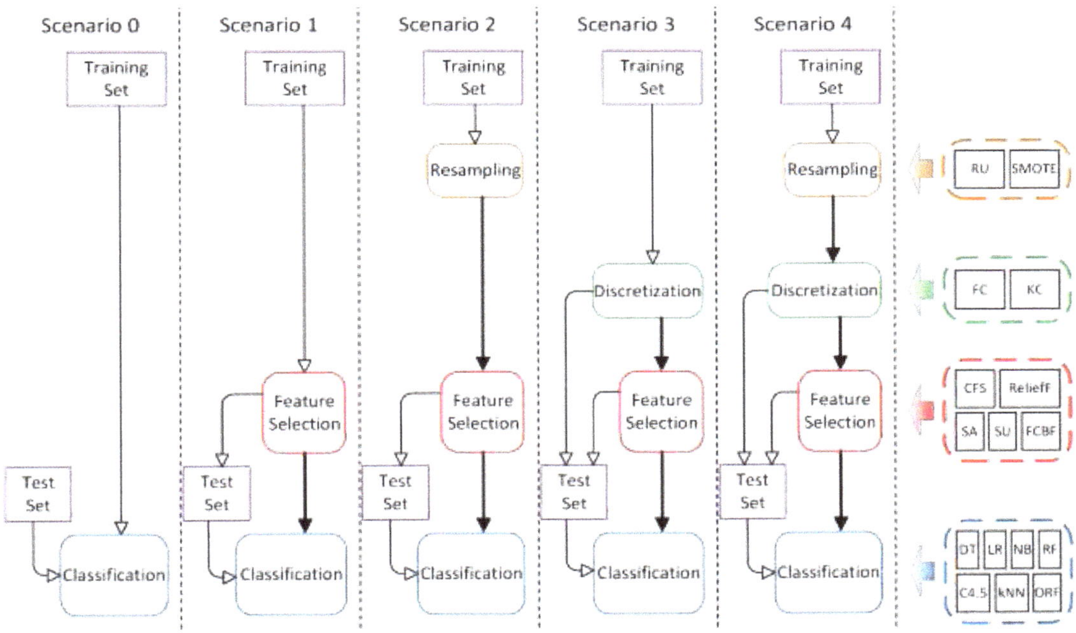

Figure 1. Scenario-based research study. Abbreviations: RU—Random undersampling, SMOTE—Synthetic Minority Over-sampling Technique, FC—Fayyad criterion-based discretization, KC—Kononenko criterion-based discretization, CFS—Correlation-based Feature Selection, SA—Significance Attribute, SU—Symmetrical Uncertainty, FCBF—Fast Correlation-Based Filter, DT—Decision table, LR—Logistic regression, NB—Naïve Bayes, RF—Random forest, C4.5—C4.5 decision tree, kNN—k-nearest neighbors, ORF—Optimized random forest.

5. Results and Discussion

Full results of conducted research study are presented in Appendices A–G, while this section shows only the best results from each considered scenario. Table 4 depicts the four top classification results from each scenario. From Table 4 it can be stated that the best classification results are obtained by RF model with possible optimization and feature selection method allowing top classification results is mainly CFS. It should be also noted that overall outstanding result was achieved by RF on full dataset of 272 features. Obviously, dimensionality reduction of such data is necessary due to the lack of ability to explain classification or need to collect great amount of information in order to classify new case. Assuming feature selection is made without resampling or discretization the best classification results were obtained by ORF. However, if both feature selection and classification accuracy are important, then RF model should be supported by data resampling, which allows to balance class distribution. Moreover, in case of RF, as well as LR and DT, undersampling provides better classification results than discretization (cf. Appendices C, E and F). On the contrary, it is opposite for NB, kNN and C4.5. Furthermore, RF and LR, both with undersampling, yield superior results than with combination of undersampling and discretization. On the other hand, above combination improves quality of classification for NB. Additionally, in order to obtain acceptable results using LR or NB, it is necessary to employ methods previously mentioned while for RF model they can be entirely omitted. Moreover, the randomness in applied undersampling algorithm also plays vital role. It has serious impact on obtained feature sets, thus, on results of classification. Nevertheless, conclusions drawn here are true for each research case performed during the study. It should be noted that in order to maximize accuracy of classification, it is recommended to carry out several draws and select set of training cases that allows to obtain the best results for the classification of testing cases.

Table 4. The best classification results from each research scenario.

Scenario	Rank (GC-Based)	Feature Selection	No of Features	Resampling	Discretization	Classifier	GC	AUPRC Negative	AUPRC Positive
0	1	-	272	-	-	ORF	0.828	0.999	0.275
	2	-	272	-	-	RF	0.76	0.998	0.276
	3	-	272	-	-	DT	0.716	0.998	0.087
	4	-	272	-	-	LR	0.7	0.997	0.096
1	1	CFS	13	-	-	ORF	0.762	0.998	0.147
	2	CFS	13	-	-	RF	0.704	0.998	0.137
	3	CFS	13	-	-	LR	0.668	0.998	0.072
	4	CFS	13	-	-	NB	0.662	0.998	0.046
2	1	CFS	35/27/37	RU	-	ORF	0.805	0.999	0.118
	2	CFS	35/27/37	RU	-	RF	0.802	0.999	0.111
	3	SU	35/27/37	RU	-	ORF	0.786	0.999	0.111
	4	SU	35/27/37	RU	-	RF	0.781	0.999	0.105
3	1	CFS	13	-	KC	NB	0.76	0.998	0.102
	2	CFS	14	-	FC	LR	0.758	0.998	0.117
	3	CFS	14	-	FC	NB	0.754	0.998	0.101
	4	CFS	13	-	KC	LR	0.752	0.998	0.116
4	1	CFS	35	RU	KC	ORF	0.794	0.999	0.121
	2	CFS	35	RU	KC	RF	0.79	0.999	0.116
	3	CFS	35	RU	KC	NB	0.768	0.999	0.112
	4	SU	35	RU	KC	LR	0.768	0.998	0.094

Classifier: NB—Naive Bayes, RF—Random Forest, DT—Decision Table, LR—Logistic Regression, ORF—Optimized Random Forest; **Resampling:** RU—Random Undersampling; **Discretization:** KC—Kononenko Criterion, FC—Fayyad Criterion; **Feature selection:** CFS—Correlation-based Feature Selection, SU—Symmetrical Uncertainty.

On the other hand, if the selection of possibly smallest feature set is of great importance, then FCBF should be used. Table 5 depicts four top classification results from each scenario where feature sets were obtained by above method. From Table 5 it can be stated that feature sets consisting in five or six features do not provide acceptable classification results. Bearing in mind that the minimum number of features and the maximum accuracy are essential, results of RF in scenario 2 and NB in scenario 4 are worth noting. DT achieves also relatively good classification results compared to other models. Main reason behind that is due to the built-in feature selection, i.e., DT automatically reduces feature space. Whether input feature set is relatively large enough, this can cause deterioration of classification compared to other models, but with low number of features additional reduction is not performed, so that there is no negative impact on final results.

Table 5. The best classification results from each research scenario using FCBF.

Scenario	Rank (GC-Based)	Feature Selection	No of Features	Resampling	Discretization	Classifier	GC	AUPRC Negative	AUPRC Positive
1	1	FCBF	6	-	-	ORF	0.626	0.997	0.086
	2		6	-	-	DT	0.624	0.997	0.074
	3		6	-	-	NB	0.584	0.997	0.035
	4		6	-	-	LR	0.572	0.997	0.044
2	1		12	RU	-	ORF	0.749	0.998	0.094
	2		12	RU	-	RF	0.743	0.998	0.089
	3		12	RU	-	DT	0.699	0.998	0.057
	4		12	RU	-	LR	0.696	0.998	0.064
3	1		5	-	FC/KC	DT	0.652	0.997	0.069
	2		5	-	FC/KC	NB	0.648	0.997	0.082
	3		5	-	FC/KC	LR	0.644	0.997	0.081
	4		5	-	FC/KC	kNN	0.626	0.997	0.074
4	1		10	RU	KC	NB	0.756	0.998	0.084
	2		10	RU	KC	LR	0.732	0.998	0.081
	3		10	RU	KC	ORF/RF	0.722	0.998	0.065
	4		10	RU	KC	kNN	0.722	0.998	0.063

Classifier: NB—Naive Bayes, RF—Random Forest, DT—Decision Table, LR—Logistic Regression, ORF—Optimized Random Forest; **Resampling:** RU—Random Undersampling; **Discretization:** KC—Kononenko Criterion, FC—Fayyad Criterion; **Feature selection:** FCBF—Fast Correlation-Based Filter.

6. Conclusions

The article deals with the problem of credit decisions based on machine learning methods. In particular, the effects of the application were verified together with classifiers of other machine learning methods in the processing of the credit data set. Summarizing results of conducted research study, it is possible to indicate premises related to use of individual methods, i.e., feature selection, binary classification, data resampling, feature discretization:

- if classification result is important, then RF will return good results over a full set of data;
- if both feature selection and classification accuracy are important, then acceptable results will be obtained by undersampling with CFS and RF;
- if both minimum number of features and classification accuracy are important, then fair results will be achieved by following approaches: (1) CFS with RF, (2) undersampling with FCBF and RF, (3) discretization with CFS and LR or NB, (4) undersampling with discretization, FCBF and NB.

Of course, above heuristics do not fulfill topic in an exhaustive way of choosing appropriate approach to credit scoring problem. In some business cases, apart from classification result and size of feature set, the ability to explain classification may be also important, which gives certain advantage. Moreover, constraining oneself only to classification accuracy, it is not possible to clearly determine whether it is better to use AUROC, AUPRC or GC. Basically, the selection of classification model will consist in seeking trade-off between inherent features of classifiers. Therefore, further research is targeted on the selection of a specific approach using a classifier for credit decisions in support of stakeholders (e.g., banks) depending on their personal needs (i.e., actual requirements and preferences). Assessment of various approaches is, here, a multi-criteria decision problem, thus, a multi-criteria decision analysis [94] will be involved.

Author Contributions: Conceptualization, P.Z. and A.B.; methodology, P.Z.; validation, J.B.; formal analysis, A.B.; investigation, P.Z.; resources, A.R.-Z.; data curation, M.P.; writing—original draft preparation, P.Z.; writing—review and editing, J.B. and A.B.; supervision, P.Z. and J.B.; project administration, A.R.-Z.; funding acquisition, D.W. All authors have read and agreed to the published version of the manuscript.

Funding: The research is partially financed through the National Centre for Research and Development, Poland (grant no. POIR.01.01.01-00-0322/18-00).

Data Availability Statement: Data available on request.

Conflicts of Interest: The authors declare no conflict of interest.

Appendix A. Results of Scenario 0

Table A1. Classification results for complete feature set.

Measure	Classifier						
	C4.5	DT	kNN	LR	NB	RF	ORF
AUROC	0.604	0.858	0.677	0.850	0.811	0.880	**0.914**
GC	0.208	0.716	0.354	0.700	0.622	0.760	**0.828**
AUPRC negative	0.991	0.998	0.993	0.997	0.997	0.998	**0.999**
AUPRC positive	0.048	0.087	0.053	0.096	0.013	**0.276**	0.275
AUPRC mean	0.982	0.989	0.984	0.988	0.987	**0.991**	**0.991**

Appendix B. Results of Scenario 1

Table A2. Classification results for feature subset selected by CFS (13 features).

Measure	Classifier						
	C4.5	DT	kNN	LR	NB	RF	ORF
AUROC	0.795	0.828	0.640	0.834	0.831	0.852	**0.881**
GC	0.590	0.656	0.280	0.668	0.662	0.704	**0.762**
AUPRC negative	0.996	0.997	0.993	0.998	0.998	0.998	**0.998**
AUPRC positive	0.073	0.063	0.029	0.072	0.046	0.137	**0.147**
AUPRC mean	0.987	0.988	0.983	0.988	0.988	0.989	**0.990**

Table A3. Classification results for feature subset selected by FCBF (six features).

Measure	Classifier						
	C4.5	DT	kNN	LR	NB	RF	ORF
AUROC	0.730	0.812	0.658	0.786	0.792	0.740	**0.813**
GC	0.460	0.624	0.316	0.572	0.584	0.480	**0.626**
AUPRC negative	0.995	0.997	0.993	0.997	0.997	0.995	**0.997**
AUPRC positive	0.067	0.074	0.037	0.044	0.035	0.071	**0.086**
AUPRC mean	0.985	0.987	0.983	0.987	0.987	0.985	**0.988**

Table A4. Classification results for feature subset selected by SU (13 features).

Measure	Classifier						
	C4.5	DT	kNN	LR	NB	RF	ORF
AUROC	0.657	0.732	0.568	**0.742**	0.719	0.660	0.729
GC	0.314	0.464	0.136	**0.484**	0.438	0.320	0.458
AUPRC negative	0.993	0.995	0.991	**0.995**	0.995	0.993	**0.995**
AUPRC positive	0.050	0.060	0.039	0.042	0.032	0.062	**0.079**
AUPRC mean	0.983	0.985	0.982	**0.986**	0.985	0.983	0.985

Table A5. Classification results for feature subset selected by SA (13 features).

Measure	Classifier						
	C4.5	DT	kNN	LR	NB	RF	ORF
AUROC	0.657	0.734	0.606	0.723	0.724	0.716	**0.756**
GC	0.314	0.468	0.212	0.446	0.448	0.432	**0.512**
AUPRC negative	0.993	0.995	0.992	0.995	0.995	0.994	**0.996**
AUPRC positive	0.056	0.046	0.047	0.044	0.041	0.106	**0.113**
AUPRC mean	0.983	0.985	0.982	0.985	0.985	0.985	**0.987**

Appendix C. Results of Scenario 2—Random Undersampling

Table A6. Classification results for feature subset selected by CFS (35/27/37 features).

Measure	Classifier						
	C4.5	DT	kNN	LR	NB	RF	ORF
AUROC	0.768	0.852	0.744	0.891	0.849	0.901	**0.902**
GC	0.537	0.704	0.488	0.781	0.699	0.802	**0.805**
AUPRC negative	0.995	0.998	0.996	0.999	0.998	0.999	**0.999**
AUPRC positive	0.031	0.057	0.022	0.082	0.049	0.111	**0.118**
AUPRC mean	0.986	0.988	0.986	0.989	0.988	0.990	**0.990**

Table A7. Classification results for feature subset selected by FCBF (12/14/11 features).

Measure	Classifier						
	C4.5	DT	kNN	LR	NB	RF	ORF
AUROC	0.780	0.849	0.740	0.848	0.819	0.872	**0.874**
GC	0.560	0.699	0.481	0.696	0.637	0.743	**0.749**
AUPRC negative	0.996	0.998	0.996	0.998	0.997	0.998	**0.998**
AUPRC positive	0.030	0.057	0.028	0.064	0.048	0.089	**0.094**
AUPRC mean	0.986	0.988	0.986	0.988	0.988	0.989	**0.989**

Table A8. Classification results for feature subset selected by SU (35/27/37 features).

Measure	Classifier						
	C4.5	DT	kNN	LR	NB	RF	ORF
AUROC	0.777	0.850	0.760	0.885	0.847	0.890	**0.893**
GC	0.555	0.701	0.519	0.769	0.693	0.781	**0.786**
AUPRC negative	0.996	0.998	0.996	0.999	0.998	0.999	**0.999**
AUPRC positive	0.032	0.061	0.024	0.081	0.048	0.105	**0.111**
AUPRC mean	0.986	0.988	0.986	0.989	0.988	0.989	**0.989**

Table A9. Classification results for feature subset selected by SA (35/27/37 features).

Measure	Classifier						
	C4.5	DT	kNN	LR	NB	RF	ORF
AUROC	0.797	0.843	0.758	0.867	0.829	0.879	**0.886**
GC	0.594	0.687	0.515	0.735	0.657	0.758	**0.772**
AUPRC negative	0.997	0.998	0.996	0.998	0.997	0.999	**0.999**
AUPRC positive	0.033	0.061	0.025	0.075	0.045	0.095	**0.100**
AUPRC mean	0.987	0.988	0.986	0.992	0.988	0.989	**0.989**

Table A10. Classification results for feature subset selected by ReliefF (35/27/37 features).

Measure	Classifier						
	C4.5	DT	kNN	LR	NB	RF	ORF
AUROC	0.767	0.843	0.725	0.734	0.838	0.852	**0.853**
GC	0.535	0.687	0.450	0.469	0.675	0.703	**0.705**
AUPRC negative	0.996	0.998	0.995	0.995	0.998	0.998	**0.998**
AUPRC positive	0.030	0.056	0.028	0.031	0.049	0.077	**0.078**
AUPRC mean	0.986	0.988	0.985	0.985	0.988	0.989	**0.989**

Appendix D. Results of Scenario 2—SMOTE

Table A11. Classification results for feature subset selected by CFS (42 features).

Measure	Classifier						
	C4.5	DT	kNN	LR	NB	RF	ORF
AUROC	0.582	0.653	0.686	0.727	0.765	**0.883**	0.875
GC	0.164	0.306	0.372	0.454	0.530	**0.766**	0.750
AUPRC negative	0.990	0.994	0.994	0.995	0.996	**0.998**	0.998
AUPRC positive	0.045	0.021	0.026	0.049	0.040	**0.158**	0.111
AUPRC mean	0.980	0.984	0.984	0.986	0.986	**0.990**	0.989

Table A12. Classification results for feature subset selected by FCBF (28 features).

Measure	Classifier						
	C4.5	DT	kNN	LR	NB	RF	ORF
AUROC	0.573	0.653	0.682	0.757	0.754	**0.869**	0.869
GC	0.146	0.306	0.364	0.514	0.508	**0.738**	0.738
AUPRC negative	0.990	0.994	0.994	0.996	0.995	**0.998**	0.998
AUPRC positive	0.034	0.021	0.027	0.054	0.040	**0.131**	0.083
AUPRC mean	0.980	0.984	0.984	0.986	0.986	**0.989**	0.989

Table A13. Classification results for feature subset selected by SU (42 features).

Measure	Classifier						
	C4.5	DT	kNN	LR	NB	RF	ORF
AUROC	0.684	0.633	0.751	**0.863**	0.775	0.846	0.849
GC	0.368	0.266	0.502	**0.726**	0.550	0.692	0.698
AUPRC negative	0.993	0.994	0.995	**0.998**	0.996	0.997	0.998
AUPRC positive	0.061	0.021	0.055	0.073	0.043	**0.116**	0.092
AUPRC mean	0.983	0.984	0.985	**0.989**	0.986	0.988	0.989

Table A14. Classification results for feature subset selected by SA (42 features).

Measure	Classifier						
	C4.5	DT	kNN	LR	NB	RF	ORF
AUROC	0.639	0.623	0.772	**0.851**	0.776	0.839	0.839
GC	0.278	0.246	0.544	**0.702**	0.552	0.678	0.678
AUPRC negative	0.992	0.993	0.996	**0.998**	0.996	0.997	0.998
AUPRC positive	0.042	0.019	0.066	0.072	0.048	**0.117**	0.099
AUPRC mean	0.982	0.983	0.986	**0.989**	0.986	0.988	0.988

Appendix E. Results of Scenario 3—Fayyad Criterion

Table A15. Classification results for feature subset selected by CFS (14 features).

Measure	Classifier						
	C4.5	DT	kNN	LR	NB	RF	ORF
AUROC	0.801	0.848	0.746	**0.879**	0.877	0.761	0.771
GC	0.602	0.696	0.492	**0.758**	0.754	0.522	0.542
AUPRC negative	0.996	0.998	0.996	**0.998**	0.998	0.995	0.995
AUPRC positive	0.084	0.082	0.083	**0.117**	0.101	0.084	0.085
AUPRC mean	0.987	0.988	0.986	**0.989**	0.989	0.985	0.986

Table A16. Classification results for feature subset selected by FCBF (five features).

Measure	Classifier						
	C4.5	DT	kNN	LR	NB	RF	ORF
AUROC	0.656	**0.826**	0.813	0.822	0.824	0.810	0.810
GC	0.312	**0.652**	0.626	0.644	0.648	0.620	0.620
AUPRC negative	0.993	**0.997**	0.997	0.997	0.997	0.997	0.996
AUPRC positive	0.045	0.069	0.074	0.081	**0.082**	0.070	0.071
AUPRC mean	0.983	0.987	0.987	0.987	**0.988**	0.987	0.987

Table A17. Classification results for feature subset selected by SU (14 features).

Measure	Classifier						
	C4.5	DT	kNN	LR	NB	RF	ORF
AUROC	0.649	**0.732**	0.678	0.717	0.720	0.653	0.647
GC	0.298	**0.464**	0.356	0.434	0.440	0.306	0.294
AUPRC negative	0.993	**0.995**	0.993	0.994	0.994	0.992	0.992
AUPRC positive	0.052	0.060	0.062	**0.072**	0.067	0.052	0.053
AUPRC mean	0.983	**0.985**	0.984	**0.985**	0.985	0.983	0.983

Table A18. Classification results for feature subset selected by SA (14 features).

Measure	Classifier						
	C4.5	DT	kNN	LR	NB	RF	ORF
AUROC	0.634	0.707	0.651	**0.721**	0.717	0.609	0.600
GC	0.268	0.414	0.302	**0.442**	0.434	0.218	0.200
AUPRC negative	0.993	0.994	0.993	**0.994**	0.994	0.992	0.991
AUPRC positive	0.056	0.060	0.071	**0.072**	0.071	0.052	0.053
AUPRC mean	0.983	0.985	0.984	**0.985**	0.985	0.982	0.982

Appendix F. Results of Scenario 3—Kononenko Criterion

Table A19. Classification results for feature subset selected by CFS (13 features).

Measure	Classifier						
	C4.5	DT	kNN	LR	NB	RF	ORF
AUROC	0.814	0.847	0.744	0.876	**0.880**	0.787	0.791
GC	0.628	0.694	0.488	0.752	**0.760**	0.574	0.582
AUPRC negative	0.997	0.997	0.995	**0.998**	**0.998**	0.995	0.996
AUPRC positive	0.086	0.083	0.087	**0.116**	0.102	0.089	0.089
AUPRC mean	0.987	0.988	0.986	**0.989**	**0.989**	0.986	0.986

Table A20. Classification results for feature subset selected by FCBF (five features).

Measure	Classifier						
	C4.5	DT	kNN	LR	NB	RF	ORF
AUROC	0.656	**0.826**	0.813	0.822	0.824	0.810	0.810
GC	0.312	**0.652**	0.626	0.644	0.648	0.620	0.620
AUPRC negative	0.993	**0.997**	0.997	0.997	**0.997**	0.997	0.996
AUPRC positive	0.045	0.069	0.074	0.081	**0.082**	0.070	0.071
AUPRC mean	0.983	0.987	0.987	0.987	**0.988**	0.987	0.987

Table A21. Classification results for feature subset selected by SU (13 features).

Measure	Classifier						
	C4.5	DT	kNN	LR	NB	RF	ORF
AUROC	0.646	0.727	0.658	0.725	**0.728**	0.644	0.643
GC	0.292	0.454	0.316	0.450	**0.456**	0.288	0.286
AUPRC negative	0.993	0.995	0.993	0.995	**0.995**	0.992	0.992
AUPRC positive	0.049	0.047	0.060	**0.069**	0.065	0.052	0.052
AUPRC mean	0.983	0.985	0.984	**0.985**	**0.985**	0.983	0.983

Table A22. Classification results for feature subset selected by SA (13 features).

Measure	Classifier						
	C4.5	DT	kNN	LR	NB	RF	ORF
AUROC	0.817	**0.825**	0.703	0.793	0.814	0.713	0.714
GC	0.634	**0.650**	0.406	0.586	0.628	0.426	0.428
AUPRC negative	**0.997**	**0.997**	0.994	0.996	0.997	0.994	0.994
AUPRC positive	**0.084**	0.063	0.067	0.075	0.079	0.066	0.066
AUPRC mean	0.987	**0.988**	0.985	0.987	0.988	0.984	0.984

Appendix G. Results of Scenario 4—Random Undersampling, Kononenko Criterion

Table A23. Classification results for feature subset selected by CFS (35 features).

Measure	Classifier						
	C4.5	DT	kNN	LR	NB	RF	ORF
AUROC	0.821	0.852	0.871	0.883	0.884	0.895	**0.897**
GC	0.642	0.704	0.742	0.766	0.768	0.790	**0.794**
AUPRC negative	0.997	0.998	0.998	0.998	0.999	0.999	**0.999**
AUPRC positive	0.036	0.049	0.091	0.098	0.112	0.116	**0.121**
AUPRC mean	0.987	0.988	0.989	0.989	0.989	0.990	**0.990**

Table A24. Classification results for feature subset selected by FCBF (10 features).

Measure	Classifier						
	C4.5	DT	kNN	LR	NB	RF	ORF
AUROC	0.840	0.843	0.861	0.866	**0.878**	0.861	0.861
GC	0.680	0.686	0.722	0.732	**0.756**	0.722	0.722
AUPRC negative	0.997	0.998	0.998	0.998	**0.998**	0.998	0.998
AUPRC positive	0.045	0.054	0.063	0.081	**0.084**	0.065	0.065
AUPRC mean	0.988	0.988	0.989	0.989	**0.989**	0.989	0.989

Table A25. Classification results for feature subset selected by SU (35 features).

Measure	Classifier						
	C4.5	DT	kNN	LR	NB	RF	ORF
AUROC	0.830	0.851	0.860	**0.884**	0.860	0.865	0.870
GC	0.660	0.702	0.720	**0.768**	0.720	0.730	0.740
AUPRC negative	0.997	0.998	0.998	**0.998**	0.998	0.998	0.998
AUPRC positive	0.039	0.056	0.072	**0.094**	0.079	0.085	0.094
AUPRC mean	0.988	0.988	0.989	**0.989**	0.989	0.989	0.989

Table A26. Classification results for feature subset selected by SA (35 features).

Measure	Classifier						
	C4.5	DT	kNN	LR	NB	RF	ORF
AUROC	0.823	0.852	0.861	0.872	0.854	0.873	**0.877**
GC	0.646	0.704	0.722	0.744	0.708	0.746	**0.754**
AUPRC negative	0.997	0.998	0.998	**0.998**	0.998	0.998	**0.998**
AUPRC positive	0.041	0.049	0.079	**0.092**	0.083	0.086	0.091
AUPRC mean	0.987	0.988	0.989	**0.989**	0.989	0.989	**0.989**

Table A27. Classification results for feature subset selected by ReliefF (35 features).

Measure	Classifier						
	C4.5	DT	kNN	LR	NB	RF	ORF
AUROC	0.784	0.851	0.842	0.828	0.844	0.876	**0.878**
GC	0.568	0.702	0.684	0.656	0.688	0.752	**0.756**
AUPRC negative	0.996	0.998	0.998	0.998	0.998	**0.998**	**0.998**
AUPRC positive	0.030	0.056	0.070	0.067	0.080	**0.105**	0.104
AUPRC mean	0.986	0.988	0.988	0.988	0.989	**0.989**	**0.989**

References

1. Koutanaei, F.N.; Sajedi, H.; Khanbabaei, M. A Hybrid Data Mining Model of Feature Selection Algorithms and Ensemble Learning Classifiers for Credit Scoring. *J. Retail. Consum. Serv.* **2015**, *27*, 11–23. [CrossRef]
2. Wang, D.; Zhang, Z.; Bai, R.; Mao, Y. A Hybrid System with Filter Approach and Multiple Population Genetic Algorithm for Feature Selection in Credit Scoring. *J. Comput. Appl. Math.* **2018**, *329*, 307–321. [CrossRef]
3. Tunç, A. Feature Selection in Credibility Study for Finance Sector. *Procedia Comput. Sci.* **2019**, *158*, 254–259. [CrossRef]
4. Tripathi, D.; Edla, D.R.; Kuppili, V.; Bablani, A.; Dharavath, R. Credit Scoring Model Based on Weighted Voting and Cluster Based Feature Selection. *Procedia Comput. Sci.* **2018**, *132*, 22–31. [CrossRef]
5. Pawlak, Z. Rough Sets and Fuzzy Sets. *Fuzzy Sets Syst.* **1985**, *17*, 99–102. [CrossRef]
6. Maldonado, S.; Peters, G.; Weber, R. Credit Scoring using Three-Way Decisions with Probabilistic Rough Sets. *Inf. Sci.* **2020**, *507*, 700–714. [CrossRef]
7. Capotorti, A.; Barbanera, E. Credit Scoring Analysis using a Fuzzy Probabilistic Rough Set Model. *Comput. Stat. Data Anal.* **2012**, *56*, 981–994. [CrossRef]
8. Zhou, X.; Zhang, D.; Jiang, Y. A New Credit Scoring Method Based on Rough Sets and Decision Tree. In *Advances in Knowledge Discovery and Data Mining*; Washio, T., Suzuki, E., Ting, K.M., Inokuchi, A., Eds.; Springer: Berlin/Heidelberg, Germany, 2008; pp. 1081–1089.
9. Zhou, J.; Tian, J. *Credit Risk Assessment Based on Rough Set Theory and Fuzzy Support Vector Machine*; Atlantis Press: Paris, France, 2007.
10. Zhou, J.; Bai, T. Credit Risk Assessment using Rough Set Theory and GA-Based SVM. In Proceedings of the 2008 the 3rd International Conference on Grid and Pervasive Computing—Workshops, Kunming, China, 25–28 May 2008; pp. 320–325. [CrossRef]
11. Ziemba, P. Multi-Criteria Fuzzy Evaluation of the Planned Offshore Wind Farm Investments in Poland. *Energies* **2021**, *14*, 978. [CrossRef]
12. López, J.; Maldonado, S. Profit-Based Credit Scoring Based on Robust Optimization and Feature Selection. *Inf. Sci.* **2019**, *500*, 190–202. [CrossRef]
13. Liu, Y.; Schumann, M. Data Mining Feature Selection for Credit Scoring Models. *J. Oper. Res. Soc.* **2005**, *56*, 1099–1108. [CrossRef]

14. Somol, P.; Baesens, B.; Pudil, P.; Vanthienen, J. Filter-versus Wrapper-Based Feature Selection for Credit Scoring. *Int. J. Intell. Syst.* **2005**, *20*, 985–999. [CrossRef]
15. Ha, S.; Nguyen, H.-N. Credit Scoring with a Feature Selection Approach Based Deep Learning. In *MATEC Web of Conferences*; EDP Sciences: Les ulis, France, 2016; Volume 54, p. 05004. [CrossRef]
16. Aryuni, M.; Madyatmadja, E. Feature Selection in Credit Scoring Model for Credit Card Applicants in XYZ Bank: A Comparative Study. *Int. J. Multimed. Ubiquitous Eng.* **2015**, *10*, 17–24. [CrossRef]
17. Boughaci, D.; Alkhawaldeh, A.A. Three Local Search-Based Methods for Feature Selection in Credit Scoring. *Vietnam J. Comput. Sci.* **2018**, *5*, 107–121. [CrossRef]
18. Van, S.H.; Ha, N.N.; Bao, H.N.T. A Hybrid Feature Selection Method for Credit Scoring. *EAI Endorsed Trans. Context-Aware Syst. Appl.* **2017**, *4*, e2.
19. Kozodoi, N.; Lessmann, S.; Papakonstantinou, K.; Gatsoulis, Y.; Baesens, B. A Multi-Objective Approach for Profit-Driven Feature Selection in Credit Scoring. *Decis. Support Syst.* **2019**, *120*, 106–117. [CrossRef]
20. Guo, X.; Yin, Y.; Dong, C.; Yang, G.; Zhou, G. On the Class Imbalance Problem. In Proceedings of the Fourth International Conference on Natural Computation, Jinan, China, 18–20 October 2008; Volume 4. [CrossRef]
21. García, S.; Luengo, J.; Sáez, J.A.; López, V.; Herrera, F. A Survey of Discretization Techniques: Taxonomy and Empirical Analysis in Supervised Learning. *IEEE Trans. Knowl. Data Eng.* **2013**, *25*, 734–750. [CrossRef]
22. Ziemba, P.; Radomska-Zalas, A.; Becker, J. Client Evaluation Decision Models in the Credit Scoring Tasks. *Procedia Comput. Sci.* **2020**, *176*, 3301–3309. [CrossRef]
23. Becker, J.; Radomska-Zalas, A.; Ziemba, P. Rough Set Theory in the Classification of Loan Applications. *Procedia Comput. Sci.* **2020**, *176*, 3235–3244. [CrossRef]
24. Andersson, F.; Mausser, H.; Rosen, D.; Uryasev, S. Credit Risk Optimization with Conditional Value-at Risk Criterion. *Math. Program.* **2001**, *89*, 273–291. [CrossRef]
25. Chen, N.; Ribeiro, B.; Chen, A. Financial Credit Risk Assessment: A Recent Review. *Artif. Intell. Rev.* **2016**, *45*, 1–23. [CrossRef]
26. Shen, G.; Jia, W. The Prediction Model of Financial Crisis Based on the Combination of Principle Component Analysis and Support Vector Machine. *Open J. Soc. Sci.* **2014**, *2*, 204–212. [CrossRef]
27. Altman, E.I. Financial Ratios, Discriminant Analysis and the Prediction of Corporate Bankruptcy. *J. Financ.* **1968**, *23*, 589–609. [CrossRef]
28. Kouki, M.; Elkhaldi, A. Toward a Predicting Model of Firm Bankruptcy: Evidence from the Tunisian Context. *Middle East. Financ. Econ.* **2011**, *14*, 26–43.
29. Kwak, W.; Shi, Y.; Kou, G. Bankruptcy Prediction for Korean Firms after the 1997 Financial Crisis: Using a Multiple Criteria Linear Programming Data Mining Approach. *Rev. Quant. Financ. Account.* **2012**, *38*, 441–453. [CrossRef]
30. Cheng, K.F.; Chu, C.K.; Hwang, R.-C. Predicting Bankruptcy using the Discrete-Time Semiparametric Hazard Model. *Quant. Financ.* **2010**, *10*, 1055–1066. [CrossRef]
31. Hwang, R.-C.; Chung, H.; Chu, C.K. Predicting Issuer Credit Ratings using a Semiparametric Method. *J. Empir. Financ.* **2010**, *17*, 120–137. [CrossRef]
32. Klein, R.; Spady, R.H. An Efficient Semiparametric Estimator for Binary Response Models. *Econometrica* **1993**, *61*, 387–421. [CrossRef]
33. Brezigar-Masten, A.; Masten, I. CART-Based Selection of Bankruptcy Predictors for the Logit Model. *Expert Syst. Appl.* **2012**, *39*, 10153–10159. [CrossRef]
34. Li, J.; Pan, L.; Chen, M.; Yang, X. Parametric and Non-Parametric Combination Model to Enhance Overall Performance on Default Prediction. *J. Syst. Sci. Complex.* **2014**, *27*, 950–969. [CrossRef]
35. Mokhatab Rafiei, F.; Manzari, S.M.; Bostanian, S. Financial Health Prediction Models using Artificial Neural Networks, Genetic Algorithm and Multivariate Discriminant Analysis: Iranian Evidence. *Expert Syst. Appl.* **2011**, *38*, 10210–10217. [CrossRef]
36. Chen, N.; Vieira, A.; Ribeiro, B.; Duarte, J.; Neves, J. A Stable Credit Rating Model Based on Learning Vector Quantization. *Intell. Data Anal.* **2011**, *15*, 237–250. [CrossRef]
37. Blanco, A.; Pino-Mejías, R.; Lara, J.; Rayo, S. Credit Scoring Models for the Microfinance Industry using Neural Networks: Evidence from Peru. *Expert Syst. Appl.* **2013**, *40*, 356–364. [CrossRef]
38. Huang, F. A Genetic Fuzzy Neural Network for Bankruptcy Prediction in Chinese Corporations. In Proceedings of the 2008 International Conference on Risk Management & Engineering Management, Beijing, China, 4–6 November 2008; pp. 542–546.
39. Yang, Z.; You, W.; Ji, G. Using Partial Least Squares and Support Vector Machines for Bankruptcy Prediction. *Expert Syst. Appl.* **2011**, *38*, 8336–8342. [CrossRef]
40. Jeganathan, J.; Joseph, K.S.; Vaishnavi, J. Bankruptcy Prediction using Svm and Hybrid Svm Survey. *Int. J. Comput. Appl.* **2011**, *34*, 39–45.
41. Li, H.; Adeli, H.; Sun, J.; Han, J.-G. Hybridizing Principles of TOPSIS with Case-Based Reasoning for Business Failure Prediction. *Comput. Oper. Res.* **2011**, *38*, 409–419. [CrossRef]
42. Wang, F.; Ding, L.; Yu, H.; Zhao, Y. Big Data Analytics on Enterprise Credit Risk Evaluation of E-Business Platform. *Inf. Syst. E-Bus. Manag.* **2020**, *18*, 311–350. [CrossRef]
43. Arora, N.; Kaur, P.D. A Bolasso Based Consistent Feature Selection Enabled Random Forest Classification Algorithm: An Application to Credit Risk Assessment. *Appl. Soft Comput.* **2020**, *86*, 105936. [CrossRef]

44. Froelich, W.; Hajek, P. IVIFCM-TOPSIS for Bank Credit Risk Assessment. In *Intelligent Decision Technologies 2019*; Czarnowski, I., Howlett, R.J., Jain, L.C., Eds.; Springer: Singapore, 2020; pp. 99–108.
45. Heidary Dahooie, J.; Razavi Hajiagha, S.H.; Farazmehr, S.; Zavadskas, E.K.; Antucheviciene, J. A Novel Dynamic Credit Risk Evaluation Method using Data Envelopment Analysis with Common Weights and Combination of Multi-Attribute Decision-Making Methods. *Comput. Oper. Res.* **2021**, *129*, 105223. [CrossRef]
46. Bellacosa, M. AI Can Transform Trade Finance through Better SME Credit Scoring. Available online: https://www.theglobaltreasurer.com/2018/06/08/ai-can-transform-trade-finance-through-better-sme-credit-scoring/ (accessed on 19 August 2021).
47. Ziemba, P.; Jankowski, J.; Wątróbski, J.; Piwowarski, M. Web Projects Evaluation using the Method of Significant Website Assessment Criteria Detection. In *Transactions on Computational Collective Intelligence XXII*; Nguyen, N.T., Kowalczyk, R., Eds.; Springer: Berlin/Heidelberg, Germany, 2016; pp. 167–188.
48. Ärje, J.; Raitoharju, J.; Iosifidis, A.; Tirronen, V.; Meissner, K.; Gabbouj, M.; Kiranyaz, S.; Kärkkäinen, S. Human Experts vs. Machines in Taxa Recognition. *Signal Process. Image Commun.* **2020**, *87*, 115917. [CrossRef]
49. Marous, J. *Retail Banking Trends and Priorities*; Temenos: Geneva, Switzerland, 2021; p. 119.
50. Sulikowski, P.; Zdziebko, T. Deep Learning-Enhanced Framework for Performance Evaluation of a Recommending Interface with Varied Recommendation Position and Intensity Based on Eye-Tracking Equipment Data Processing. *Electronics* **2020**, *9*, 266. [CrossRef]
51. Quinlan, J.R. *C4.5: Programs for Machine Learning*; Morgan Kaufmann Publishers Inc.: San Francisco, CA, USA, 1993; ISBN 978-1-55860-238-0.
52. Wang, X.; Zhou, C.; Xu, X. Application of C4.5 Decision Tree for Scholarship Evaluations. *Procedia Comput. Sci.* **2019**, *151*, 179–184. [CrossRef]
53. Breiman, L. Random Forests. *Mach. Learn.* **2001**, *45*, 5–32. [CrossRef]
54. Sulikowski, P.; Zdziebko, T.; Turzyński, D. Modeling Online User Product Interest for Recommender Systems and Ergonomics Studies. *Concurr. Comput. Pract. Exp.* **2019**, *31*, e4301. [CrossRef]
55. Demski, T. *Od Pojedynczych Drzew do Losowego Lasu*; StatSoft Polska: Kraków, Poland, 2011.
56. Breiman, L. Bagging Predictors. *Mach. Learn.* **1996**, *24*, 123–140. [CrossRef]
57. Kohavi, R. The Power of Decision Tables. In *European Conference on Machine Learning*; Springer: Berlin/Heidelberg, Germany, 1995; pp. 174–189.
58. Kalmegh, S.R. Comparative Analysis of the WEKA Classifiers Rules Conjunctiverule & Decisiontable on Indian News Dataset by using Different Test Mode. *Int. J. Eng. Sci. Invent.* **2018**, *7*, 2319–6734.
59. Perzyk, M.; Biernacki, R. Zaawansowane metody statystyczne w sterowaniu procesami produkcyjnymi. *Arch. Odlew.* **2004**, *4*, 19–28.
60. John, G.H.; Langley, P. Estimating Continuous Distributions in Bayesian Classifiers. In Proceedings of the Eleventh Conference on Uncertainty in Artificial Intelligence, Montreal, QC, Canada, 18–20 August 1995; Morgan Kaufmann Publishers Inc.: San Francisco, CA, USA, 1995; pp. 338–345.
61. StatSoft. Available online: https://www.statsoft.pl (accessed on 28 April 2021).
62. Hastie, T.; Tibshirani, R.; Friedman, J. *The Elements of Statistical Learning: Data Mining, Inference, and Prediction*, 2nd ed.; Springer Series in Statistics; Springer: New York, NY, USA, 2009; ISBN 978-0-387-84857-0.
63. Le Cessie, S.; Van Houwelingen, J.C. Ridge Estimators in Logistic Regression. *J. R. Stat. Soc. Ser. C (Appl. Stat.)* **1992**, *41*, 191–201. [CrossRef]
64. Guo, G.; Wang, H.; Bell, D.; Bi, Y.; Greer, K. KNN Model-Based Approach in Classification. In *On The Move to Meaningful Internet Systems 2003: CoopIS, DOA, and ODBASE*; Meersman, R., Tari, Z., Schmidt, D.C., Eds.; Springer: Berlin/Heidelberg, Germany, 2003; pp. 986–996.
65. Sá, J.P.M. *De Pattern Recognition: Concepts, Methods and Applications*; Springer: Berlin/Heidelberg, Germany, 2001; ISBN 978-3-642-62677-7.
66. Chizi, B.; Maimon, O. Dimension Reduction and Feature Selection. In *Data Mining and Knowledge Discovery Handbook*; Maimon, O., Rokach, L., Eds.; Springer: Boston, MA, USA, 2005; pp. 93–111. ISBN 978-0-387-25465-4.
67. Guyon, I. Practical Feature Selection: From Correlation to Causality. In *Mining Massive Data Sets for Security—Advances in Data Mining, Search, Social Networks and Text Mining, and Their Applications to Security*; IOS Press: Amsterdam, The Netherlands, 2008; pp. 27–43.
68. Ziemba, P.; Piwowarski, M.; Jankowski, J.; Wątróbski, J. Method of Criteria Selection and Weights Calculation in the Process of Web Projects Evaluation. In *Computational Collective Intelligence*; Technologies and Applications; Hwang, D., Jung, J.J., Nguyen, N.-T., Eds.; Springer International Publishing: Cham, Switzerland, 2014; pp. 684–693.
69. Biswas, S.; Bordoloi, M.; Purkayastha, B. Review on Feature Selection and Classification using Neuro-Fuzzy Approaches. *Int. J. Appl. Evol. Comput.* **2016**, *7*, 28–44. [CrossRef]
70. Liu, H.; Yu, L.; Motoda, H. Feature Extraction, Selection, and Construction. In *The Handbook of Data Mining*; Lawrence Erlbaum Associates: Mahwah, NJ, USA, 2003; pp. 409–424.
71. Witten, I.H.; Frank, E.; Hall, M.A. *Data Mining: Practical Machine Learning Tools and Techniques*; Elsevier: Amsterdam, The Netherlands, 2011; ISBN 978-0-12-374856-0.

72. Hall, M.A.; Holmes, G. Benchmarking Attribute Selection Techniques for Discrete Class Data Mining. *IEEE Trans. Knowl. Data Eng.* **2003**, *15*, 1437–1447. [CrossRef]
73. Chandrashekar, G.; Sahin, F. A Survey on Feature Selection Methods. *Comput. Electr. Eng.* **2014**, *40*, 16–28. [CrossRef]
74. Bins, J.; Draper, B. Evaluating Feature Relevance: Reducing Bias in Relief. In Proceedings of the 6th Joint Conference on Information Science, Research Triangle Park, NC, USA, 8–13 March 2002; pp. 757–760.
75. Yang, Q.; Shao, J.; Scholz, M.; Plant, C. Feature Selection Methods for Characterizing and Classifying Adaptive Sustainable Flood Retention Basins. *Water Res.* **2011**, *45*, 993–1004. [CrossRef] [PubMed]
76. Hall, M.A.; Smith, L.A. Feature Selection for Machine Learning: Comparing a Correlation-Based Filter Approach to the Wrapper. In Proceedings of the Twelfth International Florida Artificial Intelligence Research Society Conference, Orlando, FL, USA, 1–5 May 1999; AAAI Press: Palo Alto, CA, USA, 1999; pp. 235–239.
77. Yu, L.; Liu, H. Feature Selection for High-Dimensional Data: A Fast Correlation-Based Filter Solution. In Proceedings of the 20th International Conference on Machine Learning, Washington, DC, USA, 1 January 2003; Volume 2, pp. 856–863.
78. Ahmad, A.; Dey, L. A Feature Selection Technique for Classificatory Analysis. *Pattern Recognit. Lett.* **2005**, *26*, 43–56. [CrossRef]
79. Chang, C.-C. Generalized Iterative RELIEF for Supervised Distance Metric Learning. *Pattern Recognit.* **2010**, *43*, 2971–2981. [CrossRef]
80. Kononenko, I.; Hong, S.J. Attribute Selection for Modelling. *Future Gener. Comput. Syst.* **1997**, *13*, 181–195. [CrossRef]
81. Kononenko, I. Estimating Attributes: Analysis and Extensions of RELIEF. In *Machine Learning: ECML-94*; Bergadano, F., De Raedt, L., Eds.; Springer: Berlin/Heidelberg, Germany, 1994; pp. 171–182.
82. Senthamarai Kannan, S.; Ramaraj, N. A Novel Hybrid Feature Selection via Symmetrical Uncertainty Ranking Based Local Memetic Search Algorithm. *Knowl.-Based Syst.* **2010**, *23*, 580–585. [CrossRef]
83. Hall, M.A. Correlation-Based Feature Selection for Discrete and Numeric Class Machine Learning. In Proceedings of the Seventeenth International Conference on Machine Learning, Standord, CA, USA, 29 June–2 July 2000; Morgan Kaufmann Publishers Inc.: San Francisco, CA, USA, 2000; pp. 359–366.
84. Pozzolo, A.D.; Caelen, O.; Johnson, R.A.; Bontempi, G. Calibrating Probability with Undersampling for Unbalanced Classification. In Proceedings of the 2015 IEEE Symposium Series on Computational Intelligence, Cape Town, South Africa, 7–10 December 2015; pp. 159–166.
85. Chawla, N.V.; Bowyer, K.W.; Hall, L.O.; Kegelmeyer, W.P. SMOTE: Synthetic Minority Over-Sampling Technique. *J. Artif. Intell. Res.* **2002**, *16*, 321–357. [CrossRef]
86. De Sá, C.R.; Soares, C.; Knobbe, A.; Azevedo, P.; Jorge, A.M. Multi-Interval Discretization of Continuous Attributes for Label Ranking. In *Discovery Science*; Fürnkranz, J., Hüllermeier, E., Higuchi, T., Eds.; Springer: Berlin/Heidelberg, Germany, 2013; pp. 155–169.
87. Zhu, Q.; Lin, L.; Shyu, M.-L.; Chen, S.-C. Effective Supervised Discretization for Classification Based on Correlation Maximization. In Proceedings of the 2011 IEEE International Conference on Information Reuse Integration, Las Vegas, NV, USA, 3–5 August 2011; pp. 390–395.
88. Fayyad, U.M.; Irani, K.B. Multi-Interval Discretization of Continuous-Valued Attributes for Classification Learning. In Proceedings of the 13th International Joint Conference on Artificial Intelligence (IJCAI-93), Chambèry, France, 28 August–3 September 1993; pp. 1022–1027.
89. Kononenko, I. On Biases in Estimating Multi-Valued Attributes. In Proceedings of the 14th International Joint Conference on Artificial Intelligence, Montreal, QC, Canada, 20–25 August 1995; Morgan Kaufmann Publishers Inc.: San Francisco, CA, USA, 1995; Volume 2, pp. 1034–1040.
90. Duda, R.O.; Hart, P.E.; Stork, D.G. *Pattern Classification*, 2nd ed.; Wiley: Hoboken, NJ, USA, 2012. Available online: https://www.wiley.com/en-us/Pattern+Classification%2C+2nd+Edition-p-9781118586006 (accessed on 12 November 2019).
91. Boyd, K.; Eng, K.H.; Page, C.D. Area under the Precision-Recall Curve: Point Estimates and Confidence Intervals. In *Machine Learning and Knowledge Discovery in Databases*; Blockeel, H., Kersting, K., Nijssen, S., Železný, F., Eds.; Springer: Berlin/Heidelberg, Germany, 2013; pp. 451–466.
92. Saito, T.; Rehmsmeier, M. The Precision-Recall Plot is More Informative than the ROC Plot When Evaluating Binary Classifiers on Imbalanced Datasets. *PLoS ONE* **2015**, *10*, e0118432. [CrossRef]
93. Wierzba, D.; Ziemba, P.; Becker, J. Mendeley Data—Anonymized Data about Loan Repayment and Borrowers. Available online: http://dx.doi.org/10.17632/fr99jcnkxg.2 (accessed on 27 August 2021).
94. Ziemba, P. Multi-Criteria Approach to Stochastic and Fuzzy Uncertainty in the Selection of Electric Vehicles with High Social Acceptance. *Expert Syst. Appl.* **2021**, *173*, 114686. [CrossRef]

electronics

Article

Evaluation of Feature Selection Methods on Psychosocial Education Data Using Additive Ratio Assessment

Fitriani Muttakin [1], Jui-Tang Wang [2,*], Mulyanto Mulyanto [2] and Jenq-Shiou Leu [2]

1 Information Systems Department, Universitas Islam Negeri Sultan Syarif Kasim, Pakanbaru 28293, Indonesia; fitrianimuttakin@uin-suska.ac.id
2 Electronic and Computer Engineering, National Taiwan University of Science and Technology, Taipei City 106335, Taiwan; d10602813@mail.ntust.edu.tw (M.M.); jsleu@mail.ntust.edu.tw (J.-S.L.)
* Correspondence: rtwang@mail.ntust.edu.tw

Abstract: Artificial intelligence, particularly machine learning, is the fastest-growing research trend in educational fields. Machine learning shows an impressive performance in many prediction models, including psychosocial education. The capability of machine learning to discover hidden patterns in large datasets encourages researchers to invent data with high-dimensional features. In contrast, not all features are needed by machine learning, and in many cases, high-dimensional features decrease the performance of machine learning. The feature selection method is one of the appropriate approaches to reducing the features to ensure machine learning works efficiently. Various selection methods have been proposed, but research to determine the essential subset feature in psychosocial education has not been established thus far. This research investigated and proposed methods to determine the best feature selection method in the domain of psychosocial education. We used a multi-criteria decision system (MCDM) approach with Additive Ratio Assessment (ARAS) to rank seven feature selection methods. The proposed model evaluated the best feature selection method using nine criteria from the performance metrics provided by machine learning. The experimental results showed that the ARAS is promising for evaluating and recommending the best feature selection method for psychosocial education data using the teacher's psychosocial risk levels dataset.

Keywords: evaluation feature selection; evaluation model; decision model; psychosocial education

Citation: Muttakin, F.; Wang, J.-T.; Mulyanto, M.; Leu, J.-S. Evaluation of Feature Selection Methods on Psychosocial Education Data Using Additive Ratio Assessment. *Electronics* **2022**, *11*, 114. https://doi.org/10.3390/electronics11010114

Academic Editors: Agnieszka Konys and Agnieszka Nowak-Brzezińska

Received: 14 November 2021
Accepted: 28 December 2021
Published: 30 December 2021

Publisher's Note: MDPI stays neutral with regard to jurisdictional claims in published maps and institutional affiliations.

1. Introduction

Psychosocial education is multidisciplinary and covers a vast field of study. Therefore, it is not surprising that research in psychosocial education encompasses an abundance of environments and features that are logically expected to be linked to the problem-solving of educational quality improvements. Research from various perspectives, such as personal environment [1], family [2], nutrition [3], and physical activities [4], has been conducted to get an overview of the various psychosocial relationships in education. Accordingly, research linked to psychosocial education is categorized as one of the most active in education. Indeed, the search using the keyword "psychosocial education" in Google Scholar shows 212,000 results of research published between 2017 and 2021.

On the other hand, the success of artificial intelligence and big data influences decision-making perspectives, particularly those based on predictive problems. Big data can effectively handle more large-scale amounts, more complex varieties, and higher data dimensions [5]. Meanwhile, artificial intelligence, especially machine learning, significantly improves the quality of decision models [6,7]. These two factors encourage researchers to collect more data with massive features.

Theoretically, the more data that are collected, the more information that is obtained. The more information obtained, the better the prediction will be generated to be. However, the increase in the number of variables and the volume of data impacts data sparsity, especially if the data quality is poor. The increase in sparsity makes it much more difficult to find data representative of the population. Furthermore, it makes machine learning challenging to generalize to the domain problem. The vague generalization will cause machine learning to lose its ability to adapt to new problems [8,9].

Instead of thrusting all features into machine learning, performing input feature optimization is often more efficient and effective. Feature selection can eliminate all features that are irrelevant to the prediction target. There have been various methods of selecting a feature that has been proposed and proven to impact machine learning performance. With many feature selection methodologies and different approaches in each method, it is relatively easy to raise a question about which method can give the optimum and effective results in machine learning, especially regarding the psychosocial education problem.

Hence, this paper proposed a methodology to evaluate the best feature selection method in the domain of psychosocial education. The evaluation was performed using a decision model approach that utilized multi-criteria decision making (MCDM). Furthermore, additive ratio assessment (ARAS) was adopted to evaluate and rank the best feature selection method. The evaluation and ranking used the metrics from the machine learning classification performance on the teacher's psychosocial risk level dataset.

2. Related Work

Feature selection is one of the critical stages in machine learning modeling, and the relevant feature has implications for better stability, robustness, and generalization of machine learning [10]. The feature selection method can be divided into three approaches [11–13]: filtering, wrapper, and embedded method.

Moorthy and Gandhi [14] previously conducted research using the filtering method. They optimized medical data using feature selection techniques for classification problems. They combined analysis of variance (ANOVA) and whale optimization (WO) to give a better result for SVM and k-NN classifiers than the one without ANOVA-WO. Ding and Li [15] also conducted a similar study identifying mitochondrial proteins in malaria by combining ANOVA and incremental feature selection (IFS) methods to find the most optimal feature. The proposed model achieved 97.1% accuracy compared to 92.0% on the comparison model. Next, Utama [16] performed feature selection using the mutual information (MI) model to predict the airline's tweet sentiment analysis. The feature selection made contributions to the classifier improvement.

Similarly, the wrapper method also gives promising results. Richhariya et al. [17] proposed a Universum support vector machine based on the recursive feature elimination (USVM-RFE) method to diagnose Alzheimer's. Feature selection was performed on the MRI data of brain tissue, and the classification using USVM-RFE showed better results than the one using SVR-RFE. The implementation of RFE was also done in the study [18], where RFE-SVM was used to determine the best feature among the various heart rate variability (HRV) data. The study showed that RFE-SVM could identify the HRV feature and detect the stress level better.

The approach using the embedded method has been widely used. Liu et al. [19] implemented feature selection using the embedded method. The implementation was performed during a cyberattack on the Internet of Things (IoT) data. The accuracy of the proposed method was relatively comparable to that of the comparison model. However, it was better in training speed, 1000 times faster than the overall features model. The implementation of the embedded method as the feature selection was also conducted by Loscalzo et al. [20]. Feature selection was used to remove unneeded input in robotic sensors. The paper showed that the embedding methodology significantly reduces unimportant sensors. Lastly, Liu et al. [21] compared embedding methodology to the others, such as Chi-Square, F-Statistic, and Gini Index. The experiment showed that the weighted Gini Index (WGI) method was better than the other methodologies on the data with limited features.

Given the importance of choosing a suitable feature selection method for the data characteristics of domain problems, selecting the best feature selection method is quite challenging. There are various techniques for selecting feature selection methods, one of which is the decision system model approach. Kou et al. [22] conducted a study to select the best subset feature for a text classification case. The study compared several models from the MCDM, such as TOPSIS, GRA, WSM, VIKOR, and PROMOTHEE. The results showed that PROMOTHEE was better for evaluation models in the text-based classification case than other models. Hasemi et al. [23] proposed the EFS-MCDM method to determine the best feature on the computer network dataset. The features ranking in the EFS-MCDM delivered more optimal and efficient results in the measurement of accuracy, f-score, and run-time algorithm compared to other methods. Similarly, Singh [24] implemented TOPSIS to select features in the network traffic dataset. The research concluded that the classification model with the TOPSIS-based feature subset had the same accuracy yet much lower computation time.

Despite all the studies conducted on selecting existing feature selection methods so far, to the best of the authors' knowledge, there has been no study comparing and evaluating the best feature selection method to be implemented in psychosocial education. Previous studies on psychosocial education only implemented machine learning without extensive analysis of the used features.

Research Contribution

Based on the knowledge gaps derived from the previous studies, this paper would advance the body of knowledge about the feature selection method in two primary contributions:

1. This paper provides a systematic model for determining the best feature selection method using an adapted additive ratio assessment model [24]. Specifically, the selection of the feature selection method is implemented in the psychosocial education dataset.
2. This paper offers a comprehensive study and evaluation by comparing the performance of machine learning in every feature selection method. ARAS used the performance metrics from machine learning as criteria in determining the best feature selection method.

3. Methodology

3.1. Theoretical Overview

3.1.1. Artificial Intelligence Research on Psychosocial Education

Nowadays, the research in the education field focuses not only on the academic aspects, such as academic achievement, graduation level, academic grading, and teaching methods, but also on non-academic aspects, such as community relationships [25] and psychosocial. As such, the non-academic aspect also influences the quality of education [26–28].

On the other hand, the flourishing of research in artificial intelligence has made an impressive contribution to the psychosocial education field. Numerous artificial intelligence-based studies have successfully revealed psychosocial phenomena that influence education development. In a study conducted by Navarro [29], artificial intelligence was successfully used to predict the link between the condition of the environment and educators' stress levels. In addition, the research successfully interviewed and provided 4890 data points with 118 features used in predicting the level of stress on the educators. An extensive amount of data and high-dimensional features in the study indicated that psychosocial research is essential and exciting to be carried out.

3.1.2. Feature Selection Methods

Real-world problems are often represented by extensive data collection and high-dimensional features. Occasionally, existing features may not directly relate to the target problems that need to be solved [30,31]. Under such circumstances, the selection of features becomes critical. Selecting the right features makes it possible to improve model performance and efficiency in the computation process [32,33].

Three approaches are available to select features. The first approach, the filtering method, performs a selected subset of features based on the characteristics of the feature itself. The best feature is obtained from the statistical analysis of each feature with other features or target data. Next, the wrapper method uses machine learning to select the best data subsets for analysis. The wrapper method uses machine learning to reconstruct the feature subset and tests it using statistical modeling. The third approach, the embedded method, uses the same principle as the wrapper method, however, in evaluating the feature subset by analyzing the performance of machine learning.

Next, this section will briefly describe eight feature selection methods evaluated in this paper. There are three filtering methods: analyzing variance, mutual information, and chi-square; the exhaustive search feature is a wrapped method; embedding random forest, Lasso, and recursive feature elimination are embedded methods. Those methods would be compared to the models of machine learning using the baseline feature.

3.1.3. ANOVA

ANOVA is a statistical analysis used to calculate the distance of difference (variance) between two clusters [34]. ANOVA uses the *f-ratio* to calculate the magnitude of every feature and target class. Magnitude values above the *f-ratio* will be retained, and others will be discarded. In an ANOVA with class k, the variance among classes is defined as follows [35]:

$$\sigma^2_{v-all} = \frac{\sum(\overline{x}_i - \overline{x})^2 n_i}{(k-1)} \tag{1}$$

where n_i is the value discovered from the calculation on the i-th class, $\overline{x}_i$ is the mean of the i-th class, and $\overline{x}$ is the mean of all classes; the class variance is defined as follows:

$$\sigma^2_{v-class} = \frac{\left(\sum\sum(\overline{x}_{ij} - \overline{x})^2\right) - \left(\sum(\overline{x}_i - \overline{x})^2 n_i\right)}{(R-k)} \tag{2}$$

Then, the *f-ratio* is calculated based on the degree of the two variances:

$$f\text{-}ratio = \frac{\sigma^2_{v-all}}{\sigma^2_{v-class}} \tag{3}$$

3.1.4. Chi-Square

Chi-square is a statistical method that is widely used for calculating the correlation between two variables [36–38]. The implementation of Chi-square as the method to select subset features in machine learning can be done by calculating the dependency level of each feature towards the target data [39,40]. If n is the observed frequency and μ is the expected frequency, then the Chi-square (X^2) for a feature with a number of f and class C is defined as follows:

$$X^2 = \sum_{i=1}^{f} \sum_{j=1}^{C} \frac{(n_{ij} - \mu_{ij})^2}{\mu_{ij}} \tag{4}$$

3.1.5. Mutual Information (MI)

MI is used to calculate the distance of random vectors between clusters [41,42]. Mutual information looks for the similarity value between the distribution of probability $P(X, Y)$ and product of entropy $P(X)P(Y)$ [43]. Mutual information between two random vectors X and Y is defined as follows:

$$MI(X,Y) = \sum_{x \in X} \sum_{y \in Y} P(X = x,\ Y = y) \ln \frac{P(X = x,\ Y = y)}{P(X = x)P(Y = y)} \tag{5}$$

In feature selection problems, the implementation of mutual information was used to calculate the information value of how significant the contribution of a feature is towards the prediction of the target class [44–46]. Mutual information for feature set S and m feature, which have a large dependence on the target class C, is defined as follows:

$$MI(S_m, C_i) = \frac{\log(P(S_m, C_i)}{P(S_m) * P(C_i)} \tag{6}$$

3.1.6. Exhaustive Search Feature (EFS)

In the EFS method, the algorithm performance is obtained by evaluating the existing features in all possible combinations. The feature subset with the highest performance will be selected [47,48]. EFS works by finding the value of validity (P, S), assessing the entire subset of candidate feature S for a whole solution to a problem P. The result is obtained from Output (P, S), in which the entire values of S are suitable for the problem P. The EFS method is a greedy algorithm, as it uses a brute force approach to find the best possible feature subset. Due to its exhaustive nature, ESF usually requires large amounts of resources.

3.1.7. Embedding Random Forest (ERF)

ERF is an ensemble method to reconstruct the average output of an individual tree [49]. The recursive approach is needed to find the best value from the feature subset during the elimination process, especially for highly correlated features [50]. Evaluating the high correlation can be done using the mean decreasing impurity approach. The Gini Index is one of the most popular measures of mean decreasing impurities, and it is defined as follows:

$$Gini = 1 - \sum_{i=1}^{n} (P_i)^2 \tag{7}$$

3.1.8. Lasso

The least absolute shrinkage and selection operator (Lasso) is one of the shrinkage techniques. Lasso selects the variables by minimizing the number of squared errors using penalty regularization [49,51]. Shrinkage regression is carried out towards zero along with the increase in the value of the lambda (λ) parameter used to control the number of shrinkages [52]. Lasso is defined as follows:

$$L_{lasso} = (\hat{\beta}) = \sum_{i=1}^{n} (y_i - x_i' \hat{\beta})^2 + \lambda \sum_{j=1}^{m} |\hat{\beta}_j| \tag{8}$$

3.1.9. Recursive Feature Elimination (RFE)

RFE is a feature selection method that works iteratively to rank features' importance [50]. In minimizing computational resources, some approaches eliminate instead of one by one but based on a subset of features [53]. An analysis and elimination process is performed in each iteration on the feature subsets with low relevance values. Two components of RFE are the number of features and the algorithm used to analyze the performance of the feature subsets. Generally, the iteration procedure of the RFE is performed as follows [54]: (1). Train each feature subset with a classifier, (2). Regarding the ranking of the feature subsets, calculate each feature subset's ranking, (3). Removing the feature subset that has low significance.

3.1.10. ARAS: Decision System Approach for the Feature Evaluation Method

Additive ratio assessment (ARAS) is one of the MCDM modeling techniques. ARAS is a method that relies on the intuitive principle that the best solution must have the largest ratio. Ranking using the ARAS method is performed by comparing the value of each criterion on each alternative by looking at its weight to obtain the ideal alternative [55,56].

The ARAS method utilizes a function value that determines the complexity of feasible alternatives. The ARAS method was directly proportional to the values and weights of the main criteria considered to determine the best alternative. ARAS is based on the argument that complex problems can be understood simply by using relative comparisons. In ARAS, the ratio of the sum of normalized and weighted criteria values describes the possible alternatives to obtaining the optimal alternative rank. The ARAS method compares the utility functions of alternatives with optimal utility function values [57].

Like the classical MCDM approach, ARAS focuses on the ranking of criteria. Ranking with ARAS is done in several stages [55]. The first stage is forming a decision-making matrix. The matrix consists of $0 - m$ alternatives (rows) and $1 - n$ criteria (columns). If i represents the number of alternatives, j is the number of criteria. The decision-making matrix is denoted as follows:

$$X = \begin{bmatrix} x_{01} & \dots & x_{0j} & \dots & x_{0n} \\ \vdots & \ddots & \vdots & \ddots & \vdots \\ x_{i1} & \dots & x_{ij} & \dots & x_{in} \\ \vdots & \ddots & \vdots & \ddots & \vdots \\ x_{m1} & \dots & x_{mj} & \dots & x_{mn} \end{bmatrix}; i = \overline{0,\ m}; j = \overline{1,\ n}; \tag{9}$$

The x_{0j} optimal value of the criterion is the best value that can be used to represent the performance on each j criterion. In this paper, x_{0j} optimal criterion is defined as follows:

$$\begin{aligned} x_{0j} = \max_i x_{ij},\ if \max_i x_{ij}\ is\ benefit; \\ x_{0j} = \min_i x^*_{ij},\ if \min_i x^*_{ij}\ is\ cost; \end{aligned} \tag{10}$$

The next stage is normalizing all the criteria defined from $\overline{x}_{ij}$ of the matrix $\overline{X}$. The normalized decision-making matrix $\overline{X}$ is defined as follows:

$$\overline{X} = \begin{bmatrix} \overline{x}_{01} & \dots & \overline{x}_{0j} & \dots & \overline{x}_{0n} \\ \vdots & \ddots & \vdots & \ddots & \vdots \\ \overline{x}_{i1} & \dots & \overline{x}_{ij} & \dots & \overline{x}_{in} \\ \vdots & \ddots & \vdots & \ddots & \vdots \\ \overline{x}_{m1} & \dots & \overline{x}_{mj} & \dots & \overline{x}_{mn} \end{bmatrix}; i = \overline{0,\ m}; j = \overline{1,\ n}; \tag{11}$$

Normalization of benefits criteria can be done using the following formula:

$$\overline{x}_{ij} = \frac{x_{ij}}{\sum_{i=0}^{m} x_{ij}} \tag{12}$$

Meanwhile, normalization of cost criteria can be done using the normalized two-stage procedure following the notation:

$$\overline{x}_{ij} = \frac{1}{x^*_{ij}};\ \overline{x}_{ij} = \frac{x_{ij}}{\sum_{i=0}^{m} x_{ij}} \tag{13}$$

The next step is defining the normalized-weighted matrix, starting with determining the value of w_j. The sum of weights of all the criteria is 1, and the weight w_j is limited as follows:

$$\sum_{j=1}^{n} w_j = 1 \tag{14}$$

After that, the normalized-weighted matrix is calculated using the following formula:

$$\hat{X} = \begin{bmatrix} \hat{x}_{01} & \dots & \hat{x}_{0j} & \dots & \hat{x}_{0n} \\ \vdots & \ddots & \vdots & \ddots & \vdots \\ \hat{x}_{i1} & \dots & \hat{x}_{ij} & \dots & \hat{x}_{in} \\ \vdots & \ddots & \vdots & \ddots & \vdots \\ \hat{x}_{m1} & \dots & \hat{x}_{mj} & \dots & \hat{x}_{mn} \end{bmatrix}; i = \overline{0, m}; j = \overline{1, n} \quad (15)$$

Normalization using weight w_j for all criteria can be calculated using the following formula:

$$\hat{x}_{ij} = \bar{x}_{ij} * w_j;\ i = \overline{0, m}, \quad (16)$$

where w_j is the weight of criterion j, and $\hat{x}_{ij}$ is the normalized ranking of criterion j. The next step is to calculate the values of the optimality function using the following formula:

$$S_i = \sum_{j=1}^{n} \hat{x}_{ij};\ i = \overline{0, m}, \quad (17)$$

The final step in the ARAS model is to determine the ranking of the alternatives. If S_i and S_0 are optimality criterion values, then the ranking K for alternatives i follows the definition:

$$K_i = \frac{S_i}{S_0}; i = \overline{0, m}, \quad (18)$$

3.2. Experimental Design

In this section, the stages of the proposed methodology will be discussed. Three steps comprise the proposed method: preprocessing, machine learning, and the decision system. The first step is preprocessing the dataset, and the preprocessing stages aim to improve the quality of the data. Furthermore, preprocessing is performed to make the dataset more visible and is considered to improve the machine learning algorithm [58,59]. Data preprocessing concerns cleaning data, transforming categorical data to numerical form, and normalizing data.

After the preprocessing, the next step is the machine learning phase. This phase involves feature selection methods, classification, and performance evaluation. The feature selection method determines the best subset of features from the dataset. In the classification stage, a decision tree classifier is employed to generate the performance of models such as accuracy, precision, recall, f1-score, weighted precision, weighted recall, weighted f1-score, train time, and inference time using the selected feature from the previous stage.

The next stage is the decision system phase. In this step, the performance metrics are compared to determine the rank of the feature selection methods. ARAS uses the performance matrices as the ranking criteria, and this method is essential for formulating the best feature selection methods. The final result, ARAS, presented the feature selection method ranking. The stages of the proposed methodology are depicted in Figure 1.

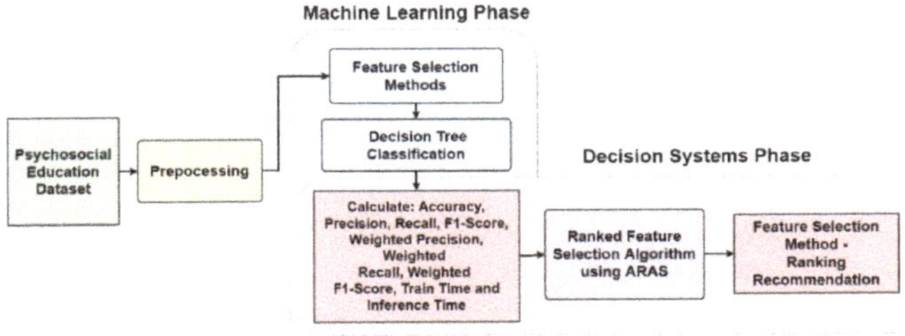

Figure 1. The proposed method of decision model to evaluate feature selection methods.

The proposed model evaluates the feature selection method using two metrics, i.e., model performance and computation performance. Model performance is a measurement of machine learning performance using selected features, and in contrast, computational performance refers to computational capabilities during the training and inference process. Experiments and evaluations are carried out on seven methods and one baseline model, which is a model that uses all features. The schematic detail of the criteria selection of the feature selection method is portrayed in Figure 2.

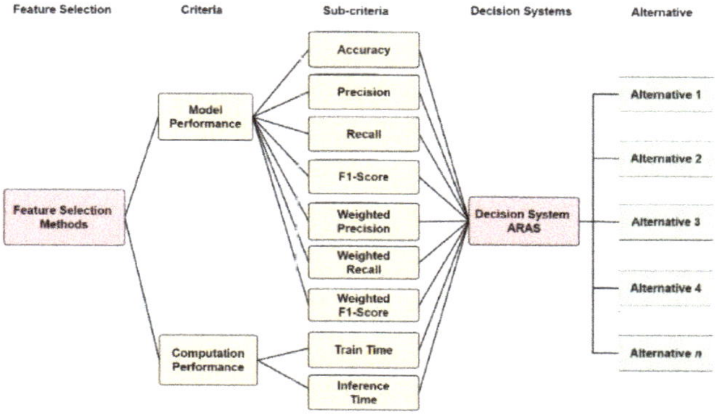

Figure 2. Schematic diagram of decision model to evaluate feature selection methods.

3.3. Dataset Description

The psychosocial education dataset used here refers to the research [29,60] to test the proposed method. It is a public dataset obtained from a psychosocial assessment to identify Colombia's teachers' stress levels. The dataset consists of 4890 instances and 118 features divided into six domains. The complete specification of the dataset can be seen in Table 1.

Table 1. Detailed specification of dataset.

No	Detail	Specification
1	Number of Features	118
2	Number of Classes	4
3	Number of Instances	4890
4	Classes Name	Low, Medium, High, Very High
5	Features Domain	Sociodemographic (S), Demands of the Job (D), Control over Work (C), Leadership and Social Relations at Work (L), Rewards (R)

3.4. Dataset Preprocessing

In a machine learning problem, the dataset is present to demonstrate the effectiveness of the proposed method. Therefore, a high-quality dataset is required to evaluate the proposed model against the existing model. Data prepossessing is a well-known technique to improve dataset quality.

The teacher's psychosocial risk level dataset is valuable and pristine, and it provides the basis for delivering research on the degree of psychosocial distress among teachers in Columbia. Several studies have been conducted using the same dataset [29,60]. Primarily, the dataset was preprocessed appropriately. It will still be necessary for us to perform several preprocessing steps to prepare a suitable dataset for the proposed methods.

The first step involves performing common preprocessing steps, such as clearing improper data and handling missing values. Then, we divided the data into two subsets by following the Pareto distribution rule [61]. In this case, 80% of the data was used for training, and 20% was used for testing. A randomly selected distribution is made to ensure fair data distribution.

The next step is to apply standardization to rescale the distribution of each dataset subset. By performing a standardization transformation of the dataset, each feature dataset will have a mean value of 0 with a standard deviation value of 1. Hopefully, a preprocessed dataset will lead machine learning to the optimal model.

3.5. Evaluation of Performance Metrics for Feature Selection Methods

Evaluation is done to measure machine learning performance. Generally, machine learning performance is measured by using a confusion matrix. A confusion matrix combines actual value and predicted value in the classifier. The confusion matrix is True Positive (TP), False Positive (FP), True Negative (TN), and False Negative (FN). The matrices for measuring accuracy, precision, recall, and f1-Score are obtained from the following calculation:

$$Accuracy = \frac{TP + TN}{TP + TN + FP + FN} \tag{19}$$

$$Precision = \frac{TP}{TP + FP} \tag{20}$$

$$Recall = \frac{TP}{TP + FN} \tag{21}$$

$$F1 = 2\,\frac{Precision \times Recall}{Precision + Recall} \tag{22}$$

The fundamental concept for calculating the confusion matrix is binary classification [62]. A single comparison is made between two classes in binary classification, while this single comparison becomes irrelevant in multi-class classification [63]. Each class's precision, recall, and f1-score are estimated as micro-averaged and macro-averaged. After that, the metrics are calculated using the one vs. all method [64]. For example, the micro-averaged scores and macro-averaged precision (PRE) scores in the k-class are defined as follows [65]:

$$PRE_{micro} = \frac{TP_1 + \cdots + TP_k}{TP_1 + \cdots + TP_k + FP_1 + \cdots + FP_k} \tag{23}$$

$$PRE_{macro} = \frac{PRE_1 + \cdots + PRE_k}{k} \tag{24}$$

4. Results and Discussion

This section reviews the performance evaluation of the proposed method. We actualized the discussion in two parts: the performance of each feature selection method on the psychosocial education dataset and the implementation of ARAS in selecting the best feature selection method. Analysis and evaluation will also be conducted by comparing performance against a single criterion. In this case, accuracy criteria are used as a comparison.

4.1. Performance Analysis of the Feature Selection Method

This section discusses the performance measures of the feature selection method. The feature selection reduces the dimension by eliminating the least important features and retaining the important ones. It is expected that, by reducing the dimensions, the model and computational performance will increase. If the baseline used all 118 features, the other methods only performed the subset features according to the algorithm. Table 2 shows the selected feature for each method.

Table 2. Selected features in each method.

No	Models	Σ Feature	Features Selected
1	**Baseline** [29,60]	118	All Features
2	**ANOVA** [35]	11	S2, S3, S4, S5, S7, S8, S10, D3, D28, L1, R8
3	**Chi-Square** [39,40]	11	S2, S4, S5, S10, D3, D6, D28, C16, L1, R5, R8
4	**MI** [45,46]	11	S2, S3, S7, S8, S10, D9, D27, D37, C21, L1, L4
5	**EFS** [47,48]	24	S1, S2, S3, D2, D5, D8, D14, D17, D19, D25, D27, D31, D34, D38, C2, C9, L3, L9, L10, L12, L30, R3, R4, M1
6	**ERF** [50]	9	S2, S3, S4, S5, S7, S8, S10, D3, L1
7	**Lasso** [52]	10	S1, S3, S4, S5, S7, S8, S10, M1, M2, M3
8	**RFE** [54]	11	S2, S3, S8, D2, D18, D21, D26, D28, D38, L17, R5

The performance measure of the feature selection method is carried out to obtain performance parameters that will be used as the criteria for ARAS in the future. The measurement consists of performance models such as accuracy, precision, recall, f1-score, weighted precision, weighted recall, and weighted f1-score. The computation performance consists of train time and inference time. From a series of experiments conducted, what is interesting is that the baseline model requires the longest training time (34.3910 s) compared to other feature selection methods. It is decent because the baseline model used all the features in the psychosocial education datasets. However, the baseline model produced lower results than other models with far fewer selection features in the accuracy metric. Details of performance matrices for each feature selection method can be seen in Table 3.

Table 3. The performance metrics of feature selection methods.

Model	Accuracy	Precision	Recall	F1-Score	Weighted Prec.	Weighted Recall	Weighted F1-Score	Train Time (s)	Inference Time (s)
Baseline	0.9725	0.9581	0.9427	0.9501	0.9721	0.9725	0.9722	34.3910	0.9917
ANOVA	0.9734	0.9585	0.9453	0.9517	0.9730	0.9734	0.9731	3.9885	0.9933
Chi-Square	0.9752	0.9614	0.9484	0.9547	0.9749	0.9752	0.9750	3.9892	0.9950
MI	0.9757	0.9647	0.9479	0.9569	0.9754	0.9757	0.9753	3.9893	0.9636
EFS	0.9265	0.9324	0.9257	0.9267	0.9344	0.9265	0.9273	8.4677	0.9770
ERF	0.9770	0.9719	0.9478	0.9591	0.9769	0.9770	0.9766	2.0112	0.9823
Lasso	0.9770	0.9684	0.9513	0.9597	0.9768	0.9770	0.9768	2.9948	0.9646
RFE	0.9706	0.9537	0.9400	0.9470	0.9703	0.9706	0.9704	3.9878	0.9646

4.2. Evaluation Feature Selection Method Using ARAS

At this stage, choosing the best feature selection method is performed. ARAS determines the ranking using performance metrics from each feature selection method. The first step is to initialize the decision-making matrices for each alternative and their respective criteria pairs. By assigning each feature selection method as an alternative, and assignment performance matrices, i.e., accuracy (A), precision (P), recall (R), f1-score (FS), weighted precision (WP), weighted recall (WR), weighted f1-score (WFS), train time (TT), and inference time (IT) as criteria x_n. Based on the analysis, it is determined that the value of the criteria x_1–x_7 are the benefit, while x_8 and x_9 are as the cost. In addition, it is also determined that the weighted value (w) of criteria x_1 is 0.2 and criteria x_2–x_9 is 0.1, with the sum of their weighted values of 1. Criteria x_1 gets a higher weight because, in real problems, accuracy is one of the most important performance matrices that is widely used as a benchmark for machine learning [66,67]. The initial decision-making matrix's complete formation with each criterion's weight and optimization is shown in Table 4.

After the initial decision matrix is completed, the next step is to normalize the decision matrix. The step is finding the optimal value of A_0 value. The max operator is used for criteria with the benefit value, and the min operator is used for criteria with the cost value using Equation (10):

Table 4. Initial decision-making matrix X.

Alternative	Criteria								
	A	P	R	FS	WP	WR	WFS	TT	IT
	x_1	x_2	x_3	x_4	x_5	x_6	x_7	x_8	x_9
Optimization	Benefit	Benefit	Benefit	Benefit	Benefit	Benefit	Benefit	Cost	Cost
Weight (w)	0.2	0.1	0.1	0.1	0.1	0.1	0.1	0.1	0.1
Baseline	0.9725	0.9581	0.9427	0.9501	0.9721	0.9725	0.9722	34.3910	0.9917
ANOVA	0.9734	0.9585	0.9453	0.9517	0.9730	0.9734	0.9731	3.9885	0.9933
Chi-Square	0.9752	0.9614	0.9484	0.9547	0.9749	0.9752	0.9750	3.9892	0.9950
MI	0.9757	0.9647	0.9479	0.9569	0.9754	0.9757	0.9753	3.9893	0.9636
EFS	0.9265	0.9324	0.9257	0.9267	0.9344	0.9265	0.9273	8.4677	0.9770
ERF	0.9770	0.9719	0.9478	0.9591	0.9769	0.9770	0.9766	2.0112	0.9823
Lasso	0.9770	0.9684	0.9513	0.9597	0.9768	0.9770	0.9768	2.9948	0.9646
RFE	0.9706	0.9537	0.9400	0.9470	0.9703	0.9706	0.9704	3.9878	0.9646

After obtaining the value A_{0_j}, all of the criteria in the matrix are normalized. The decision matrix is normalized using Equation (12) for benefit and Equation (13) for cost. The formation of the normalization of the decision matrix $\overline{X}$ is shown in detail in Table 5, and for example of calculating the values of $\overline{x}_{1\ (A_0)}$ and $\overline{x}_{1\ (Baseline)}$ are as follows:

$$\overline{x}_{1\ (A_0)} = \frac{0.9770}{0.9770+0.9725+0.9734+0.9752+0.9757+0.9265+0.9770+0.9770+0.9706}$$

$$\overline{x}_{1\ (A_0)} = 0.1120$$

$$\overline{x}_{1\ (Baseline)} = \frac{0.9725}{0.9770+0.9725+0.9734+0.9752+0.9757+0.9265+0.9770+0.9770+0.9706}$$

$$\overline{x}_{1\ (Baseline)} = 0.1115$$

$$x_{A0j} = \left\{ \begin{array}{c} [\max(0.9725,\ 0.9734,\ 0.9752,\ 0.9757,\ 0.9265,\ 0.9770,\ 0.9770,\ 0.9706)], \\ [\max(0.9581,\ 0.9585,\ 0.9614,\ 0.9647,\ 0.9324,\ 0.9719,\ 0.9684,\ 0.9537)], \\ [\max(0.9427,\ 0.9453,\ 0.9484,\ 0.9479,\ 0.9257,\ 0.9478,\ 0.9513,\ 0.9400)], \\ [\max(0.9501,\ 0.9517,\ 0.9547,\ 0.9569,\ 0.9267,\ 0.9591,\ 0.9597,\ 0.9400)], \\ [\max(0.9721,\ 0.9730,\ 0.9749,\ 0.9754,\ 0.9344,\ 0.9769,\ 0.9768,\ 0.9703)], \\ [\max(0.9725,\ 0.9734,\ 0.9752,\ 0.9757,\ 0.9265,\ 0.9770,\ 0.9770,\ 0.9706)], \\ [\max(0.9722,\ 0.9731,\ 0.9750,\ 0.9753,\ 0.9273,\ 0.9766,\ 0.9768,\ 0.9704)], \\ [\min(35.942,\ 3.9885,\ 3.9892,\ 3.9893,\ 8.4677,\ 2.0112,\ 2.9948,\ 3.9878)], \\ [\min(0.9668,\ 0.9933,\ 0.9950,\ 0.9636,\ 0.9770,\ 0.9823,\ 0.9646,\ 0.9646)] \end{array} \right\}$$

$$x_{A0j} = \{0.9770,\ 0.9719,\ 0.9513,\ 0.9597,\ 0.9770,\ 0.9770,\ 0.9766,\ 2.0112,\ 0.9646\}$$

After the normalization of the matrix $\overline{X}$ is obtained, the next step is to perform the weighted normalization by multiplying the criteria weight by the normalized weighted matrix according to the formula (16). The results of weighted normalization are presented in detail in Table 6.

Table 5. Normalized decision-making matrix $\overline{X}$ on each criterion.

Alternative	Criteria								
	$\bar{x}_1$	$\bar{x}_2$	$\bar{x}_3$	$\bar{x}_4$	$\bar{x}_5$	$\bar{x}_6$	$\bar{x}_7$	$\bar{x}_8$	$\bar{x}_9$
Weight (w)	0.2	0.1	0.1	0.1	0.1	0.1	0.1	0.1	0.1
A_0	0.1120	0.1125	0.1119	0.1120	0.1119	0.1120	0.1120	0.2007	0.1124
Baseline	0.1115	0.1109	0.1109	0.1109	0.1113	0.1115	0.1114	0.0112	0.1120
ANOVA	0.1116	0.1109	0.1112	0.1111	0.1114	0.1116	0.1116	0.1012	0.1090
Chi-Square	0.1118	0.1113	0.1116	0.1115	0.1117	0.1118	0.1118	0.1012	0.1088
MI	0.1118	0.1116	0.1115	0.1117	0.1117	0.1118	0.1118	0.1012	0.1124
EFS	0.1062	0.1079	0.1089	0.1082	0.1070	0.1062	0.1063	0.0477	0.1108
ERF	0.1120	0.1125	0.1115	0.1120	0.1119	0.1120	0.1120	0.2007	0.1102
Lasso	0.1120	0.1121	0.1119	0.1120	0.1119	0.1120	0.1120	0.1348	0.1122
RFE	0.1112	0.1104	0.1106	0.1106	0.1111	0.1112	0.1112	0.1012	0.1122

Table 6. Normalization-weighted decision-making matrix $\hat{X}$ of each criterion.

Alternative	Criteria								
	$\hat{x}_1$	$\hat{x}_2$	$\hat{x}_3$	$\hat{x}_4$	$\hat{x}_5$	$\hat{x}_6$	$\hat{x}_7$	$\hat{x}_8$	$\hat{x}_9$
A_0	0.0224	0.0113	0.0112	0.0112	0.0112	0.0112	0.0112	0.0201	0.0112
Baseline	0.0223	0.0111	0.0111	0.0111	0.0111	0.0112	0.0111	0.0011	0.0112
ANOVA	0.0223	0.0111	0.0111	0.0111	0.0111	0.0112	0.0112	0.0101	0.0109
Chi-Square	0.0224	0.0111	0.0112	0.0112	0.0112	0.0112	0.0112	0.0101	0.0109
MI	0.0224	0.0112	0.0112	0.0112	0.0112	0.0112	0.0112	0.0101	0.0112
EFS	0.0212	0.0108	0.0109	0.0108	0.0107	0.0106	0.0106	0.0048	0.0111
ERF	0.0224	0.0113	0.0112	0.0112	0.0112	0.0112	0.0112	0.0201	0.0110
Lasso	0.0224	0.0112	0.0112	0.0112	0.0112	0.0112	0.0112	0.0135	0.0112
RFE	0.0223	0.0110	0.0111	0.0111	0.0111	0.0111	0.0111	0.0101	0.0112

Next, the optimal value S_i is calculated, where S_i is the value of the ideal function of alternative i. After that, the criteria K_i is ranked using Equations (17) and (18). Meanwhile, the value K_i is calculated by dividing the value S_i by the value S_0. For the values S_0, $K_{Baseline}$ and K_{ANOVA} can be computed as follows:

$$S_0 = 0.0224 + 0.0113 + 0.0112 + 0.0112 + 0.0112 + 0.0112 + 0.0112 + 0.0201 + 0.0112$$

$$S_0 = 0.1210$$

$$K_{(Baseline)} = \frac{0.1013}{0.1210}\ ; \quad K_{(Baseline)} = 0.8372$$

$$K_{(ANOVA)} = \frac{0.1101}{0.1210}\ ; \quad K_{(ANOVA)} = 0.9099$$

In detail, the calculation of the optimal value K is presented in Table 7. Then, based on the results of the K_i, the final results of the rankings are shown in Table 8.

Table 7. The result of optimality value on the feature selection methods.

Alternative	i	S	K	Rank
A_0	0	0.1210	-	-
Baseline	1	0.1013	0.8372	8
ANOVA	2	0.1101	0.9099	5
Chi-Square	3	0.1105	0.9132	4
MI	4	0.1109	0.9165	3
EFS	5	0.1015	0.8388	7
ERF	6	0.1208	0.9983	1
Lasso	7	0.1143	0.9446	2

Table 8. Final results of the ARAS rank for feature selection methods.

Model	Rank
ERF	1
Lasso	2
MI	3
Chi-Square	4
ANOVA	5
RFE	6
EFS	7
Baseline	8

The decision results using ARAS show that the ERF method is top-ranking, and the baseline is at the lowest rank. ERF with 11 features gives better results than the baseline, which uses 118 features. It shows that selecting the best subset features is still relevant to machine learning problems.

We compare the ARAS rank with single machine learning measurements such as accuracy. In that case, the results obtained tend to be the same while on ARAS: ERF > Lasso > MI > Chi-square > Anova > RFE > EFS > Baseline, and on the other hand, using accuracy, the performance order is obtained as follows: ERF > Lasso > MI > Chi-square > Anova > Baseline > RFE > EFS. It happens because the overall performance produced by feature selection methods is mostly stable, so there are no models with cross-dominating criteria. To consider the dominating performance result, Figure 3 shows the comparative performance of every model.

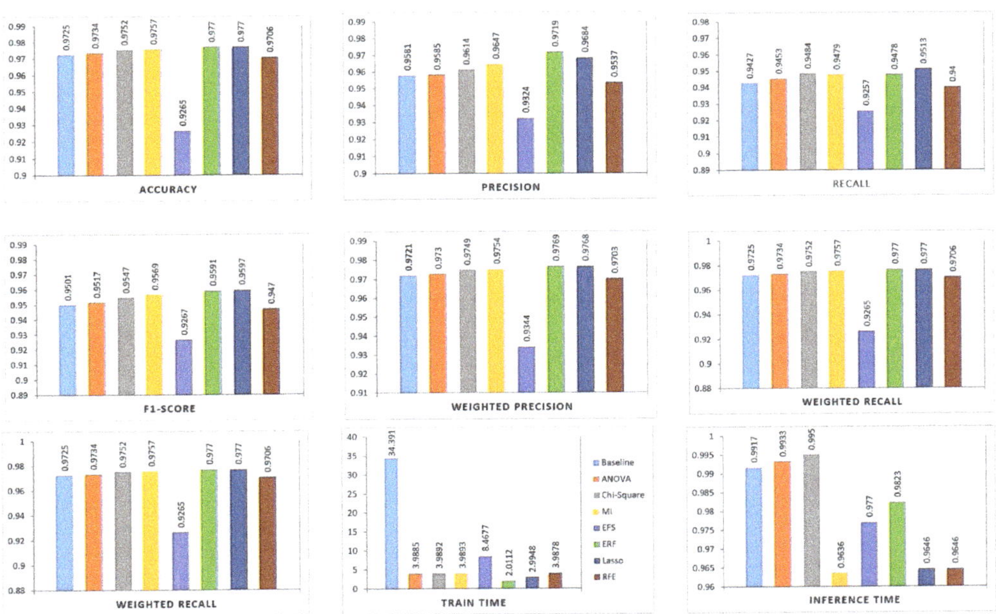

Figure 3. Performance comparison of feature selection methods.

The experiment shows that the machine learning phase accomplished the model's performance analysis. By selecting specific metrics, the aim of the performance of machine learning can be defined. For example, the accuracy metric can be used as a benchmark metric to find the best accuracy model. Nevertheless, a decision model to measure and evaluate the overall performance metrics of feature selection methods is still necessary.

Finally, the goal of the proposed method is to show that the proposed model can resolve the problem formulation. Theoretically, this methodology is relevant and should be proposed. ARAS can perform a fair mapping in ranking the feature selection methods in the psychosocial education domain, especially to identify Colombia's teachers' stress level problems. However, this methodology has not fully demonstrated the significance of performance evaluation in the current dataset case, where several dominant criteria ultimately dictate the ranking results. More experience is necessary to provide a robust comparison and conclusion, and more experience based on a similar dataset might provide better results.

5. Conclusions

ARAS has proven effective and can be implemented as an evaluation model to determine the best feature selection method in the psychosocial education dataset. The evaluation used performance matrices to rank the feature selection methods. From the evaluation that has been accomplished, the determination of weight and optimization value plays an essential role in the ARAS model. Giving subjective weights affects the overall ARAS ranking.

Regarding future research directions, we recommend further investigation on the proposed method on different datasets with conditions where each criterion contradicts and does not predominate the other. The problem associated with imbalanced datasets that show uneven and contradictory performance matrices can be challenging. This problem is expected to measure the extent of ARAS's ability to provide an optimal ranking.

Author Contributions: Conceptualization, F.M. and J.-T.W.; Methodology, J.-T.W. and F.M.; Software, M.M.; Visualization, F.M. and M.M.; Project administration, J.-S.L. and J.-T.W.; Supervision, J.-S.L. and J.-T.W.; Writing—original draft, F.M. and M.M.; Writing—review and editing, J.-T.W. and J.-S.L. All authors have read and agreed to the published version of the manuscript.

Funding: This work was supported by the Ministry of Science and Technology, Taiwan, under Grant MOST-110-2637-E-011-003-.

Institutional Review Board Statement: Not applicable.

Informed Consent Statement: Not applicable.

Data Availability Statement: Not applicable.

Acknowledgments: The authors gratefully acknowledge the support extended by the Ministry of Science and Technology, Taiwan, under Grant MOST-110-2637-E-011-003-.

Conflicts of Interest: The authors declare no conflict of interest.

References

1. Hoti, A.H.; Heinzmann, S.; Müller, M.; Buholzer, A. Psychosocial Adaptation and School Success of Italian, Portuguese and Albanian Students in Switzerland: Disentangling Migration Background, Acculturation and the School Context. *J. Int. Migr. Integr.* **2015**, *18*, 85–106. [CrossRef]
2. Wong, R.S.M.; Ho, F.; Wong, W.H.S.; Tung, K.T.S.; Chow, C.B.; Rao, N.; Chan, K.L.; Ip, P. Parental Involvement in Primary School Education: Its Relationship with Children's Academic Performance and Psychosocial Competence through Engaging Children with School. *J. Child Fam. Stud.* **2018**, *27*, 1544–1555. [CrossRef]
3. Raskind, I.G.; Haardörfer, R.; Berg, C.J. Food insecurity, psychosocial health and academic performance among college and university students in Georgia, USA. *Public Health Nutr.* **2019**, *22*, 476–485. [CrossRef]
4. Sierra-Díaz, M.J.; González-Víllora, S.; Pastor-Vicedo, J.C.; Sánchez, G.F.L. Can We Motivate Students to Practice Physical Activities and Sports Through Models-Based Practice? A Systematic Review and Meta-Analysis of Psychosocial Factors Related to Physical Education. *Front. Psychol.* **2019**, *10*, 2115. [CrossRef]
5. Souravlas, S.; Anastasiadou, S. Pipelined Dynamic Scheduling of Big Data Streams. *Appl. Sci.* **2020**, *10*, 4796. [CrossRef]
6. López-Belmonte, J.; Segura-Robles, A.; Moreno-Guerrero, A.-J.; Parra-González, M.E. Machine Learning and Big Data in the Impact Literature. A Bibliometric Review with Scientific Mapping in Web of Science. *Symmetry* **2020**, *12*, 495. [CrossRef]
7. Al-Jarrah, O.Y.; Yoo, P.; Muhaidat, S.; Karagiannidis, G.K.; Taha, K. Efficient Machine Learning for Big Data: A Review. *Big Data Res.* **2015**, *2*, 87–93. [CrossRef]
8. Altman, N.; Krzywinski, M. The curse(s) of dimensionality. *Nat. Methods* **2018**, *15*, 399–400. [CrossRef]

9. Köppen, M. The curse of dimensionality. In Proceedings of the 5th Online World Conference on Soft Computing in Industrial Applications (WSC5), Online, 4–8 September 2000; Volume 1, pp. 4–8.
10. Khaire, U.M.; Dhanalakshmi, R. Stability of feature selection algorithm: A review. *J. King Saud Univ.Comput. Inf. Sci.* **2019**, *34*. [CrossRef]
11. Jović, A.; Brkić, K.; Bogunović, N. A review of feature selection methods with applications. In Proceedings of the 2015 38th International Convention on Information and Communication Technology, Electronics and Microelectronics (MIPRO), Opatija, Croatia, 25–29 May 2015; pp. 1200–1205.
12. Chandrashekar, G.; Sahin, F. A survey on feature selection methods. *Comput. Electr. Eng.* **2014**, *40*, 16–28. [CrossRef]
13. Cai, J.; Luo, J.; Wang, S.; Yang, S. Feature selection in machine learning: A new perspective. *Neurocomputing* **2018**, *300*, 70–79. [CrossRef]
14. Moorthy, U.; Gandhi, U.D. A novel optimal feature selection technique for medical data classification using ANOVA based whale optimization. *J. Ambient. Intell. Humaniz. Comput.* **2020**, *12*, 3527–3538. [CrossRef]
15. Ding, H.; Li, D. Identification of mitochondrial proteins of malaria parasite using analysis of variance. *Amino Acids* **2015**, *47*, 329–333. [CrossRef]
16. Utama, H. Sentiment analysis in airline tweets using mutual information for feature selection. In Proceedings of the 2019 4th International Conference on Information Technology, Information Systems and Electrical Engineering (ICITISEE), Yogyakarta, Indonesia, 20–21 November 2019; pp. 295–300.
17. Richhariya, B.; Tanveer, M.; Rashid, A. Diagnosis of Alzheimer's disease using universum support vector machine based recursive feature elimination (USVM-RFE). *Biomed. Signal Process. Control.* **2020**, *59*, 101903. [CrossRef]
18. Park, D.; Lee, M.; Park, S.E.; Seong, J.-K.; Youn, I. Determination of Optimal Heart Rate Variability Features Based on SVM-Recursive Feature Elimination for Cumulative Stress Monitoring Using ECG Sensor. *Sensors* **2018**, *18*, 2387. [CrossRef]
19. ZLiu, Z.; Thapa, N.; Shaver, A.; Roy, K.; Siddula, M.; Yuan, X.; Yu, A. Using Embedded Feature Selection and CNN for Classification on CCD-INID-V1—A New IoT Dataset. *Sensors* **2021**, *21*, 4834.
20. Loscalzo, S.; Wright, R.; Acunto, K.; Yu, L. Sample aware embedded feature selection for reinforcement learning. In Proceedings of the 14th Annual Conference on Genetic and Evolutionary Computation, Philadelphia, PA, USA, 7–11 July 2012; pp. 887–894.
21. Liu, H.; Zhou, M.; Liu, Q. An embedded feature selection method for imbalanced data classification. *IEEE/CAA J. Autom. Sin.* **2019**, *6*, 703–715. [CrossRef]
22. Kou, G.; Yang, P.; Peng, Y.; Xiao, F.; Chen, Y.; Alsaadi, F.E. Evaluation of feature selection methods for text classification with small datasets using multiple criteria decision-making methods. *Appl. Soft Comput.* **2020**, *86*, 10583. [CrossRef]
23. Hashemi, A.; Dowlatshahi, M.B.; Nezamabadi-Pour, H. Ensemble of feature selection algorithms: A multi-criteria decision-making approach. *Int. J. Mach. Learn. Cybern.* **2021**, 1–21. [CrossRef]
24. Singh, R.; Kumar, H.; Singla, R.K. TOPSIS based multi-criteria decision making of feature selection techniques for network traffic dataset. *Int. J. Eng. Technol.* **2014**, *5*, 4598–4604.
25. Souravlas, S.; Anastasiadou, S.; Katsavounis, S. A Survey on the Recent Advances of Deep Community Detection. *Appl. Sci.* **2021**, *11*, 7179. [CrossRef]
26. Acosta, D.; Fujii, Y.; Joyce-Beaulieu, D.; Jacobs, K.D.; Maurelli, A.T.; Nelson, E.J.; McKune, S.L. Psychosocial Health of K-12 Students Engaged in Emergency Remote Education and In-Person Schooling: A Cross-Sectional Study. *Int. J. Environ. Res. Public Health* **2021**, *18*, 8564. [CrossRef]
27. Carreon, A.D.V.; Manansala, M.M. Addressing the psychosocial needs of students attending online classes during this COVID-19 pandemic. *J. Public Health* **2021**, *43*, e385–e386. [CrossRef]
28. Mahapatra, A.; Sharma, P. Education in times of COVID-19 pandemic: Academic stress and its psychosocial impact on children and adolescents in India. *Int. J. Soc. Psychiatry* **2021**, *67*, 397–399. [CrossRef]
29. Navarro, R.M.; Castrillón, O.D.; Osorio, L.P.; Oliveira, T.; Novais, P.; Valencia, J.F. Improving classification based on physical surface tension-neural net for the prediction of psychosocial-risk level in public school teachers. *PeerJ. Comput. Sci.* **2021**, *7*, e511. [CrossRef]
30. Kira, K.; Rendell, L.A. A practical approach to feature selection. In *Machine Learning Proceedings 1992*; Sleeman, D., Edwards, P., Eds.; Morgan Kaufmann: San Francisco, CA, USA, 1992; pp. 249–256.
31. Dash, M.; Liu, H. Feature selection for classification. *Intell. Data Anal.* **1997**, *1*, 131–156. [CrossRef]
32. Urbanowicz, R.J.; Meeker, M.; La Cava, W.; Olson, R.S.; Moore, J.H. Relief-based feature selection: Introduction and review. *J. Biomed. Inform.* **2018**, *85*, 189–203. [CrossRef] [PubMed]
33. Bommert, A.; Sun, X.; Bischl, B.; Rahnenführer, J.; Lang, M. Benchmark for filter methods for feature selection in high-dimensional classification data. *Comput. Stat. Data Anal.* **2020**, *143*, 106839. [CrossRef]
34. Ashik, M.; Jyothish, A.; Anandaram, S.; Vinod, P.; Mercaldo, F.; Martinelli, F.; Santone, A. Detection of Malicious Software by Analyzing Distinct Artifacts Using Machine Learning and Deep Learning Algorithms. *Electronics* **2021**, *10*, 1694. [CrossRef]
35. Johnson, K.J.; Synovec, E.R. Pattern recognition of jet fuels: Comprehensive GC×GC with ANOVA-based feature selection and principal component analysis. *Chemom. Intell. Lab. Syst.* **2002**, *60*, 225–237. [CrossRef]
36. Vora, S.; Yang, H. A comprehensive study of eleven feature selection algorithms and their impact on text classification. In Proceedings of the 2017 Computing Conference, London, UK, 18–20 July 2017; pp. 440–449.

37. Ghosh, M.; Sanyal, G. Performance Assessment of Multiple Classifiers Based on Ensemble Feature Selection Scheme for Sentiment Analysis. *Appl. Comput. Intell. Soft Comput.* **2018**, *2018*, 8909357. [CrossRef]
38. Alazab, M. Automated Malware Detection in Mobile App Stores Based on Robust Feature Generation. *Electronics* **2020**, *9*, 435. [CrossRef]
39. Cilia, N.D.; De Stefano, C.; Fontanella, F.; di Freca, A.S. A ranking-based feature selection approach for handwritten character recognition. *Pattern Recognit. Lett.* **2019**, *121*, 77–86. [CrossRef]
40. Bahassine, S.; Madani, A.; Al-Sarem, M.; Kissi, M. Feature selection using an improved Chi-square for Arabic text classification. *J. King Saud Univ. Comput. Inf. Sci.* **2020**, *32*, 225–231. [CrossRef]
41. Thejas, G.S.; Joshi, S.R.; Iyengar, S.S.; Sunitha, N.R.; Badrinath, P. Mini-Batch Normalized Mutual Information: A Hybrid Feature Selection Method. *IEEE Access* **2019**, *7*, 116875–116885. [CrossRef]
42. Macedo, F.; Oliveira, M.R.; Pacheco, A.; Valadas, R. Theoretical foundations of forward feature selection methods based on mutual information. *Neurocomputing* **2019**, *325*, 67–89. [CrossRef]
43. Guyon, I.; Elisseeff, A. An Introduction to Variable and Feature Selection. *J. Mach. Learn. Res.* **2003**, *3*, 1157–1182.
44. Peng, H.; Long, F.; Ding, C. Feature selection based on mutual information criteria of max-dependency, max-relevance, and min-redundancy. *IEEE Trans. Pattern Anal. Mach. Intell.* **2005**, *27*, 1226–1238. [CrossRef] [PubMed]
45. Gonzalez-Lopez, J.; Ventura, S.; Cano, A. Distributed multi-label feature selection using individual mutual information measures. *Knowl.-Based Syst.* **2020**, *188*, 105052. [CrossRef]
46. Zhou, H.; Zhang, Y.; Zhang, Y.; Liu, H. Feature selection based on conditional mutual information: Minimum conditional relevance and minimum conditional redundancy. *Appl. Intell.* **2019**, *49*, 883–896. [CrossRef]
47. Ruggieri, S. Complete Search for Feature Selection in Decision Trees. *J. Mach. Learn. Res.* **2019**, *20*, 1–34.
48. Igarashi, Y.; Ichikawa, H.; Nakanishi-Ohno, Y.; Takenaka, H.; Kawabata, D.; Eifuku, S.; Tamura, R.; Nagata, K.; Okada, M. ES-DoS: Exhaustive search and density-of-states estimation as a general framework for sparse variable selection. *J. Phys. Conf. Ser.* **2018**, *1036*, 012001. [CrossRef]
49. Lee, C.-Y.; Chen, B.-S. Mutually-exclusive-and-collectively-exhaustive feature selection scheme. *Appl. Soft Comput.* **2018**, *68*, 961–971. [CrossRef]
50. Granitto, P.; Furlanello, C.; Biasioli, F.; Gasperi, F. Recursive feature elimination with random forest for PTR-MS analysis of agroindustrial products. *Chemom. Intell. Lab. Syst.* **2006**, *83*, 83–90. [CrossRef]
51. Tibshirani, R. Regression Shrinkage and Selection Via the Lasso. *J. R. Stat. Soc. Ser. B Methodol.* **1996**, *58*, 267–288. [CrossRef]
52. Hesterberg, T.; Choi, N.H.; Meier, L.; Fraley, C. Least angle and $\ell 1$ penalized regression: A review. *Stat. Surv.* **2008**, *2*, 61–93. [CrossRef]
53. Abdulsalam, S.O.; Mohammed, A.A.; Ajao, J.F.; Babatunde, R.S.; Ogundokun, R.O.; Nnodim, C.T.; Arowolo, M.O. Performance Evaluation of ANOVA and RFE Algorithms for Classifying Microarray Dataset Using SVM. *Lect. Notes Bus. Inf. Process.* **2020**, 480–492. [CrossRef]
54. Guyon, I.; Weston, J.; Barnhill, S.; Vapnik, V. Gene Selection for Cancer Classification using Support Vector Machines. *Mach. Learn.* **2002**, *46*, 389–422. [CrossRef]
55. Zavadskas, E.K.; Turskis, Z. A new additive ratio assessment (ARAS) method in multi-criteria decision-making. *Technol. Econ. Dev. Econ.* **2010**, *16*, 159–172. [CrossRef]
56. Radović, D.; Stević, Ž.; Pamučar, D.; Zavadskas, E.K.; Badi, I.; Antuchevičiene, J.; Turskis, Z. Measuring Performance in Transportation Companies in Developing Countries: A Novel Rough ARAS Model. *Symmetry* **2018**, *10*, 434. [CrossRef]
57. Maulana, C.; Hendrawan, A.; Pinem, A.P.R. Pemodelan Penentuan Kredit Simpan Pinjam Menggunakan Metode Additive Ratio Assessment (Aras). *J. Pengemb. Rekayasa Teknol.* **2019**, *15*, 7–11. [CrossRef]
58. García, S.; Luengo, J.; Herrera, F. Data preparation basic models. In *Data Preprocessing in Data Mining*; International Publishing; Springer: Cham, Switzerland, 2015; pp. 39–57.
59. Kotsiantis, S.B.; Kanellopoulos, D.; Pintelas, P.E. Data preprocessing for supervised leaning. *Int. J. Comput. Sci.* **2006**, *1*, 111–117.
60. Mosquera, R.; Castrillón, O.D.; Parra, L. Prediction of Psychosocial Risks in Colombian Teachers of Public Schools using Machine Learning Techniques. *Inf. Tecnol.* **2018**, *29*, 267–280. [CrossRef]
61. Newman, M.E.J. Power laws, Pareto distributions and Zipf's law. *Contemp. Phys.* **2005**, *46*, 323–351. [CrossRef]
62. Luque, A.; Carrasco, A.; Martín, A.; de las Heras, A. The impact of class imbalance in classification performance metrics based on the binary confusion matrix. *Pattern Recognit.* **2019**, *91*, 216–231. [CrossRef]
63. Takahashi, K.; Yamamoto, K.; Kuchiba, A.; Koyama, T. Confidence interval for micro-averaged F1 and macro-averaged F1 scores. *Appl. Intell.* **2021**, 1–12. [CrossRef]
64. Pillai, I.; Fumera, G.; Roli, F. F-measure optimisation in multi-label classifiers. In Proceedings of the 21st International Conference on Pattern Recognition (ICPR2012), Tsukuba, Japan, 11–15 November 2012; pp. 2424–2427.
65. Van Asch, V. Macro- and Micro-Averaged Evaluation Measures. 2013, pp. 1–27. Available online: https://www.semanticscholar.org/paper/Macro-and-micro-averaged-evaluation-measures-%5B-%5B-%5D-Asch/1d106a2730801b6210a67f7622e4d192bb309303 (accessed on 14 November 2021).

66. Sokolova, M.; Japkowicz, N.; Szpakowicz, S. Beyond accuracy, F-score and ROC: A family of discriminant measures for performance evaluation. In *AI 2006: Advances in Artificial Intelligence*; Springer: Berlin/Heidelberg, Germany, 2006; pp. 1015–1021.
67. Yin, M.; Vaughan, J.W.; Wallach, H. Understanding the effect of accuracy on trust in machine learning models. In Proceedings of the 2019 CHI Conference on Human Factors in Computing Systems, Glasgow, UK, 4–9 May 2019; pp. 1–12.

electronics

Article

Aggregation of Rankings Using Metaheuristics in Recommendation Systems

Michał Bałchanowski *,† and Urszula Boryczka †

Institute of Computer Science, Faculty of Science and Technology, University of Silesia in Katowice, Będzińska 39, 41-200 Sosnowiec, Poland; urszula.boryczka@us.edu.pl
* Correspondence: michal.balchanowski@us.edu.pl
† These authors contributed equally to this work.

Abstract: Recommendation systems are a powerful tool that is an integral part of a great many websites. Most often, recommendations are presented in the form of a list that is generated by using various recommendation methods. Typically, however, these methods do not generate identical recommendations, and their effectiveness varies between users. In order to solve this problem, the application of aggregation techniques was suggested, the aim of which is to combine several lists into one, which, in theory, should improve the overall quality of the generated recommendations. For this reason, we suggest using the Differential Evolution algorithm, the aim of which will be to aggregate individual lists generated by the recommendation algorithms and to create a single list that will be fine-tuned to the user's preferences. Additionally, based on our previous research, we present suggestions to speed up this process.

Keywords: recommendation systems; rank aggregation; differential evolution; supervised learning; matrix factorization; metaheuristic

Citation: Bałchanowski, M.; Boryczka, U. Aggregation of Rankings Using Metaheuristics in Recommendation Systems. *Electronics* **2022**, *11*, 369. https://doi.org/10.3390/electronics11030369

Academic Editor: Stefano Ferilli

Received: 30 November 2021
Accepted: 22 January 2022
Published: 26 January 2022

1. Introduction

In today's world where the amount of information available is overwhelming for a common user, the use of systems designed to support the user in making decisions is becoming more apparent. This role is taken on by recommendation systems, which are more commonly used in various areas of our life. From buying items on auction sites through selecting a movie to adding new friends on social networks. The growing popularity of this type of website means that there is a real demand for recommendation systems that work efficiently and not only increase the quality of the generated recommendations but also ensure their novelty and diversity [1].

Within the recommendation systems, we can distinguish two main approaches to creating a recommendation. They can be based on an attempt to predict what rating (e.g., on a scale from 1 to 5) the user would give to an item in the system. They can also attempt to predict a certain set of items, most often presented in the form of a list that would be recommended to the user [2] (this problem is also called the top-N recommendations problem). Additionally, we can rely on data entered directly by the user or we can infer their preferences by observing how they use the system.

This article will also discuss the problem of rank aggregation, which has been described thoroughly in the literature, especially in the context of information retrieval systems [3–5] and proven to be NP-hard [6] even for small collections of ranks (e.g., 4 or more). However, according to some researchers [7], this topic has not yet been sufficiently studied in the context of recommended systems. Depending on the dataset used, individual recommendation algorithms can generate different recommendations, and choosing one particular algorithm over others can decrease the quality of recommendations for some of the users.

Therefore, the use of aggregation techniques has been proposed also in this context where the aim is to combine the individual lists generated by different recommendation techniques in order to create one "super" list.

Additionally, due to the fact that we will be optimizing the average precision (AP) measure, the Differential Evolution (DE) algorithm will be used, which is a metaheuristic that makes the direct optimization of this measure possible [8]. Our method is universal, and thus any metaheuristic algorithm that is used for real-valued optimization can be used here (e.g., PSO [9]). We chose the DE to conduct our research, due to the fact that it is well-suited for this type of optimization [10–12]. DE is arguably one of the most versatile and stable population-based search algorithms that exhibits robustness to many different optimization problems [13]. Additionally, it is relatively simple to implement and has a small number of control parameters, which makes this algorithm easy to tune.

The main contribution of this paper is to present how the DE algorithm can be applied to the problem of rank aggregation in recommendation systems, which will be supported by tests performed on the MovieLens 100k data set [14]. We will also present, based on our previous work [15], how to accelerate this algorithm while generating ranking lists of items using a dedicated fitness function. This function can also be successfully used in other metaheuristics that use real-valued representations of individuals in a population. In addition, we will present research that will show that the use of metaheuristic algorithms in the context of the problem of rank aggregation can be additionally justified due to the resistance of these techniques to algorithms that generate low-quality recommendations.

The article is divided into six chapters. Section 2 constitutes a literature review with information about the current literature. Section 3 presents a formal definition of a recommendation system, an explanation of the ranking aggregation problem and the Differential Evolution algorithm. Section 4 presents a description of our algorithm along with the system architecture and a figure showing a simple example regarding how the matrix fitness function is calculated. Section 5 discusses how the test environment was prepared for conducting the experiments and presents the results with commentary. The final Section 6 discusses our conclusions and research proposals for the future.

2. Literature Overview

The problem of recommendations can be presented as the problem of predicting how a user would rate a given item (e.g., on a scale from 1 to 5) [16], or as the problem of creating a list of suggested items and is referred to as the Top-N recommendation problem [17]. In fact, the latter is more similar to the real-life scenario when working with recommendation systems [18], where the recommendations are most often presented in the form of a list of suggested items in which the elements at the beginning are more important than the ones at the end.

There have been many works describing this approach in the context of recommendation systems [2,17]. In order to evaluate the quality of such recommended lists, measures that take into account the order in which the items appear on the list are used, e.g., Mean Average Precision (MAP) and Normalized Discounted Cumulative Gain (NDCG). Due to the fact that these measures are usually difficult to directly optimize, metaheuristic algorithms can be applied here [8,19]. A good review of evolutionary algorithms in recommendation systems is the paper [20], in which the authors presented an overview of the current research in this area and suggestions for research in the future.

In this article, we also pay attention to the problem of rank aggregation. A great deal of work has been done on this subject, especially in the context of information retrieval systems [21]. We generally divide the algorithms used for rank aggregation into two categories: permutation-based and score-based. There are many suggested techniques in the literature, for example: Borda Count [6], COMB* [22] (e.g., COMBSUM and COMBMNZ), or OutRank [23]. Within the context of recommendation systems, there have also been several works addressing this problem. In [24], a system for creating recommendations for the entire group of users was suggested, instead of as usually done for one user only.

In the work [25], the authors suggested creating a multi-criteria recommendation system, which, in addition to the quality of the generated recommendations, also took into account measures, such as novelty and diversity. In [26], the authors used genetic programming to create a recommendation system that generated recommendations by optimizing the MAP measure. It is also worth paying attention to [7], in which the researchers asked themselves whether the problem of rank aggregation in the context of recommendation systems is worth looking into. They performed extensive experiments and suggested the direction in which future work in this area should go.

3. Background of the Research

This chapter explains the basic information and the definitions used in this article. At first, the definition of the recommendation system, methods of obtaining feedback from users and the problem of matrix factorization will be discussed. Then, we will present the problem of rank aggregation in the context of recommendation systems. Finally, we will present a metaheuristic algorithm that will be used during our research.

3.1. Recommender System

In a recommendation system, we distinguish a certain set of users $U = u_1, \ldots, u_{|U|}$ and a certain set of items $I = i_1, \ldots, i_{|I|}$. Each of the users $u \in U$ has interacted with some of the items $i \in I$. The task of the recommendation system for the Top-N recommendation problem is to, on the basis of the historical data collected in the system, predict the user's next choices and create a list of items that are likely to interest the user. High-quality recommendations contribute to user satisfaction, which can translate into an overall good impression when using the platform. Depending on what kind of feedback is obtained from the user, recommendation techniques can be based on data from:

- Implicit feedback—This feedback is obtained by analyzing the user's behavior in the system, e.g., clicking on a specific product, page views and adding an item to the basket [27]. This type of feedback is easier to obtain as there is no need to ask the user to interact with the system (e.g., commenting and rating items). The main disadvantage of this approach is the lack of information on whether the interaction with the object was positive or negative [28]. For example, the user may have accidentally added an item to the basket and later removed it, and the mere fact of opening a page does not mean that the user likes the item. For this reason, the implementation of systems based on this type of data is associated with a number of challenges and has been described in many works [29,30].
- Explicit feedback—Feedback is obtained from the user in a direct way, for example the system asks the user to rate a given item [31]. The main advantage of this type of feedback is that it is easier to determine whether the interaction with the system was positive or negative. For example, if the user can enter a rating on a scale from 1 to 5 and selects a rating of 5, then, with a high probability, it can be assumed that this is an item that the user likes.
- Hybrid feedback—This is a combination of the two previously discussed techniques [32].

It should also be noted that recommendation systems often do not have good quality features for users and items. For this reason, various methods of obtaining them have been proposed, and one of the most popular techniques is to factorize the user–item matrix. With this, we can obtain features that are also called latent features. More on the subject can be found in [33].

3.2. Rank Aggregation Problem

This section describes the problem of rank aggregation in the context of recommendation systems. We define a ranking as an ordered list of items $\tau = [i_j >= i_h >= \cdots >= i_z]$, where the items at the beginning of the list (first position) are more significant than those at the end (last position). Item positions i_j in ranking τ, we define as $\tau(i_j)$. Two items $i_j \in \tau$

and $i_h \in \tau$ can be compared by checking their position in the list τ. If the item i_j is ranked higher in the τ in comparison to the item i_h, it is defined as $\tau(i_j) > \tau(i_h)$.

In recommendation systems, aggregations are generated through various algorithms, where a single algorithm will be defined as a_h, and a set of n recommendation algorithms will be defined as $A = \{a_1, a_2, \ldots, a_n\}$. Each of the algorithms $a_h \in A$ generates a ranking τ, and the set of all n created rankings is defined as $T = \{\tau^1, \tau^2, \ldots, \tau^n\}$. In addition, all algorithms that generate recommendations take, as input, matrix $M_{m\times n}$. Each row in this matrix represents a user $u_i \in U$, and each column represents an item $i_j \in I$. The value of this matrix $M_{i,j}$ corresponds to the rating given by the user u_i to the item i_j. Note that users rate only a small fraction of the items appearing in such a matrix; therefore, such a matrix is very sparse.

The problem of rank aggregation can be defined as the problem of finding such a combination of rankings in T generated by a set of recommendation algorithms A for each user $u_i \in U$, to create a single list ("super-list") that will optimize a given criterion (in our case, the average precision) to the greatest extent. Such a list should, in theory, be "better" than individual lists.

3.3. *Differential Evolution*

In order to optimize the AP measure, the Differential Evolution algorithm was used, which is a metaheuristic developed by K. Price and R. Storn [10]. It is based on individuals, which are represented as vectors of real numbers. For this reason, it is primarily suitable for the optimization of continuous functions, although there are papers that have suggested modifications to the algorithm and its adaptation to the optimization of discrete problems [30].

There is a population P of individuals, where each individual is a solution to an optimization problem, often represented as a d dimensional vector of real-valued numbers. The initial population P can be initialized randomly and should cover the entire search space. In the classic version of the algorithm, this is assumed to have a uniform probability distribution. In order to determine how good a given individual is in the population, it is necessary to define the fitness function, which assigns a certain value to each individual in the population.

This value is later used in the selection process, which is the process of choosing which individuals should go to the next generation. With each iteration, the algorithm attempts to improve the population of individuals until the stopping criterion is reached (e.g., a certain number of iterations). Owing to the use of crossover and mutation operators [34], the population of individuals changes and the algorithm attempts to find a better solution. Mutation creates a new individual by combining three randomly selected individuals and can be expressed with the following formula:

$$\vec{v_i} = \vec{x_{r_1}} + F(\vec{x_{r_2}} - \vec{x_{r_3}}), \tag{1}$$

where r_1, r_2 and r_3 are random unique individuals ($r_1 \neq r_2 \neq r_3$). The F parameter is the parameter responsible for amplification and usually takes a value in the range $[0, 1]$. After creating a new individual $\vec{v_i}$ using the mutation operator, we use the crossover operator according to Formula (2). The CR parameter is the parameter that determines the crossover probability. Additionally, there is a $rand$ function that generates a random number between [0, 1].

$$u_{i,j} = \begin{cases} v_{i,j} & \text{if } (rand(j) \leq CR \text{ } or \text{ } i = i_{rand}) \\ x_{i,j} & otherwise. \end{cases} \tag{2}$$

4. Suggested AggRankDE Method

Our AggRankDE method is designed based on the values issued by the individual recommender algorithms for each item i in the set of all items I to find a vector of the weight W that achieves the largest AP value on the training set TS. It should be noted that this vector is created for each user $u_i \in U$ separately, since each user has their own individual recommendation preferences. Additionally, based on our previous research, we suggest a matrix representation for the scores given by individual algorithms and the population of individuals of the DE algorithm.

Details of this representation can be found in our previous work [15], and a simple example is presented in Figure 1. As a result it is easier to parallelize the process of learning user preferences and, thus, to reduce the computation time that is needed to find the particular preference vector W.

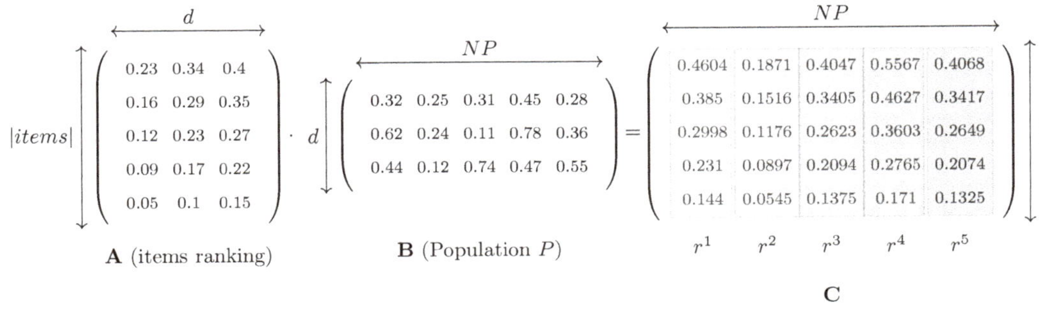

Figure 1. Toy example of the multiplication of two matrices. Matrix A represents scores assigned by the recommendation algorithms to each item $i \in I$ and some population P (real value vectors) of the metaheuristic algorithm represented by matrix B. Matrix product C represents new scores for each item $i \in I$, which, after sorting, create new rankings τ^n where $n \in \{1, 2, \ldots, NP\}$.

The hybridization technique was taken from [25] and is based on assigning weights $W = \{w_{a_1}, w_{a_2}, \ldots, w_{a_n}\}$ for each algorithm a_h, from the set of algorithms $A = \{a_1, a_2, \ldots, a_n\}$. The aggregated value for each item is calculated according to the formula:

$$\hat{p}(i_j|u_i) = \sum_{h=1}^{n} \hat{p}_{a_h}(i_j|u_i)\, w_{a_h} \tag{3}$$

where w_{a_h} is the weight assigned to the algorithm $a_h \in A$, with each algorithm assigning a value of $\hat{p}_{a_h}(i_j|u_i)$ to each item i_j, which determines the degree of potential interest of user u_i in this item. We should also remember to use the normalization technique so that all the algorithms in A can operate on the same scale.

The use of the metaheuristic algorithm based on evolution is associated with the need to define the fitness function so that, in subsequent iterations, the algorithm can reward individuals who are better adapted, i.e., with a greater value of the fitness function. In our case, this will be the average precision (AP) measure calculated for the active user u_A as follows:

$$Fitness = AP@k(R, S) \tag{4}$$

where S is the set of items recommended by the system and R is the set of items that user u_A rated in TS. According to our experiments, the value of k in AP during the learning process should be defined as the number of items that the user u_A rated in his TS. In our opinion, such a value is most appropriate due to the fact that it does not cause the algorithm to overfit. The details for how to calculate AP, especially in the context of recommendation systems, can be found in our paper [35]. The architecture of our system is presented below Figure 2.

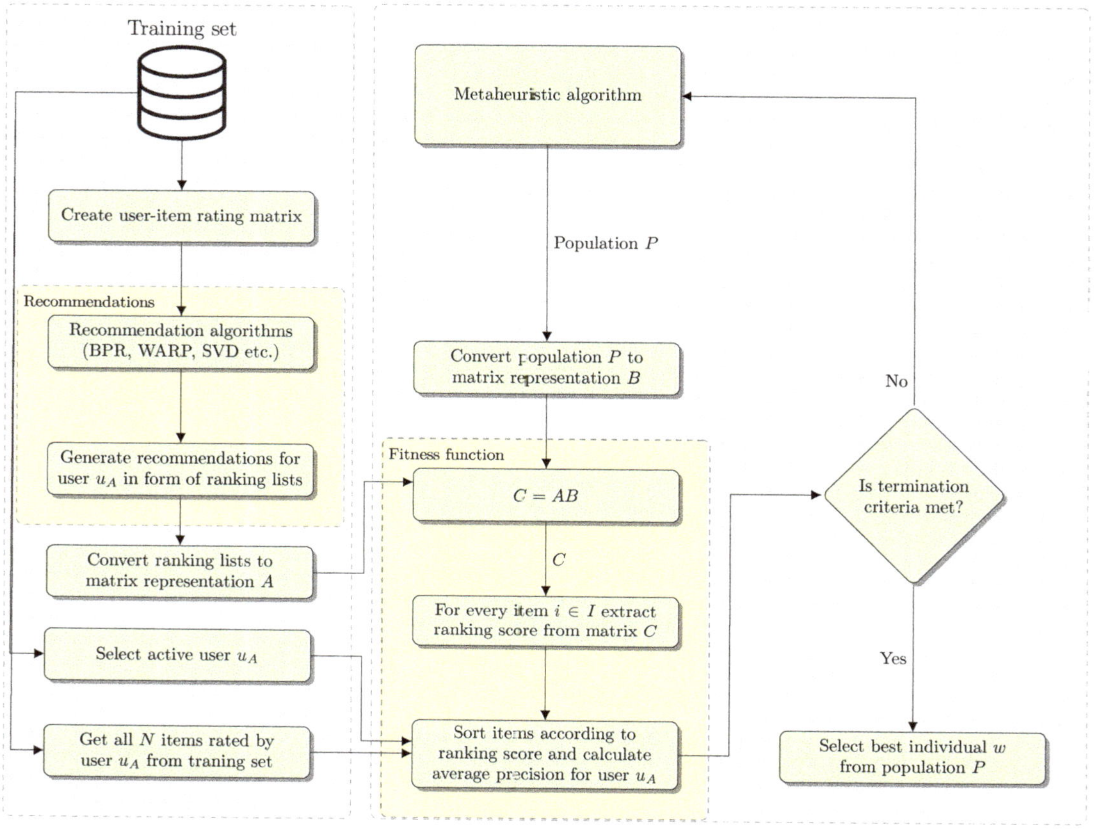

Figure 2. System architecture. The recommendation process is divided into two phases. In the first phase, recommendation algorithms generate recommendations in the form of lists, and active user u_A is selected with all his N items from the training set. In the second phase, a metaheuristic algorithm works (in our case DE) with the dedicated fitness function, which allows for faster calculation of item scores, on the basis of which, new rankings will be created.

5. Experimental Evaluation

Due to the fact that recommendations are most often presented to users in the form of a list, in our experiments, we used the average precision measure (AP) and the mean average precision measure (MAP). The *AP* measure is used in the context of a specific (one) user, and, in our research, it was used to compare the list of items recommended to the user with the list of items available in the test set for a given user. This allowed us to calculate the quality of the generated recommendations.

In addition, it should be noted that this measure also takes into account where the relevant items are located on the list. If the relevant items are higher (closer to the first position), then the *AP* value is also higher. Due to the fact that metaheuristics are computationally expensive, we chose only a certain subset of users for the experiments. We randomly selected 50 users who rated at least 150 movies in the dataset. The experiments carried out as part of this paper were performed using the popular MovieLens 100k dataset. The AggRankDE algorithm adopts four algorithms as the input: SVD, WMF, BPR and WARP. All of them are based on matrix factorization, and thus features are generated for each item and for each user on the basis of the user–item matrix.

These features are called latent features due to the fact that their meaning cannot be explained. In addition, these algorithms are considered to be the current state-of-the-art and are often used to compare research results in recommendation systems for the Top-N recommendation problem. The research environment was implemented in Python and C#, and the research was carried out on a computer with an Intel Core i5-7600 (3.50 GHz) with 16 GB RAM.

5.1. Parameters Tuning

Before creating an aggregation, the parameters of the algorithms that are included must be tuned. To this end, experiments were conducted to tune their values so that they could achieve the best possible MAP measure on the set of users used for the experiments. This is an important step, due to the fact that improper tuning of the parameters can result in the generation of poor quality recommendations. Table 1, presented below, shows the parameter values used during the tuning process.

This process consisted of first setting all parameters to the default values and then changing only one parameter that was selected for the tuning. After the process was completed, the best values were saved in the ("*Best values*" column in Table 1). The detailed MAP@10 values obtained during this process for various parameters are presented in tables: Table 2 (learning rate), Table 3 (regularization) and Table 4 (latent features).

The process of tuning the CR and F parameters for the DE algorithm was also performed, and the results of these experiments are presented in Tables 5 and 6. In addition, in article [10], the authors indicated that a good value for the parameter NP is a value between $5 \cdot d$ and $10 \cdot d$, where d is the number of dimensions. The authors also point out that the parameter F, equal to 0.5, is usually a good initial value and this parameter typically takes a value in the range $[0.4, 1]$. The final values of the Differential Evolution algorithm that were used during the experiments are presented in Table 7.

Table 1. The recommendation algorithms parameters that were used during the tuning process.

Algorithm Name	Parameter Name	Values Used in Tuning Process	Default Values	Best Values
BPR	Regularization	{0.0, 0.005, 0.01, 0.05, 0.1, 0.15, 0.2}	0.0	0.0
	Learning rate	{0.005, 0.01, 0.25, 0.5, 0.1, 0.15, 0.2, 0.25, 0.3}	0.05	0.025
	Latent factors	{4, 5, 6, 7, 8, 9, 10, 20, 30, 40, 50, 60, 70, 80, 90, 100}	10	10
WARP	Regularization	{0.0, 0.005, 0.01, 0.05, 0.1, 0.15, 0.2}	0.0	0.0
	Learning rate	{0.005, 0.01, 0.25, 0.5, 0.1, 0.15, 0.2, 0.25, 0.3}	0.05	0.15
	Latent factors	{4, 5, 6, 7, 8, 9, 10, 20, 30, 40, 50, 60, 70, 80, 90, 100}	10	50
WMF	Latent factors	{4, 5, 6, 7, 8, 9, 10, 20, 30, 40, 50, 60, 70, 80, 90, 100}	10	10
SVD	Latent factors	{4, 5, 6, 7, 8, 9, 10, 20, 30, 40, 50, 60, 70, 80, 90, 100}	10	6

Table 2. Learning rate parameter tuning. This table presents MAP@10 for different parameter values. The remaining parameters are set to the default values according to Table 1.

Learning Rate	BPR_MAP	WARP_MAP
0.005	0.176	0.158
0.01	0.208	0.185
0.025	0.259	0.198
0.05	0.228	0.214
0.1	0.209	0.220
0.15	0.191	0.244
0.2	0.174	0.230
0.25	0.149	0.122
0.3	0.137	0.079

Table 3. Regularization parameter tuning. This table presents MAP@10 for different parameter values. The remaining parameters are set to the default values according to Table 1.

Regularization	BPR_MAP	WARP_MAP
0	0.228	0.214
0.005	0.170	0.173
0.01	0.171	0.147
0.05	0.170	0.073
0.1	0.166	0.002
0.15	0.166	0.005
0.2	0.009	0.000

Table 4. Latent features (dimensions) parameter tuning. This table presents MAP@10 for different parameter values. The remaining parameters are set to the default values according to Table 1.

Latent Features	BPR_MAP	WARP_MAP	WMF_MAP	SVD_MAP
4	0.20	0.21	0.19	0.18
5	0.18	0.21	0.15	0.20
6	0.17	0.21	0.20	0.21
7	0.17	0.20	0.20	0.21
8	0.19	0.21	0.20	0.19
9	0.20	0.20	0.20	0.20
10	0.23	0.21	0.22	0.19
20	0.22	0.21	0.21	0.13
30	0.21	0.25	0.18	0.08
40	0.23	0.24	0.17	0.07
50	0.22	0.27	0.17	0.06
60	0.21	0.26	0.16	0.05
70	0.21	0.25	0.13	0.06
80	0.21	0.23	0.15	0.05
90	0.21	0.24	0.14	0.04
100	0.20	0.22	0.13	0.05

Table 5. F parameter tuning. This table presents MAP@10 for different parameter values. The remaining parameters are set to the default values according to Table 1.

Amplification Factor F	DE_MAP
0.3	0.42
0.4	0.43
0.5	0.46
0.6	0.45
0.7	0.44
0.8	0.42
0.9	0.43
1	0.44

Table 6. CR parameter tuning. This table presents MAP@10 for different parameter values. The remaining parameters are set to the default values according to Table 1.

Crossover's Probability CR	DE_MAP
0.3	0.45
0.4	0.43
0.5	0.46
0.6	0.44
0.7	0.41
0.8	0.41
0.9	0.49
1	0.43

Table 7. The differential evolution parameters used in the experiments.

Parameter Name	Value
Population	50
Number of Iterations	500
Crossover's Probability	0.9
Amplification Factor F	0.5

5.2. Experimental Setup

In order to prepare the environment for testing, first, the data was prepared in an appropriate way. User ratings were sorted by the time in which a given rating was issued and then divided into two sets: training (80%) and test (20%). Owing to this approach, our algorithm attempts to predict the user's future preferences based on the user's previous activity. The task is not trivial due to the number of items from which we can choose items and which will later be presented to the user.

Fifty users were randomly selected for the study, where a recommendation was generated for each user, and then the results of the suggested recommendations were compared with the test sets of each user. The AP measure was used to calculate the quality of the generated recommendations, and then its value was averaged for all users selected for testing; thus, the tables show the results given using the MAP measure. In order to show that our algorithm gives good results, we compared it with other algorithms used for the rank aggregation problem, such as the Borda Count, Majority Judgement, Pairwise Method (Copeland's) and Score Voting (mean).

In the research, we additionally took into account the quality of recommendations that was achieved through algorithms that participated in the creation of aggregation. These included the Bayesian Personal Ranking (BPR) and Weighted Approximate-Rank Pairwise (WARP) algorithms, the implementation of which is available in the LightFM library [36]. In addition, the usual SVD algorithm marked in the results as "SVD" and a weighted matrix factorization (WMF) algorithm were implemented.

5.3. Results

In Section 5.1, we presented the process of tuning the parameters for the various algorithms used to create aggregations. This is an important step, due to the fact that the quality of the generated recommendations by the different recommendation techniques can largely depend on the parameters that are set. For example, by analyzing Table 4, it can be seen that the MAP value obtained was highly dependent on the number of latent features. Additionally, the research presented in Table 2 showed that the parameter "Learning rate", which is characteristic for the BPR and WARP techniques, also required tuning as opposed to the parameter "Regularization" (Table 3) where the default value (0) generated the best quality of the recommendations.

While analyzing the results presented in Table 8, it can be seen that the AggRankDE algorithm aggregated the recommendation algorithms and improved the overall quality of the generated recommendations even compared to other aggregation techniques. This is an important observation because it shows that one "super" list can be created from several lists to improve the quality of recommendations, which is consistent with the experimental results by [7].

Looking at the quality of the recommendations generated by the different recommendation algorithms, we can see that, depending on *MAP@*, the quality of the recommendations varies. In general, as the number of items based on which the MAP@ measure is calculated increases, it can be seen that the quality of the recommendations decreases, although the AggRankDE algorithm improved the quality of the generated recommendations in all cases.

Additionally, after the introduction of the "Random" method (Table 9), which purposefully generated poor quality recommendations, in the case of the AggRankDE, this did not significantly degrade the quality of the produced aggregation in contrast with, for example, the Borda Count method. This indicates that the AggRankDE has some resistance to weak algorithms that are used in the aggregation.

Table 10 presents the improvement in the speed (in seconds) of the generated recommendations after implementing the matrix fitness function. Time is measured for a single user in the system and depends on the number of iterations. Looking at this table, it can be seen that the improvement in speed is significant, and this is due to the fact that the operation on entire matrices can be easily parallelized. This is particularly important in the context of metaheuristic algorithms due to the fact that computing the fitness function is the most costly step in this type of algorithm.

Table 8. The quality of the generated recommendations (MAP) for different *MAP@* values for the best parameters presented in Table 1.

MAP@	Bpr	Warp	WMF	SVD	Borda	Majority	Pairwise	Score	AggRankDE
1	0.46	0.44	0.46	0.44	0.5	0.46	0.48	0.5	0.58
2	0.39	0.36	0.43	0.37	0.4	0.38	0.39	0.41	0.43
3	0.34	0.31	0.36	0.34	0.36	0.36	0.36	0.35	0.4
4	0.32	0.28	0.28	0.32	0.32	0.32	0.33	0.33	0.36
5	0.30	0.26	0.27	0.29	0.31	0.30	0.30	0.31	0.32
6	0.27	0.25	0.24	0.26	0.28	0.28	0.28	0.28	0.29
7	0.24	0.24	0.23	0.24	0.26	0.26	0.26	0.27	0.28
8	0.23	0.23	0.22	0.23	0.24	0.24	0.25	0.25	0.26
9	0.23	0.21	0.20	0.22	0.23	0.23	0.23	0.23	0.24
10	0.22	0.21	0.20	0.21	0.22	0.23	0.21	0.21	0.24

Table 9. The quality of the generated recommendations (MAP) for different *MAP@* values for the best parameters presented in Table 1 with the additional *RANDOM* algorithm.

MAP@	Bpr	Warp	WMF	SVD	Random	Borda	Majority	Pairwise	Score	AggRankDE
1	0.46	0.44	0.46	0.44	0.04	0.26	0.42	0.5	0.2	0.56
2	0.39	0.36	0.38	0.37	0.06	0.22	0.39	0.39	0.16	0.48
3	0.34	0.31	0.34	0.34	0.01	0.19	0.34	0.36	0.13	0.42
4	0.32	0.28	0.30	0.32	0.02	0.16	0.30	0.32	0.10	0.35
5	0.30	0.26	0.29	0.29	0.02	0.15	0.28	0.29	0.09	0.32
6	0.27	0.25	0.24	0.26	0.02	0.14	0.26	0.28	0.08	0.28
7	0.24	0.24	0.24	0.24	0.02	0.14	0.23	0.26	0.07	0.27
8	0.23	0.23	0.22	0.23	0.01	0.13	0.22	0.24	0.06	0.25
9	0.23	0.21	0.21	0.22	0.01	0.12	0.21	0.23	0.06	0.24
10	0.22	0.21	0.21	0.21	0.01	0.12	0.20	0.21	0.05	0.23

Table 10. The average time (in seconds) depending on the number of iterations. The remaining parameters are according to Table 7.

Iterations	DE	AggRankDE
100	0.89 s	0.45 s
200	1.59 s	0.83 s
300	2.28 s	1.2 s
400	3.0 s	1.51 s
500	3.72 s	1.87 s
600	4.46 s	2.28 s
700	5.19 s	2.66 s
800	5.92 s	3.01 s
900	7.03 s	3.45 s
1000	7.5 s	3.86 s

When analyzing the experimental results, the application of the DE algorithm with the hybridization technique presented in [25] produced good results. However, in our paper, we suggested how to improve it by using a dedicated fitness function to directly optimize the average precision measure and to speed up its calculation process. By assigning different weights to the different algorithms included in the aggregation, the DE algorithm optimizes the average precision measure using a weighted hybridization technique in order to obtain the highest possible value of the average precision measure on the training set.

During the testing phase, this translated into an increase in the quality of the generated recommendations. However, this process is computationally very expensive; therefore, we suggested using the matrix representation in the fitness function, which significantly accelerated the process of calculating the values for each item by the hybridization technique on the basis of which the ranking was created.

6. Conclusions

In this article, we presented how the Differential Evolution algorithm can be used to optimize the problem of rank aggregation in recommendation systems. The experiments were conducted on the database MovieLens 100k, and they showed that our algorithm improved the quality of the recommendations expressed by the MAP measure by 5% compared to other algorithms used for this purpose. Our research showed that, even using simple aggregation techniques, we could improve the quality of the generated recommendations.

In addition, in analyzing the research results, it can be seen that the AggRankDE algorithm is resistant to algorithms that generate poor-quality recommendations. We believe that this is due to the fact that, through the presence of a training phase in which the DE algorithm optimizes the AP measure, it is able to detect algorithms that generate low-quality recommendations and assign them correspondingly low weights, which results in them participating least in the creation of the list of recommended items.

Based on our previous work, we also suggested the use of matrix representation for the population of the DE algorithm and the values of coefficients calculated by individual aggregation algorithms for each item in the system. Such a representation makes it much easier to parallelize the process of calculating the values for individual items in the training phase on the basis of which new rankings (recommendations) are created. The calculation of the fitness function is the most expensive operation in the metaheuristic algorithms. In the context of the recommendation systems, this is particularly important, due to the relatively large data sets that are processed.

In following papers, we will increase the number of algorithms that are part of the aggregation, add more aggregation techniques and increase the number of data sets on the basis of which the research is carried out. We will also conduct a more detailed analysis of the effectiveness of our algorithm, taking into account a larger number of users, and conduct a more detailed analysis of how the parameters of the individual algorithms included in the aggregation and the model itself affect the quality of the generated recommendations.

Another interesting direction of research would be to take a closer look at the quality of the generated recommendations by particular algorithms in relation to individual users. Although the AggRankDE algorithm is more robust to algorithms that generate poor recommendations, the decrease in the quality is noticeable. Presumably, eliminating the weaker quality algorithms would generally improve the quality of the aggregation produced. We believe that the problem of rank aggregation within the context of the recommendation systems has not yet been sufficiently studied, and this will likely be the direction of our future work.

Author Contributions: Conceptualization, U.B. and M.B.; Formal analysis, U.B. and M.B.; Investigation, M.B.; Methodology, U.B. and M.B.; Project administration, M.B.; Software, M.B.; Validation, U.B. and M.B.; Visualization, M.B.; Writing—original draft, U.B. and M.B.; Writing—review and editing, U.B. and M.B. All authors have read and agreed to the published version of the manuscript.

Funding: This research received no external funding.

Institutional Review Board Statement: Not applicable.

Informed Consent Statement: Not applicable.

Data Availability Statement: Data is available at https://grouplens.org/datasets/movielens/100k/ accessed on 29 November 2021.

Conflicts of Interest: The authors declare no conflict of interest.

Abbreviations

The following abbreviations are used in this manuscript:

u	Generic user
u_i	Specific user
u_A	Active user in system for which recommendations are generated
U	The set of all users
i	Generic item
i_j	Specific item
I	Set of all items
a_h	Specific recommendation algorithm
A	Set of n recommendation algorithms $A = \{a_1, a_2, \ldots, a_n\}$
τ	Generic ranking
τ_i^r	Ranking recommended to user u_i by algorithm a_r where $r \in \{1, 2, \ldots, n\}$
$\tau(i_j)$	The position of item i_j in ranking τ
T	Set of n rankings $T = \{\tau^1, \tau^2, \ldots, \tau^n\}$
w_{a_h}	Weight assigned to recommendation algorithm a_h where $h \in \{1, 2, \ldots, n\}$
W	Set of n weights $W = \{w_{a_1}, w_{a_2}, \ldots, w_{a_n}\}$
R	Set of items that user u_A rated in his training set
S	Set of items recommended to user u_A
P	Population of metaheuristic algorithm
NP	Number of individuals in population
TS	Training set

References

1. Castells, P.; Hurley, N.; Vargas, S. *Novelty and Diversity in Recommender Systems*; Springer: Boston, MA, USA, 2015 ; pp. 881–918. [CrossRef]
2. Cremonesi, P.; Koren, Y.; Turrin, R. Performance of recommender algorithms on top-N recommendation tasks. In Proceedings of the Fourth ACM Conference on Recommender Systems, Barcelona, Spain, 26–30 September 2010; pp. 39–46. [CrossRef]
3. Dwork, C.; Naor, M.; Sivakumar, D. Rank Aggregation Revisited. 2003 . Available online: http://www.cse.msu.edu/~cse960/Papers/games/rank.pdf (accessed on 29 November 2021).
4. Vanderpooten, D.; Farah, M. An Outranking Approach for Rank Aggregation in Information Retrieval. 2007 . Available online: https://dl.acm.org/doi/10.1145/1277741.1277843 (accessed on 29 November 2021). [CrossRef]
5. Dourado, Í.C.; Pedronette, D.C.G.; da Silva Torres, R. Unsupervised Graph-based Rank Aggregation for Improved Retrieval. *CoRR* **2019** , 56, 1260–1279. [CrossRef]
6. Dwork, C.; Kumar, R.; Naor, M.; Sivakumar, D. Rank Aggregation Methods for the Web. In Proceedings of the 10th International Conference on World Wide Web WWW'01, Hong Kong, China, 1–5 May 2001; pp. 613–622. [CrossRef]
7. Oliveira, S.E.L.; Diniz, V.; Lacerda, A.; Merschmanm, L.; Pappa, G.L. Is Rank Aggregation Effective in Recommender Systems? An Experimental Analysis. *ACM Trans. Intell. Syst. Technol. (TIST)* **2020**, *11*, 16. [CrossRef]
8. Bollegala, D.; Noman, N.; Iba, H. RankDE: Learning a ranking function for information retrieval using differential evolution. In Proceedings of the 13th Annual Conference on Genetic and Evolutionary Computation, Dublin, Ireland, 12–16 July 2011; pp. 1771–1778. [CrossRef]
9. Kennedy, J.; Eberhart, R. Particle swarm optimization. In Proceedings of the ICNN'95-International Conference on Neural Networks, Perth, WA, Australia, 27 November–1 December 1995; Volume 4, pp. 1942–1948. [CrossRef]
10. Storn, R.; Price, K. Differential Evolution—A Simple and Efficient Heuristic for Global Optimization over Continuous Spaces. *J. Glob. Optim.* **1997**, *11*, 341–359. [CrossRef]
11. Bilal.; Pant, M.; Zaheer, H.; Garcia-Hernandez, L.; Abraham, A. Differential Evolution: A review of more than two decades of research. *Eng. Appl. Artif. Intell.* **2020**, *90*, 103479.
12. Ronkkonen, J.; Kukkonen, S.; Price, K. Real-parameter optimization with differential evolution. In Proceedings of the 2005 IEEE Congress on Evolutionary Computation, Edinburgh, UK, 2–5 September 2005; Volume 1, pp. 506–513. [CrossRef]
13. Feoktistov, V. *Differential Evolution*; Springer: Boston, MA, USA, 2006.
14. Harper, F.M.; Konstan, J.A. The MovieLens Datasets History and Context. *ACM Trans. Interact. Intell. Syst.* **2015**, *5*, 19 . [CrossRef]

15. Boryczka, U.; Bałchanowski, M. Speed up Differential Evolution for ranking of items in recommendation systems. *Procedia Comput. Sci.* **2021**, *192*, 2229–2238. [CrossRef]
16. Bennett, J.; Lanning, S.; Netflix, N. The Netflix Prize. In *KDD Cup and Workshop in Conjunction with KDD*; 2007 . Available online: https://www.cs.uic.edu/~liub/KDD-cup-2007/NetflixPrize-description.pdf (accessed on 29 November 2021).
17. Deshpande, M.; Karypis, G. Item-Based Top-N Recommendation Algorithms. *ACM Trans. Inf. Syst.* **2004**, *22*, 143–177. [CrossRef]
18. Karatzoglou, A.; Baltrunas, L.; Shi, Y. Learning to rank for recommender systems. In Proceedings of the 7th ACM Conference on Recommender Systems, Hong Kong, China, 12–16 October 2013; pp. 493–494. [CrossRef]
19. Diaz-Aviles, E.; Nejdl, W.; Schmidt-Thieme, L. Swarming to Rank for Information Retrieval. In Proceedings of the 11th Annual Conference on Genetic and Evolutionary Computation GECCO'09, Montreal, QC, Canada, 8–12 July 2009; Association for Computing Machinery: New York, NY, USA, 2009; pp. 9–16. [CrossRef]
20. Horvath, T.; de Carvalho, A. Evolutionary computing in recommender systems: A review of recent research. *Nat. Comput.* **2016** , 16, 441–462. [CrossRef]
21. Klementiev, A.; Roth, D.; Small, K. *Unsupervised Rank Aggregation with Distance-Based Models ICML'08*; Association for Computing Machinery: New York, NY, USA, 2008; pp. 472–479. [CrossRef]
22. Shaw, J.A.; Fox, E.A. Combination of Multiple Searches. In Proceedings of the Second Text Retrieval Conference (TREC-2), Plainsboro, NJ, USA, 8–11 March 1994; pp. 243–252.
23. Farah, M.; Vanderpooten, D. An Outranking Approach for Rank Aggregation in Information Retrieval. In Proceedings of the 30th Annual International ACM SIGIR Conference on Research and Development in Information Retrieval SIGIR'07, Amsterdam, The Netherlands, 23–27 July 2007; Association for Computing Machinery: New York, NY, USA, 2007; pp. 591–598. [CrossRef]
24. Baltrunas, L.; Makcinskas, T.; Ricci, F. Group Recommendations with Rank Aggregation and Collaborative Filtering. In Proceedings of the Fourth ACM Conference on Recommender Systems RecSys'10, Barcelona, Spain, 26–30 September 2010; Association for Computing Machinery: New York, NY, USA, 2010; pp. 119–126. [CrossRef]
25. Ribeiro, M.T.; Ziviani, N.; Moura, E.S.D.; Hata, I.; Lacerda, A.; Veloso, A. Multiobjective Pareto-Efficient Approaches for Recommender Systems. *ACM Trans. Intell. Syst. Technol.* **2015**, *5*, 53 . [CrossRef]
26. Oliveira, S.; Diniz, V.; Lacerda, A.; Pappa, G.L. Evolutionary rank aggregation for recommender systems. In Proceedings of the 2016 IEEE Congress on Evolutionary Computation (CEC), Vancouver, BC, Canada, 24–29 July 2016; pp. 255–262. [CrossRef]
27. Oard, D.; Kim, J. Implicit Feedback for Recommender System. In Proceedings of the AAAI Workshop on Recommender Systems 2000 . Available online: https://www.aaai.org/Papers/Workshops/1998/WS-98-08/WS98-08-021.pdf (accessed on 29 November 2021).
28. Hu, Y.; Koren, Y.; Volinsky, C. Collaborative Filtering for Implicit Feedback Datasets. In Proceedings of the 2008 Eighth IEEE International Conference on Data Mining, Pisa, Italy, 15–19 December 2008; pp. 263–272. [CrossRef]
29. Rendle, S.; Freudenthaler, C.; Gantner, Z.; Schmidt-Thieme, L. BPR: Bayesian Personalized Ranking from Implicit Feedback. In Proceedings of the Twenty-Fifth Conference on Uncertainty in Artificial Intelligence UAI'09, Montreal, QC, Canada, 18–21 June 2009; AUAI Press: Arlington, VA, USA, 2009; pp. 452–461.
30. Pan, R.; Zhou, Y.; Cao, B.; Liu, N.N.; Lukose, R.; Scholz, M.; Yang, Q. One-Class Collaborative Filtering. In Proceedings of the 2008 Eighth IEEE International Conference on Data Mining, Pisa, Italy, 15–19 December 2008; pp. 502–511. [CrossRef]
31. Jawaheer, G.; Szomszor, M.; Kostkova, P. Comparison of implicit and explicit feedback from an online music recommendation service. In Proceedings of the 1st International Workshop on Information Heterogeneity and Fusion in Recommender Systems, Barcelona, Spain, 26–30 September 2010; [CrossRef]
32. Chou, C.L.; Lu, T.Y. A hybrid-feedback recommender system for employment websites. *J. Ambient. Intell. Humaniz. Comput.* **2020** . Available online: https://link.springer.com/article/10.1007/s12652-020-01772-y (accessed on 29 November 2021). [CrossRef]
33. Koren, Y.; Bell, R.; Volinsky, C. Matrix Factorization Techniques for Recommender Systems. *Computer* **2009**, *42*, 30–37. [CrossRef]
34. Boryczka, U.; Juszczuk, P.; Kłosowicz, L. A Comparative Study of Various Strategies in Differential Evolution. In *Evolutionary Computing and Global Optimization KAEiOG'09*; 2009; pp. 19–26. Available online: https://www.researchgate.net/publication/230788075_A_Comparative_Study_of_Various_Strategies_in_Differential_Evolution (accessed on 29 November 2021).
35. Boryczka, U.; Bałchanowski, M. Using Differential Evolution in order to create a personalized list of recommended items. *Procedia Comput. Sci.* **2020**, *176*, 1940–1949. [CrossRef]
36. Kula, M. Metadata Embeddings for User and Item Cold-start Recommendations. In Proceedings of the 2nd Workshop on New Trends on Content-Based Recommender Systems Co-Located with 9th ACM Conference on Recommender Systems (RecSys 2015), Vienna, Austria, 16–20 September 2015; Volume 1448, pp. 14–21.

electronics

MDPI

Article

Integration Strategy and Tool between Formal Ontology and Graph Database Technology

Stefano Ferilli †

Department of Computer Science, University of Bari, 70125 Bari, Italy; stefano.ferilli@uniba.it;
Tel.: +39-080-5442293
† Current address: Via E. Orabona 4, 70125 Bari, Italy.

Abstract: Ontologies, and especially formal ones, have traditionally been investigated as a means to formalize an application domain so as to carry out automated reasoning on it. The union of the terminological part of an ontology and the corresponding assertional part is known as a Knowledge Graph. On the other hand, database technology has often focused on the optimal organization of data so as to boost efficiency in their storage, management and retrieval. Graph databases are a recent technology specifically focusing on element-driven data browsing rather than on batch processing. While the complementarity and connections between these technologies are patent and intuitive, little exists to bring them to full integration and cooperation. This paper aims at bridging this gap, by proposing an intermediate format that can be easily mapped onto the formal ontology on one hand, so as to allow complex reasoning, and onto the graph database on the other, so as to benefit from efficient data handling.

Keywords: knowledge representation; formal ontologies; graph databases

Citation: Ferilli, S. Integration Strategy and Tool between Formal Ontology and Graph Database Technology. *Electronics* **2021**, *10*, 2616. https://doi.org/10.3390/electronics10212616

Academic Editor: Agnieszka Konys

Received: 24 September 2021
Accepted: 24 October 2021
Published: 26 October 2021

1. Introduction

Two main perspectives, very different from each other, have been adopted in Computer Science for information storage and handling. The 'Knowledge Base' (KB) perspective is interested in high-level reasoning on the available information, so as to infer implicit information or check the consistency of the information with respect to the reference domain. It is pursued by the Knowledge Representation (KR) branch of Artificial Intelligence (AI) and includes the research field of formal ontologies. The 'Data Base' (DB) perspective is a traditional branch of research in Computer Science interested in developing optimal data organizations aimed at efficient storage, management and retrieval. While clearly complementary, these two perspectives have traditionally been investigated separately. However, due to the increasingly pervasive use of AI solutions in many applications, it would be extremely relevant to take advantage of both.

A new opportunity for cooperation comes from the recent development of *Graph Databases*, a kind of NoSQL DB aimed at optimizing element-driven data browsing rather than batch processing as in traditional relational DBs. Another difference between graph and relational DBs is that the former do not have a pre-defined schema to describe and organize the data, which obviously affects the interpretability and accessibility of the data by the applications and their interoperability. A *graph* is a data structure consisting of nodes (usually representing things) and arcs connecting these nodes (usually representing relationships between things). The arcs may be directed, if they have a direction, and may have attributes or labels qualifying or quantifying the relationship. Interestingly, when the terminological part of an ontology (Tbox, reporting definitions and axioms) is considered in conjunction with the assertional part (Abox, specifying individuals or instances) the result is a so-called *Knowledge Graph* (KG, a kind of KB) [1]. Whilst the literature on ontologies often defines them as encompassing both parts, the relevant literature adopts this very definition for KGs, equating the ontology to the data model only:

- "A knowledge graph is created when you apply an ontology (the data model) to a dataset of individual data points (the [...] data). In other words:

  ontology + data = knowledge graph" [2].
- "Ontologies represent the backbone of the formal semantics of a knowledge graph. They can be seen as the data schema of the graph" [3].
- Knowledge graphs derive from "the core idea of using graphs to represent data, often enhanced with some way to explicitly represent knowledge" [4].
- "In general, a knowledge graph describes objects of interest and connections between them. [...] Many practical implementations impose constraints on the links in knowledge graphs by defining a schema or ontology' [5].

It is clear that graph representation can be the missing link to join the two perspectives/technologies and take the best from each. Unfortunately, formal ontologies and graph DBs refer to different graph models which cannot straightforwardly be combined together. This paper proposes a technology, called *GraphBRAIN*, aimed at bridging the gap between them through the following contributions:

- Defining a formalism for expressing graph DB schemas, so as to allow data interpretability and applications interoperability;
- Defining the mapping between the graph DB model expressed by this formalism and the standard ontological model adopted in the literature;
- Defining the basics for the operational connection between graph DBs and ontologies, through the two mentioned standards;
- Implementing a software library (intended to act as a wrapper for the DB, permitting only interactions that are compliant to the schema) and tools for the practical exploitation of the proposed formalisms and methodologies.

It would allow graph DB developers to carry out high-level reasoning on their data. Indeed, formal, automated reasoning is much more powerful than the DB's query language, e.g., using ontological reasoning one may check consistency, correctness or completeness of the data. Using rule-based reasoning one may infer information that is not explicitly expressed in the data, possibly defined by complex patterns (as expressible in Logic Programming). Even more, multiple inference strategies (e.g., abduction, argumentation, etc.), not just deduction, can be carried out.

We already developed prototypes of the library, of a tool for building and maintaining the schema and of a tool for handling and consulting the DB based on the schema. This preliminary implementation of GraphBRAIN [6] is currently in use as part of a larger ongoing project [7], aimed at building an integrated system for AI-supported tourism, providing advanced support to end-users, entrepreneurs and institutions involved in touristic activities. It currently includes schemas describing the inter-related domains of 'tourism' (concerning history, cultural heritage items, points of interest, logistics and services, etc.), 'food' (concerning typical dishes and beverages from specific regions), 'computing' (concerning computing devices and their history) [8] and 'lam' (concerning libraries, archives and museums) [9].

Original contributions of this paper are:

- For the first time, a detailed specification of the proposed formalism with a complete account and explanation of its components;
- An extension and refinement of the formalism's components proposed in the previous papers;
- A description of its use as a schema for graph DBs;
- A full mapping of it on a standard ontological format.

This paper is organized as follows. After discussing in Section 2 the basic concepts and related works about formal ontologies and graph databases, Section 3 describes our proposed formalism for interfacing the two technologies. Then, Section 4 shows how schemas expressed in our formalism can be mapped onto graph DBs on one hand and onto a standard ontological format on the other. Finally, Section 5 concludes the paper.

2. Basics and Related Work

According to one of its many definitions in Computer Science, an *ontology* is "a formal, explicit specification of a shared conceptualization" [10]. Therefore, building an ontology requires a conceptualization step, by which: (1) the relevant entities, relationships and their attributes in a domain of interest are identified; (2) names are defined for them; (3) possibly (in the case of *formal* ontologies) axioms are stated expressing what is mandatory, permitted or prohibited in that domain. Explicit or implicit ontology building is pervasive in Computer Science (e.g., when designing E-R diagrams in DBs, or class diagrams in Object-Oriented systems, or predicates, functions and constants in KBs), to determine what can be represented in a (family of) application(s) and to define the rules driving their operation. Indeed, ontologies are key to improving communication among agents, foster systems interoperability and support reuse. Formal Ontologies specifically focus on automated reasoning aimed at making inferences on the available knowledge (concerning both the concepts and their instances) expressed according to the ontology. The main reasoning tasks include KB satisfiability, axiom entailment, concept satisfiability, instance retrieval, classification, query answering [11].

A standard formalism for expressing ontologies and KGs is the Web Ontology Language (OWL) [12]. In fact, a number of reasoners based on OWL are available [13] that provide implementations for all or part of the inferences. OWL is based on the Resource Definition Framework (RDF) [14], originally developed for describing resources on the Web but amenable to knowledge representation in general. RDF graphs are based on a directed graph data model in which nodes are Uniform Resource Identifiers (URIs). A Named Graph is an RDF graph named by a graph URI. An RDF Graph is a collection of RDF Triples, representing arcs, i.e., units of RDF Data of the form:

(Subject, Predicate, Object)

where the Subject and Predicate are URIs and the Object may be a URI or a literal value. Triplestores (or 'Semantic Graph Databases') are DB Management Systems (DBMSs) specifically focusing on RDF Data. Sometimes they need to extend Triples to store extra information, thus actually becoming Column Stores. A common extension are Quads, useful to add context or provenance to triples. Another NoSQL semantic graph database is GraphDB, which may work schema-free or exploiting an RDF ontological schema. Triplestores are specialized for RDF knowledge graphs and thus not optimized for generic data handling, like standard DBMSs. Since data representation constrained to using URIs does not necessarily make sense out of the Automated Reasoning applications (e.g., the Semantic Web), we aim at working with 'normal' DBs but still adopting the graph approach and still being able to carry out formal reasoning on their contents.

A more general structure than Triplestores is provided by graph DBs, based on the Labeled Property Graphs (LPGs) model [15]. In LPGs, both nodes and arcs may have names (called *labels* for nodes and *types* for arcs) and can store *properties* represented as key/value maps. Many arcs, possibly labeled with the same type, may exist between the same pair of nodes. Operationally, nodes and arcs are associated with unique identifiers. The most relevant differences between RDF graphs and LPGs are [16]:

- Nodes are atomic in RDF graphs while they carry information in LPGs; this ensures a much more compact structure in the latter (the estimated decrease in number of nodes is of up to one order of magnitude), which means that not only the former are much less readable, they also cause a significant decay in efficiency, especially in browsing-intensive tasks such as Social Network Analysis or Graph Mining algorithms;
- RDF cannot distinguish different occurrences of the same relationship between the same pair of entities; this is possible in LPGs thanks to the unique identifiers of relationships instances;
- RDF cannot attach properties to instances of relationships; the reification solution (transforming a relationship instance into an object which has relationships to the

original Subject and Object and to the additional properties) worsens readability; another partial solution is via annotations;

- RDF admits multivalued properties (triples with same subject and predicate but different object); these are recovered in LPGs by using arrays as property values;
- The notion of Quad has no equivalent in LPGs, but LPGs have labels, types and properties to carry additional information;
- There is only one kind of node in LPGs, but two in RDF graphs (URIs or literal values for objects of triples).

Whilst not directly related to data storage and management, and seemingly irrelevant, readability may be important for exploitation purposes when a portion of the graph is to be graphically displayed for humans—one of the main strengths of graphs. For the reader's reference, Table 1 provides a comparison of the different terms used to denote the same concepts in the DB, KR and LPG communities. In the following we will use them interchangeably, depending on the needs and context.

Table 1. Alignment between DB, Ontology and LPG terminology.

DBs	KR	LPGs
Entity	Class	Node (label)
Relationship	Object Property	Arc (type)
Attribute	Data Property	Property
Value	Datatype	Value
Instance	Individual	Node/Arc

The relevance of the graph-based approach to DB technology nowadays is witnessed by many big players in the industry developing their own solutions: just consider Google's 'Knowledge Graph', Facebook's 'Social Graph' and Twitter's 'Interest Graph'. All these solutions are proprietary and specifically intended for use in the products of such companies. As a more general-purpose solution we may mention Microsoft Research's 'Graph Engine' (previously known as 'Trinity') [17], a project started in 2010 and released as open source in 2017; however, no recent news is available for it, nor any particular success has been reported for it. In the following we will refer to Neo4j [18], the most popular graph DB according to DB-Engines, a platform that ranks DBMSs according to their popularity [19]. It is currently ranked #17, gaining 4 places in the past year [20]. It has been adopted by many big companies and governmental organizations for several different and relevant use cases, including Recommendation, Biology, Artificial Intelligence and Data Analytics, Social Networks, Data Science and Knowledge Graphs [21].

In Neo4j labels usually represent classes, nodes represent class instances, types represent relationships and arcs represent relationship instances. Each node may be associated with many labels, while each arc may have at most one type. Neo4j comes with a powerful query language (Cypher) and extensive libraries for advanced data manipulation (APOC). However, Neo4j (as most graph DBs) is schema-free: the user may apply any label/type or property to each single node or arc. Only simple 'constraints' may be defined to bias the DB content; while ensuring great flexibility, this causes the lack of a clear semantics for the graph contents. This motivated this work, aimed at proposing a schema formalism for graph DBs. In particular we believe the schema must be in the form of an ontology, so as to enable high-level reasoning on the available knowledge and still benefit from the advantages provided by graph DBs and LPGs. Specifically, we may leverage the advantages of DBMSs (scalability, storage optimization, efficient handling, mining and browsing of the data, etc.) and LPGs (flexibility, expressive power) for handling individuals, and exploit the high-level functionalities of ontological reasoners (allowing formal reasoning on, and consistency or correctness checks of, the data) on the ontological part.

On the methodological side, a few theoretical works analyze the possibilities of cooperation between ontologies and graph DBs, e.g., ref. [22] recognizes the need, but limited adoption, of logic-based KR for the development of KGs and summarizes some attempts to tackle this issue. Ref. [23] uses Neo4j to show how ontological schemas can be applied to Multilayer graphs (graphs whose labeled edges belong to a number of predetermined classes) and their algebraic counterpart, ontological tensors, also elaborating on complexity.

Other approaches are more practical, aimed at mapping ontologies or KGs to graph DBs. Ref. [24] stores the Freebase KG in Neo4j. As opposed to our proposal, it is not interested in developing ontologies as schemas for the graph DB; actually, it focuses on simple 'querying', not on 'reasoning', and the power or the proposed queries is incomparable to what can be obtained using automated reasoning techniques from AI. Most other works specifically focus on the mapping between OWL and LPGs. G2GML [25] maps OWL (RDF graphs) to PGs to overcome the limitations of SPARQL in implementing traversal or analytics algorithms. It proposed an exchangeable serialization format to support different graph DBMSs and their interoperability, but redefined the PG model. OWL2LPG [26] maps OWL 2 ontologies to an LPG representation, and vice versa, identifying specific kinds of queries that in Neo4j should be both easily expressible and more performant than in WebProtégé 4.0. Since the queries concern the ontology axioms and their revisions, it translates the ontology, not the data. In our approach the ontology stays apart from the DB, where only the data are stored and queried. SciGraph [27] aims at representing OWL ontologies and data as Neo4j graphs. It is strictly 'OWL-centric' and implementation dependant: it reads only formats available to the OWLAPI [28]—an API for OWL which is fully compliant with the official OWL specifications by W3C—and ignores the rest. It is clearly stated that creating ontologies based on the graph and supporting reasoning are not goals of this work. Therefore, it is exactly opposite to our work. VirtualFlyBrain [29] aims at translating only "a well defined subset" of OWL 2 EL ontologies into Neo4j and back in such a way that entailments and annotations (not the syntactic structure) are preserved after the round-trip. Differences from other mappings, such as SciGraph, are quite technical, e.g., having to do with the treatment of blank nodes or with the use of 'safe labels' for typing relations (a safe label is basically the URI with all non-alphanumeric characters being replaced by underscores). The authors point out some 'idiosyncrasies' of the approach, again very technical. Like us they only support datatypes that are supported by both Neo4j and OWL. As opposed to us, they label individuals with their most direct class, while we label them with their top-level class. All these approaches adopted a perspective biased towards ontologies and on their mapping on the graph DB. Since LPGs are more structured than RDF graphs, this direction seems quite obvious, at least syntactically. Since we believe that the DB technology is more mature and widely exploited than the ontology one, we take the opposite perspective and aim at preserving the DB structure and organization, superimposing the ontology on it only so far as it can be easily done.

OWLStar [30] exports Neo4j to OWL but specifying ontological semantics (e.g., OWL-DL interpretations), to be converted to OWL, in edge properties, so the driving perspective is again OWL-centric. It uses RDF* (and its query language SPARQL* that extends SPARQL), in an attempt to bring PGs into RDF by adding syntax to attach properties to edges. Ref. [31] proposes a formal mapping between LPGs and RDF* that can be leveraged to keep the data in the DB and render them in RDF*. However, RDF* is an extension of RDF and thus not compliant with standard reasoners, which prevents immediate reuse of the many reasoners available in the literature for performing ontological reasoning that involves instances. To overcome this limitation we developed a mapping of LPGs onto standard RDF. This required reconciling the differences between the two models and notably the inability of RDF to express datatype properties on relationships.

Some discussions and practical proposals can be found in the Neo4j community blog. The mainstream approach [32] proposes solutions for interoperability of Neo4j data and automated reasoning on them. The former is obtained by exporting Neo4j instances to RDF, e.g., upon request of an ontological reasoner. One way to do this is exporting Neo4j data in JSON using Cypher and the APOC libraries [33] and then further translating the result into other ontological formats (e.g., using libraries such as [34]). The latter is obtained by importing an RDF ontology into Neo4j, e.g., using the tool provided by the 'official' Neo4j library [35]. The RDF triples specifying the ontology are just transposed into nodes and arcs in the graph, so that the graph DB includes the schema, almost like schemas are stored in relational DBs as tables within the DB itself. On this representation, some (simple) kinds of ontological reasoning (e.g., navigation of the subclass hierachy) are translated into DB queries using Cypher. This solution has several drawbacks. First, the graph would include two disjoint parts, the ontology and the data, to be handled in totally different ways albeit coexisting in the same graph (in relational DBs they would be stored in different schemas, while in graph DBs there is a single overall graph). Second, no formal discussion is provided about what kinds of reasoning can be mapped onto graph DB queries. We expect them to be quite limited if compared to the power of state-of-the-art ontological reasoners. Furthermore, implementing these reasoning facilities is still in charge of the applications accessing the DB. Finally, it does not prevent data that are not compliant with the intended ontology to be inserted into the DB.

Instead, we propose an API, to be exploited by all applications accessing the DB, that wraps the DB and enforces compliance of the data with the intended schemas in both building and consulting the DB. In our vision KB designers must provide pre-specified data schemas, expressed in the form of ontologies for LPGs, that this API will interpret and use to drive all subsequent accesses to the DB. By referring to a schema, the applications will commit to be compliant with it, as in traditional databases. Just like in Triplestores and RDF* this will ensure a tight integration between the data and the schema. As opposed to Triplestores, RDF* and most of the cited works, where the ontology is ingested in the graph, the data/instances (stored in the graph DB) are kept apart from the schema/ontology (specified in a file external to the DB, using an ontological representation format). As discussed in Section 4, we leverage this separation between the data repository and the data schema to obtain the additional opportunity of applying different (but compatible) schemas to the same DB. Indeed, each schema may represent a different, partial view on the same data, allowing to limit or expand the possible interactions depending on specific needs and adding flexibility to our solution. Again, this is not even thinkable in Triplestores.

Proposing an ontological format brings the need for tools to comfortably build, browse and edit the ontologies expressed in this format. Several tools have been proposed, in the literature and practice, for the current standard ontology representations (notably OWL). Each pursues different objectives as regards the construction, editing, annotation and merging of ontologies [36]. Protégé [37,38], based on the OWLAPI, is the most popular and mature. Different versions, extensions and plugins for Protégé have been proposed (e.g., [39,40]), including an online version. Since sometimes they are not completely compatible with the original tool, we will take the OWLAPI as the standard reference in the rest of this paper. Since the ontological format for LPGs we propose in this paper has different features than those available for the RDF graph model, we also developed a corresponding tool for ontology definition and handling. In particular, it allows the ontology designer to specify attributes also for relationships and to specify labels for nodes and types for arcs, which is not allowed by extant ontological standards and tools. Therefore, our starting point was the need to define a schema for the graph DB, and the tool was developed so as to allow the users to comfortably define a schema to be used for building the KB. Then, in order to enable OWL reasoning capabilities, the translation in standard ontology format was a consequential objective. The various approaches proposed in the literature to assess the quality of tools for the construction of ontologies [41] can provide useful hints for improving and extending our tool with advanced features.

3. GraphBRAIN Graph Database Scheme Format

The *GraphBRAIN Schema* (GBS) format we propose to define graph DB schemas consists of an XML file whose tags allow us to exploit the representational features provided for by the LPG model (we developed a DTD for automated syntax checking of GBS files). In the following, when specifying the GBS file structure, we will adopt the usual notation of square brackets [...] to denote optional elements, curly brackets {...} to denote repeated elements and pipes in parentheses (...|...) to denote choices. Furthermore, we write XML tag names in boldface, XML tag attribute names in italics and entity or relationship names in smallcaps. Text in plain typeface reports comments useful to understand the various elements and their behavior.

The main structure of the XML with the tags and their nesting is reported in Table 2, where the universal entity ENTITY and the universal relationship RELATIONSHIP, acting resp. as the roots of the entity and relationship hierarchies, are implicitly assumed (remember that in ontological terminology entities correspond to classes and relationships correspond to object properties). Therefore, entities and relationships are to be specified only starting from the first level of specialization, which we will call *top-level*. Since each node (resp., arc) in the graph must be associated with one top-level entity (resp., relationship), the top-level entities (resp., relationships) are to be considered as disjoint. They may be the roots of specialization hierarchies of sub-entities (resp., sub-relationships). The set of direct specializations of a (sub-)entity or (sub-)relationship are in turn disjoint and are not to be intended as a partition: instances that do not fit any of the specializations of a parent (sub-)entity or (sub-)relationship may be directly associated with the parent. Therefore, also the root and intermediate levels of each hierarchy admit instances in the knowledge base. This design choice prevents multiple inheritance (associating an instance to many classes belonging to different branches in the hierarchy). We partially recover this at the level of instances: when two instances of different (sub-)entities represent the same object, we link them using an ALIASOF relationship. The single reference object represented by all these instances takes the union of their attributes.

Table 2. Main structure of GBS files.

domain // tag enclosing the overall ontology
  [**imports**]
  entities // tag enclosing the classes
    {**entity**} // see **(*)**
  relationships // tag enclosing the relationships
    {**relationship**} // see **(*)**

Entities and relationships are specified using the structure shown in Table 3. **Reference** is used only in relationships to specify their possible domain-range pairs, **taxonomy** is optional (used only if the entity or relationship has sub-entities or sub-relationships) and allows us to conveniently represent the specialization-type assertions; all other object properties are to be specified in the **relationships** section. **Attributes** is mandatory for entities (an entity instance must be described by some attribute) and optional for relationships (a relationship may carry information in its very linking two instances). **Specialization** is a recursive tag, allowing to define hierarchies of sub-entities or sub-relationships. In addition to its own attributes each specialization inherits all the attributes of the (sub-)entities (resp., (sub-)relationships) on the hierarchy path from its specific **specialization** section up to the corresponding top-level entity (resp., relationship).

Table 3. Structure for describing entity and relationship hierarchies in GBS files.

(*) (**entity** | **relationship** | **specialization**) tag
  [**references**]
    {**reference**}
  [**taxonomy**]
    {**specialization**} // see **(*)** (recursive)
  [**attributes**] specifying the data properties
    {**attribute**}

Some tags have XML attributes that specify the details of the item they represent in the schema:

- **domain** tag:

  name the unique identifier for the domain being described
  author the author of the schema
  version the version of the schema

- **entity** tag:

  name the unique identifier for the entity

- **relationship** tag:

  name the unique identifier for the relationship
  inverse the unique identifier for the inverse relationship of *name*

- **reference** tag:

  subject the identifier of the entity that is the domain of the (sub-)relationship
  object the identifier of the entity that is the range of the (sub-)relationship

- **specialization** tag:

  name the unique identifier for the specialization (sub-entity or sub-relationship)
  [*inverse*] the unique identifier for the inverse sub-relationship of *name* (not used for sub-entities)

- **attribute** tag:

  name an identifier for the attribute
  mandatory = (**true** | **false**)
    whether the attribute must take a value in each instance
  distinguishing = (**true** | **false**)
    whether the attribute may concur in distinguish instances having the same values for mandatory attributes
  display = (**true** | **false**)
    whether the attribute represents interesting additional information with respect to mandatory and distinguishing attributes, to be possibly displayed
  datatype = (**integer** | **real** | **boolean** | **string** | **text** | **select** | **tree** | **date** | **entity**)
  [*length*] the maximum allowed number of characters (used only when datatype = string)
  [*target*] an entity name (used only when datatype = entity)

Therefore, the union of mandatory and distinguishing attributes of an entity or relationship can be used to specify a key for uniquely identifying its instances. The union of mandatory, distinguishing and display attributes of an entity or relationship can be used to build and display a summary reporting the most relevant information about the instances.

Regarding datatypes, attributes of type *integer, real, boolean, string, text* take an atomic value of the corresponding type, where *text* is intended for free text of any length, differently from *string* which has a limited maximum length that can be specified in the 'length' attribute. Attributes of type *date* take values in one of the following forms:

- Year;
- Year/month;
- Year/month/day.

where year is any integer, month $\in \{01, \ldots, 12\}$ and day $\in \{01, \ldots, 31\}$. Attributes of type *select* denote a choice in an enumeration of values, described using the substructure reported in Table 4; attributes of type *tree* denote a choice in a tree of values, described using the recursive substructure shown in Table 5. Attributes of type *entity* denote 1:1 relationships between an instance of the current entity and an instance of another entity (specified in the 'target' attribute of the tag), e.g., the birthplace of an entity Person would be modeled as an attribute of type *entity* with target='Place':

```
<entity name="Person">
   <attributes>
      <attribute name="birthplace" datatype="entity" target="Place"/>
   </attributes>
</entity>
```

Table 4. Structure for describing enumerative attribute values in GBS files.

attribute ... datatype="select" tag
values
{**value**}

Table 5. Structure for describing enumerative attribute values in GBS files.

(**) (**attribute ... datatype="tree"** | **values**) tag
values
{**value**} // see **(**)** (recursive)

As a conventional notation we propose identifiers made up of uppercase letters, lowercase letters or decimal digits only. They should start with an uppercase letter for entity names and enumeration or tree values, or with a lowercase letter for domain, relationship and attribute names. Multi-word names are built by juxtaposing their constituent words, using an uppercase letter for the first letter of each word (except for the first one, as prescribed above). When writing documentation, a relationship 'rel' between an entity 'Subj' and an entity 'Obj' can be represented using the dot notation

Subj.rel.Obj

which is not ambiguous since dots are not allowed in our entity and relationship names.

Tables 6 and 7 show a fragment of a GBS file concerning the domain of computing. We see entity 'Component', representing an electronic component and including a taxonomy of sub-classes, some of which have specific attributes of various type, e.g., sub-class 'Memory' has attributes 'capacity' and 'speed' in addition to those inherited by 'Component' ('name', 'description', 'originalPrice' and 'announcementDate'). In the relationships section we see that relationship 'wasIn' may be established between a 'Component' and an 'Event' (to signify that the component was on show at the event), or between a 'Person' and a 'Place' (meaning that the person was in that place), etc.

Table 6. Sample fragment of ontology in GBS format (part 1).

```
<!--  <!DOCTYPE domain SYSTEM "graphbrain.dtd">  -->
<domain name="retrocomputing" author="stefano" version="1">
  <entities>
    <entity name="Component">
      <attributes>
        <attribute name="name" mandatory="true" datatype="string"/>
        <attribute name="description" mandatory="false" datatype="text"/>
        <attribute name="originalPrice" mandatory="false" datatype="real"/>
        <attribute name="announcementDate" mandatory="false" datatype="date"/>
      </attributes>
      <taxonomy>
        <specialization name="Chip">
          <taxonomy>
            <specialization name="Logic">
              <taxonomy>
                <specialization name="FlipFlop">
                  <attributes>
                    <attribute name="type"
                          mandatory="false" datatype="select">
                      <values>
                        <value name="D"/>
                        <value name="FK"/>
                        <value name="JK"/>
                        <value name="T"/>
                      </values>
                    </attribute>
                  </attributes>
                </specialization>
                <specialization name="Memory">
                  <attributes>
                    <attribute name="capacity"
                          mandatory="false" datatype="string"/>
                    <attribute name="speed"
                          mandatory="false" datatype="string"/>
                  </attributes>
                  <taxonomy>
                    <specialization name="EPROM"/>
                    <specialization name="PROM"/>
                    <specialization name="RAM"/>
                    <specialization name="ROM">
                      <attributes>
                        <attribute name="content"
                              mandatory="false" datatype="string"/>
                      </attributes>
                    </specialization>
                  </taxonomy>
                </specialization>
              </taxonomy>
            </specialization>
            <specialization name="MicroProcessor">
              <attributes>
                <attribute name="speed" mandatory="false" datatype="string"/>
                <attribute name="bits" mandatory="false" datatype="integer"/>
              </attributes>
            </specialization>
            <specialization name="PLA"/>
            <specialization name="RRIOT"/>
          </taxonomy>
        </specialization>
        [...]
      </taxonomy>
    </entity>
    [...]
  </entities>
```

Table 7. Sample fragment of ontology in GBS format (part 2).

```
    <relationships>
      <relationship name="wasIn" inverse="hosted">
        <references>
          <reference subject="Company" object="Event"/>
          <reference subject="Company" object="Place"/>
          <reference subject="Component" object="Event"/>
          <reference subject="Event" object="Place"/>
          <reference subject="Person" object="Company"/>
          <reference subject="Person" object="Event"/>
          <reference subject="Person" object="Place"/>
          [...]
        </references>
        <attributes>
          <attribute name="reason" mandatory="false" datatype="string"/>
          <attribute name="position" mandatory="false" datatype="string"/>
        </attributes>
      </relationship>
      [...]
    </relationships>
</domain>
```

Each GBS schema is intended to describe one domain. However, sometimes wider domains involve ontological elements that are already described in more 'basic' schemas (e.g., the schemas for Cultural Heritage, Food and Transportations might be exploited in the ontology aimed at supporting a touristic application) and it might be useful to reuse such schemas, both for standardization of the definitions and for building on existing knowledge. Actually, the combination of many schemas is more powerful a representation than the simple juxtaposition of their elements. Indeed, their shared entities act as bridges that allow, through the relationships available in those domains, to connect proprietary entities of each domain that would not otherwise have a chance to be related with each other. In the GBS framework, classes and relationships in different ontologies are considered the same (and thus are shared) if they have the same name. They may have, however, different attributes, reflecting the different perspectives associated with the different domains. If an attribute is present in different domains it must have the same type in all of them. Moreover, additional cross-schema relationships (and entities) may be defined in the overall ontology, building on the existing ones. GBS schemas support such opportunity by providing for an optional section in which existing schemas can be imported. The structure of this section (delimited by tag **imports** and placed at the beginning of the schema, before the entities and relationships) is as shown in Table 8. The tag attributes are:

- **import** tag:
  schema: the name of a schema to be imported
- **delete** tag:
  elementtype = (entity | relationship)
  elementname: the name of the element to be deleted

Table 8. Structure for describing imported schemas in GBS files.

imports tag
  {**import**}
  [{**delete**}]

Schemas are imported in the same order as specified by the sequence of **import** tags. Definitions of top-level elements (entities or relationships) in an imported schema having the same name as elements defined in previous imported schemas override the previous definitions. Finally, elements defined in the **entities** or **relationships** sections of the importing schema override elements with the same name in all imported schemas. Since it may happen that some elements of the imported schemas are not needed in the current domain, **delete** tags allow to remove them from the overall ontology.

In addition to the API for GBS-based handling of Neo4j, we developed tools for GBS schema/ontology editing and for data management. They were implemented as Web Applications based on the Java Server Faces technology and the PrimeFaces library. JavaScript was used for handling interactive browsing of the graph. A connection to Prolog allows it to carry out rule-based reasoning on selected portions of the data. Obviously Neo4j was used to store the knowledge graph, while Postgres was used to store user and usage data (roles, access rights, change log, etc.). A demo of the tools can be found at http://193.204.187.73:8088/GraphBRAIN/ in the form of a general-purpose system for the collaborative development, management and (personalized) fruition of a KB, in the same spirit as Freebase [42]. After logging into the system, the user may choose a domain and all subsequent interaction is driven by the corresponding GBS schema. Screenshots of the current online prototypes are shown in Figures 1–3.

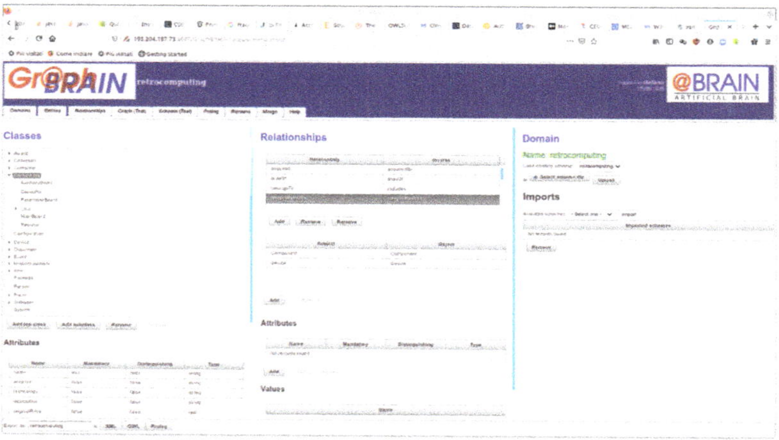

Figure 1. Online editor for GBS schemas/ontologies.

Figure 1 shows the interface for building, editing and browsing GBS schemas/ontologies. In the left-hand-side section the entity hierarchy, with entity attributes and attribute types and values, can be handled. In the center section the same can be done for relationships, also including inverse relationships and references. On the right-hand-side section imports can be handled and existing schemas can be loaded. On the bottom several save and export buttons are available. Figure 2 shows the interactive interface to feed and consult information in the knowledge base by direct interaction. It consists of two form-based tabs, one for entities (Figure 2a) and one for relationships (Figure 2b), allowing the user to insert, update, remove or query instances. The forms are automatically generated by the system from the GBS specification of a schema and interact with the graph DB using our API to enforce consistency with the selected schema. Let us first describe the entity tab. In the left-hand-side section (sub-)entities and corresponding instances can be selected. In the center section a form with the attributes of the selected (sub-)entity is shown, possibly filled with the values from the selected instance. Regarding the relationships tab, the center section allows to choose a relationship, for which subject and object (sub-)entities and corresponding instances can be selected in the left- and right-hand-side sections, respectively. When a triple (subject, relationship, object) is selected, the center section also shows a form with the attributes of the selected (sub-)relationship. If subject and object instances are also selected, a drop-down menu allows selecting a specific relationship instance, in which case the attribute form is filled with the corresponding values. More functions are available (e.g., handling of attachments to the selected instances, or search and collaborative evaluation facilities) but their description is beyond the scope of this paper.

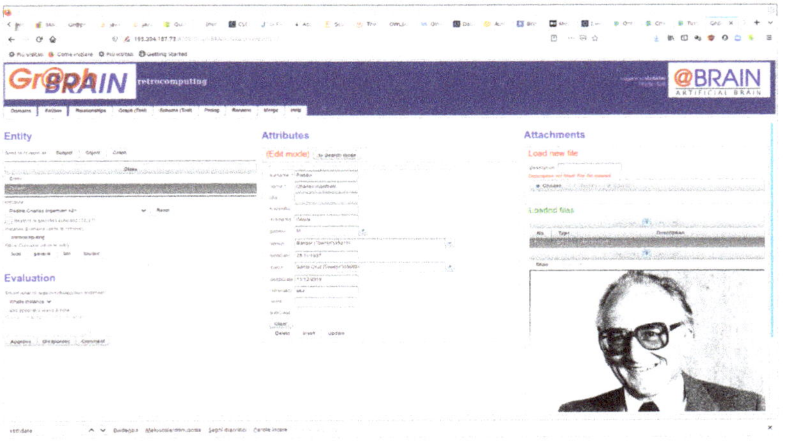

(**a**)

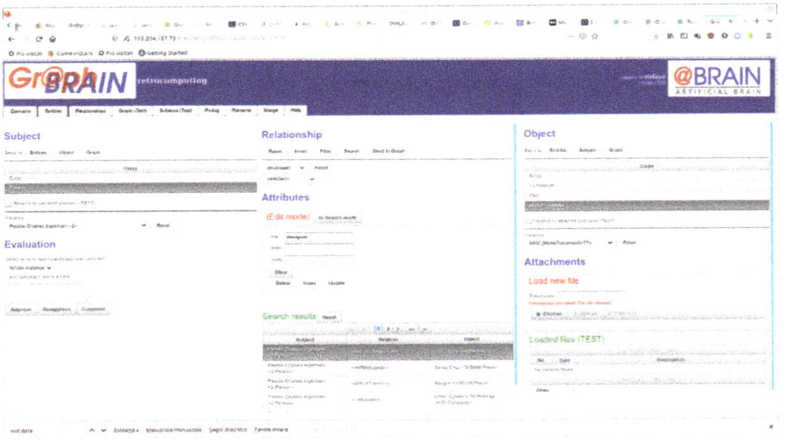

(**b**)

Figure 2. Online interfaces for managing and consulting GBS knowledge bases: (**a**) entities, (**b**) relationships.

Figure 3 shows the tab in which users can display and manually browse the graph. Since the whole KB would be too large to be readable, only a portion thereof is shown in this tab. The portion is dynamically generated so as to focus on the portion of graph of interest to the user based on their profile, optionally starting from selected nodes specified by him. In the figure, the graph was generated for user 'stefano' starting from nodes representing Chuck Peddle (a pioneer in microprocessor design) and the 6502 (one of the earliest and most successful microprocessors on the market), identified by a thicker node border. Different colors of nodes denote different classes (e.g., light blue for Person, yellow for Component, etc.). At a glance, it is possible to see clusters of nodes that represent possibly relevant aggregates of information to be investigated or explored. Note that the nodes and arcs in this view may belong to different schemas, not only to the schema selected for the form-based interaction. Therefore, here the user may discover connections that are beyond the starting domain. The user may pan and zoom on the graph, drag nodes, dynamically follow links, read attributes of nodes and/or arcs, further expand the graph around nodes of interest and run analytics and mining algorithms from menus on the right-hand-side and contextual menus that appear by clicking on the graph. The information on a node or arc in this view is the complete set of properties for that node or arc, gathered from all domains in which it is involved.

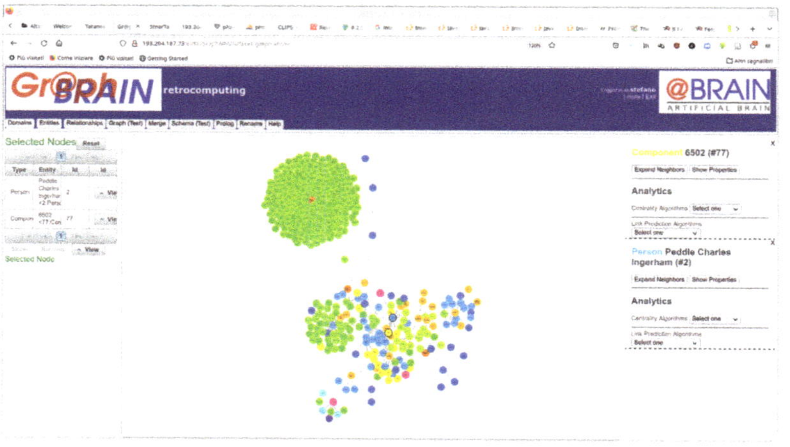

Figure 3. Online interface for browsing GBS knowledge bases.

4. Mapping onto DB and Ontology

Since graph DBs are naturally suited to express knowledge graphs, i.e., knowledge bases underlying given ontologies, a fundamental requirement of our approach is that our schemas can be mapped onto both the DB and to an OWL representation which can then be processed by a reasoner. In this section, we report in detail how these two mappings work in practice.

4.1. Use as a Graph DB Schema

As said, part of the main motivation for defining GBS schemas is to endow LPG-based graph DBs with a schema that ensures a clear semantics to the information pieces they contain and provides directions for their management and interpretation. According to this perspective the DB users will be required to work according to pre-specified data schemas expressed in the form of ontologies. Operationally, the DB will be wrapped into a layer, e.g., in the form of an API (see the previous section), that takes as input a GBS schema specifying the desired domain ontology and controls all interactions, allowing the external applications to manipulate and consult only information items that are compliant with the ontology.

In our approach we also provide an additional opportunity. Specifically, we allow a single graph DB to underlie several domains (schemas), provided that their elements (entities and relationships) are compatible. By *compatible* we mean that for elements having the same name in the different schemas, attributes having the same name must have the same datatype, too. The other attributes, or non-shared elements, can be freely defined. Therefore, using any of such schemas on the DB would provide a partial view of its contents, perhaps representing a different perspective or aimed at limiting access to the DB contents for some users or applications.

Let us now show how the GBS elements are implemented using LPG features. For easy reference, Table 9 summarizes the mapping.

Table 9. Correspondence between GBS elements and LPG features.

GBS Element	LPG Feature
entity instance	node
relationship instance	arc
entity name	label
relationship name	type
domain name	label
entity attribute	node property
relationship attribute	arc property

4.1.1. Entities and Relationships

Leveraging the possibility of using many labels for nodes, each node is labeled with the top-level entity it belongs to and with all the domains for which it is relevant (e.g., 'Herbert Simon' would be labeled with 'Person' for the entity and with 'economy' and 'computing' for the domains). When the same DB underlies several domains, this allows to select only the instances actually involved in a domain of interest. On the other hand, since each arc may take at most one type, we use it for specifying the relationship it expresses. The domains for which a relationship instance is relevant may be inferred from the domain labels of the nodes it connects by considering all the domain labels that are present in both its subject and its object.

4.1.2. Attributes

Concerning attributes, we propose to reserve an attribute name (*'specialization'*) to store which is the specific sub-entity (resp., sub-relationship) the entity (resp., relation) instance belongs to. Given the top class (resp., relationship) specified in the labels (resp., types) and the specific sub-entity specified in the 'specialization' property, the path of specializations between these two may be easily recovered bottom-up starting from the latter and climbing the specialization hierarchy in the ontology up to the former (since nodes admit many labels, one might specify all the sub-entities in such a specialization path as labels; for the sake of uniformity with arcs, where this is not possible, we propose the above solution). We also propose to implicitly assume another reserved attribute *'notes'* for both nodes and arcs, that allows to add information not considered by the other, domain-specific attributes.

4.1.3. Attribute Types and Values

Attribute values of types *integer*, *real*, *boolean*, *string* and *text* are stored as literal values for the corresponding DB types, e.g., Neo4j provides the following types matching GBS types: Integer and Float (both subtypes of an abstract type Number), Boolean, and String.

For types *select* and *tree* the string corresponding to the selected value in the list or tree is stored.

An attribute of type *entity* actually corresponds to a relationship between the current instance and an instance of the target entity and thus it is stored in the DB as an arc, connecting the nodes corresponding to these two instances and having the attribute name as type. Note that in our proposed naming policy attribute names start with a lowercase letter, just like relationship names.

Finally, albeit Neo4j provides for temporal types, including 'Date', following [18] we propose to model attributes of type *date* as relationships, as well. We assume the ontology implicitly defines four entities, as shown in Table 10:

DAY representing a specific day of a specific year, with integer attributes *day*, *month*, *year*;

MONTH representing a specific month of a specific year, with integer attributes *month*, *year*;

YEAR representing a year, with a single integer attribute *year*.

TIMELINE representing the overall timeline.

This allows to specify dates at different granularity, differently from the Date type available in Neo4j. Neo4j provides functions for Date truncation to Month or Year, but such truncations actually correspond to the first day of the month or year and thus there is no way to distinguish whether a date like 2020/01/01 actually refers to the specific day or is a truncation for the month (2020/01) or year (2020). A single TIMELINE node is automatically added to the DB. DAY, MONTH or YEAR nodes are automatically added to the DB for each year/month/day, year/month or year value, resp., in date attributes of instances. The DB will also automatically link, using arcs of type BELONGSTO, each DAY node with the corresponding MONTH node, each MONTH node with the corresponding YEAR node and finally all YEAR nodes with the TIMELINE node. This will allow collecting all instances referring to the same date at different levels of granularity. Furthermore, arcs of type FOLLOWS may be added and maintained between adjacent days, months or years in the DB. This will allow to easily extract from the DB time intervals and associated information.

Table 10. Implicit entities and relationships for time handling.

```
<entities>
   <entity name="Timeline"/>
   <entity name="Year">
      <attributes>
         <attribute name="year" mandatory=""true" datatype="integer"/>
      </attributes>
   </entity>
   <entity name="Month">
         <attribute name="month" mandatory=""true" datatype="integer"/>
         <attribute name="year" mandatory=""true" datatype="integer"/>
   </entity>
   <entity name="Day">
         <attribute name="day" mandatory=""true" datatype="integer"/>
         <attribute name="month" mandatory=""true" datatype="integer"/>
         <attribute name="year" mandatory=""true" datatype="integer"/>
   </entity>
</entities>
<relationships>
   <relationship name="belongsTo" inverse="includes">
      <references>
         <reference subject="Day" object="Month"/>
         <reference subject="Month" object="Year"/>
         <reference subject="Year" object="Timeline"/>
      </references>
   </relationship>
   <relationship name="follows" inverse="precedes">
      <references>
         <reference subject="Day" object="Day"/>
         <reference subject="Month" object="Month"/>
         <reference subject="Year" object="Year"/>
      </references>
   </relationship>
</relationships>
```

4.2. Mapping to OWL Format

The other part of our motivation for this work was using the ontology level not only as a DB schema, but also to carry out formal reasoning and consistency or correctness checks on the individuals. As noted in Section 2, a widespread standard for representing ontologies is OWL, based on a different model than LPGs, on which GraphBRAIN ontologies are based. While of course new reasoners may be purposely developed for GBS ontologies, it would be desirable to translate GBS ontologies into OWL, so as to allow immediate reuse of the many existing tools for OWL ontologies. This section provides a strategy for this translation, aimed at overcoming and reconciling the differences in concepts, perspectives and expressive power between the two ontological models. For compliance with existing tools and reasoners, our implementation of GraphBRAIN adopted the same OWL-API as Protégé for its ontology export functionality, so that the generated ontologies are fully compliant with the standard and may be edited using Protégé. So, in the following, we will use the OWL-RDF syntax accepted by Protégé.

When serializing GBS ontologies to OWL format we propose to use prefix **gbs** in the namespaces, so that they can be easily recognized.

Note that here we just provide the translation for the basic GBS format, expressing the DB schema. Additional tags/features can be added to this basic format to express information intended for use by the ontological level (e.g., transitivity of relationships, etc.), but this is a wide path of investigation and will be developed in future work.

As a reference for the subsequent discussion, we provide in Figures 4–6 some screenshots of a sample GBS ontology (concerning the domain of 'computing') exported in OWL using our API and opened with Protégé.

4.2.1. Entities

Entities in GBSs correspond to Classes in OWL. Each (sub-)entity is declared in OWL using the **owl:Class** statement. Specializations are associated with their immediate superclass using the **rdfs:subClassOf** statement. The implicit universal entity ENTITY, generalizing all (sub-)entities defined in the schema, corresponds to the 'Thing' class in OWL. Since classes are to be considered as disjoint (see Section 3), the axioms for classes in the top level and the specializations of each (sub-)class also include (many) **owl:disjointWith** statements to all of their sibling (sub-)classes, e.g., the following fragment of taxonomy for entity DOCUMENT:

```
<entity name="Document">
   <taxonomy>
      <value name="Printable">
         <taxonomy>
            <value name="Book"/>
            <value name="Letter"/>
         </taxonomy>
      </value>
   </taxonomy>
</entity>
```

translates into the following OWL fragment:

```
<owl:Class rdf:about="http://owl.api.ontology#Document">
   <owl:disjointWith rdf:resource="http://owl.api.ontology#Component"/>
   <owl:disjointWith rdf:resource="http://owl.api.ontology#Device"/>
   <owl:disjointWith rdf:resource="http://owl.api.ontology#Person"/>
   <owl:disjointWith rdf:resource="http://owl.api.ontology#Place"/>
</owl:Class>

<owl:Class rdf:about="http://owl.api.ontology#Printable">
   <rdfs:subClassOf rdf:resource="http://owl.api.ontology#Document"/>
</owl:Class>

<owl:Class rdf:about="http://owl.api.ontology#Book">
   <rdfs:subClassOf rdf:resource="http://owl.api.ontology#Printable"/>
   <owl:disjointWith rdf:resource="http://owl.api.ontology#Letter"/>
</owl:Class>
```

```
<owl:Class rdf:about="http://owl.api.ontology#Letter">
   <rdfs:subClassOf rdf:resource="http://owl.api.ontology#Printable"/>
   <owl:disjointWith rdf:resource="http://owl.api.ontology#Book"/>
</owl:Class>
```

In the OWL translation, each entity instance is associated with the sub-class specified by its 'specialization' attribute of the top-level class specified in its labels.

In Figure 4, in the left-hand-side area of the window we see the class hierarchy, in which class 'Computer' (a sub-class of 'Device') has been selected and corresponding details are shown in the right-hand-side area. We may notice that Computer has in turn several sub-classes.

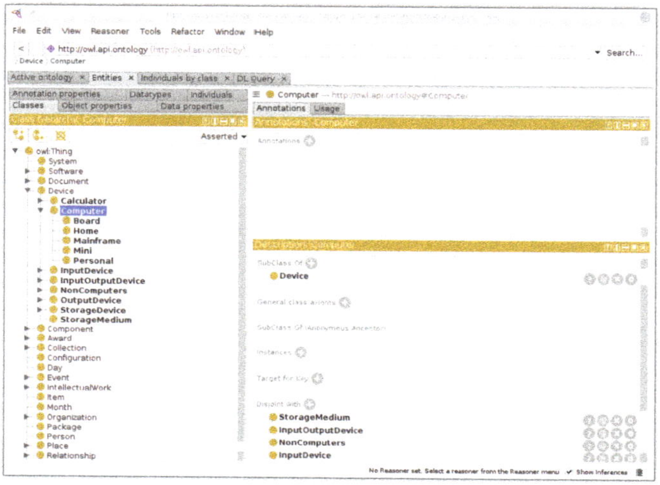

Figure 4. OWL translation of a sample GBS ontology loaded in Protègè: classes.

4.2.2. Relationships

Relationships in GBSs correspond to Object Properties in OWL. Each (sub-)relationship is declared in OWL using the **owl:ObjectProperty** construct. Specializations are associated with their immediate super-relationship using the **rdfs:subPropertyOf** construct. The implicit universal relationship RELATIONSHIP, generalizing all (sub-)relationships defined in the schema, corresponds to the 'topObjectProperty' object property in OWL. Subject and Object entities acting as references of a relationship in GBSs correspond to Domain and Range of the Object Property in OWL, expressed by constructs **rdfs:domain** and **rdfs:range**, respectively. The name for the inverse of a relationship in GBS is translated into OWL using the **owl:inverseOf** construct.

GBSs may use the same relationship name applied to possibly many Subject–Object pairs as references. This cannot be expressed directly in OWL. Adding all the Subject (resp., Object) entities as domain (resp., range) classes to the corresponding OWL object property would be interpreted in OWL as the intersection of the Subject (resp., Object) classes as the domain (resp., range) of the OWL object property.

When the subject (resp., object) of all references in a relationship is the same, the logical disjunction (OR) operator of the classes in the object (resp., subject) would solve the problem, e.g., the following relationship:

```
<relationship name="produced" inverse="producedBy">
   <references>
      <reference subject="Company" object="Device"/>
      <reference subject="Company" object="Software"/>
   </references>
</relationship>
```

meaning that companies may produce devices or software (but a specific company might produce both, or either, or none of them), might be represented as a single object property

Company.produced.(Device OR Software)

and the following relationship:

```
<relationship name="belongsTo" inverse="includes">
<references>
      <reference subject="Device" object="Collection"/>
      <reference subject="Document" object="Collection"/>
   </references>
</relationship>
```

meaning that devices or documents may belong to collections, might be represented as a single object property

(Device OR Document).belongsTo.Collection

However, in general, when the subjects and objects both involve many classes, adding the logical disjunction (OR) of the Subject entities as the domain and of the Object entities as the range would be a wrong translation, because it would not prevent OWL from accepting instances from incompatible Subject–Object pairs, e.g., if relationship WASIN can be applied to reference pairs COMPANY-EVENT and PERSON-PLACE:

```
<relationship name="wasIn" inverse="hosted">
   <references>
      <reference subject="Company" object="Event"/>
      <reference subject="Person" object="Place"/>
   </references>
</relationship>
```

using '(Company OR Person)' as the domain and '(Event OR Place)' as the range:

(Company OR Person).wasIn.(Event OR Place)

would admit relating an instance of Company to an instance of Place, which was not intended by the GBS ontology. We reconcile this by introducing in OWL one object property for each GBS relationship, using the same name and the disjunction (OR) of the Subject entities as the domain and the disjunction (OR) of the Object entities as the range. Then, for each Subject–Object reference pair for a relationship 'rel' in GBS, in OWL we define a new relationship 'rel_Subject_Object' with domain Subject and range Object, as a subObjectProperty (OWL feature **rdfs:subPropertyOf**) of 'rel' (not ambiguous since underscores are not allowed in GBS entity and relationship names).

The OWL translation of the previous example would be:

```
<owl:ObjectProperty rdf:about="http://owl.api.ontology#hosted"/>

<owl:ObjectProperty rdf:about="http://owl.api.ontology#wasIn">
   <owl:inverseOf rdf:resource="http://owl.api.ontology#hosted"/>
</owl:ObjectProperty>
```

```
<owl:ObjectProperty rdf:about="http://owl.api.ontology#wasIn_Company_Event">
   <rdfs:subPropertyOf rdf:resource="http://owl.api.ontology#wasIn"/>
   <rdfs:domain rdf:resource="http://owl.api.ontology#Company"/>
   <rdfs:range rdf:resource="http://owl.api.ontology#Event"/>
</owl:ObjectProperty>

<owl:ObjectProperty rdf:about="http://owl.api.ontology#wasIn_Person_Place">
   <rdfs:subPropertyOf rdf:resource="http://owl.api.ontology#wasIn"/>
   <rdfs:domain rdf:resource="http://owl.api.ontology#Person"/>
   <rdfs:range rdf:resource="http://owl.api.ontology#Place"/>
</owl:ObjectProperty>
```

In principle, we should add some constraint telling OWL that 'rel' is an 'abstract' relationship, i.e., it does not admit direct instances (any instances must belong to a subObjectProperty of 'rel'), but unfortunately this cannot be expressed in OWL [43]. However, since the OWL functionality will be applied only to the instances in the DB, which are controlled by the GBS ontology, in practice this constraint will be implicitly enforced for explicit instances. Only the reasoning might identify individuals belonging to 'rel'. Another option would be defining only the subObjectProperties, but semantically we would miss the information that they express the same concept declined for different references and operationally we would miss the opportunity of defining in 'rel' a core set of properties that apply to all of its sub-relationships. On the other hand, defining attributes (Datatype Properties) on Object Properties is forbidden by OWL and must be handled appropriately in the translation, as we will see in the next sections.

When the *name* of a relationship and its *inverse* in GBS are the same, instead of adding the inverse object property, the object property is labeled as symmetric, using the **owl:SymmetricProperty** construct, e.g., ALIASOF:

```
<relationship name="aliasOf" inverse="aliasOf">
   <references>
      <reference subject="Company" object="Company"/>
      <reference subject="Person" object="Person"/>
   </references>
</relationship>
```

is translated as:

```
<owl:ObjectProperty rdf:about="http://owl.api.ontology#aliasOf">
   <rdf:type rdf:resource="http://www.w3.org/2002/07/owl#SymmetricProperty"/>
</owl:ObjectProperty>

<owl:ObjectProperty rdf:about="http://owl.api.ontology#aliasOf_Company_Company">
   <rdfs:subPropertyOf rdf:resource="http://owl.api.ontology#aliasOf"/>
   <rdf:type rdf:resource="http://www.w3.org/2002/07/owl#SymmetricProperty"/>
   <rdfs:domain rdf:resource="http://owl.api.ontology#Company"/>
   <rdfs:range rdf:resource="http://owl.api.ontology#Company"/>
</owl:ObjectProperty>

<owl:ObjectProperty rdf:about="http://owl.api.ontology#aliasOf_Person_Person">
   <rdfs:subPropertyOf rdf:resource="http://owl.api.ontology#aliasOf"/>
   <rdf:type rdf:resource="http://www.w3.org/2002/07/owl#SymmetricProperty"/>
   <rdfs:domain rdf:resource="http://owl.api.ontology#Person"/>
   <rdfs:range rdf:resource="http://owl.api.ontology#Person"/>
</owl:ObjectProperty>
```

In Figure 5, the left-hand-side area reports the hierarchy of object properties corresponding to GBS relationships, all depending from the universal class 'topObjectProperty'. Object properties 'aliasOf' and 'belongsTo' have been expanded, showing the subproperties generated by the corresponding references. 'belongsTo_Award_Collection' is selected, whose details are reported on the right-hand-side area. Specifically, we see that its domain is class 'Award' and its range is class 'Collection' and that it is a subPropertyOf class 'belongsTo'.

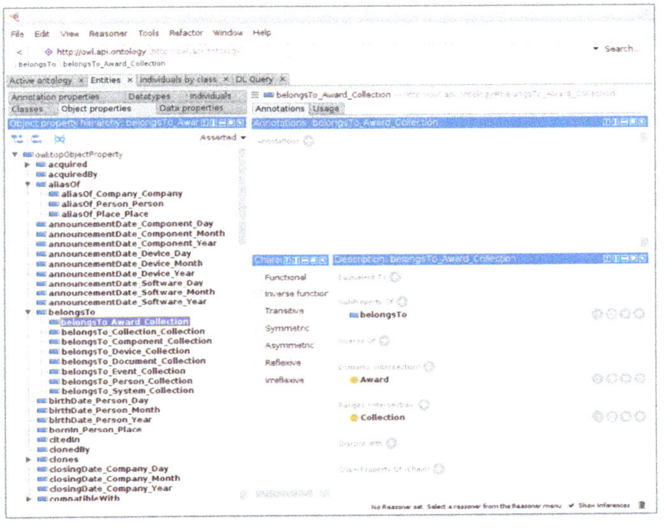

Figure 5. OWL translation of a sample GBS ontology loaded in Protègè: object properties.

4.2.3. Data Types

Attributes of data types *integer*, *real*, *boolean*, *string* and *text* are translated into OWL using the corresponding datatypes **xsd:integer**, **xsd:decimal**, **xsd:boolean**, **xsd:string** (for both string and text). Note that OWL provides several versions of some datatypes.

For types *select* and *tree*, we define in OWL an Enumerated datatype specifying the values in the list or tree. We do not need to store the tree structure, so we can flatten the tree values into a list, because (a) in GBS the tree is just a conceptual aid to the users, in order to build the interfaces to the DB; and (b) there are no duplicate values in the tree, e.g., the values for attribute 'gender' of entity 'Person' in this GBS fragement:

```
<entity name="Person">
   <attributes>
      <attribute datatype="select" mandatory="false" name="gender">
         <values>
            <value name="M"/>
            <value name="F"/>
         </values>
      </attribute>
</entity>
```

would be specifies as the range of the datatype property 'gender_Person' made up of the list of string values {M, F}:

```
<owl:DatatypeProperty rdf:ID="gender_Person">
  <rdfs:range>
    <owl:DataRange>
      <owl:oneOf>
        <rdf:List>
          <rdf:first rdf:datatype="&xsd;integer">M</rdf:first>
          <rdf:rest>
            <rdf:List>
              <rdf:first rdf:datatype="&xsd;integer">F</rdf:first>
              <rdf:rest rdf:resource="&rdf;nil" />
            </rdf:List>
          </rdf:rest>
        </rdf:List>
      </owl:oneOf>
    </owl:DataRange>
  </rdfs:range>
</owl:DatatypeProperty>
```

Attributes of type *entity* actually correspond to a relationship between the current instance and an instance of the *target* entity and thus they have as values the individuals of the corresponding *target* class.

Finally, OWL provides several datatypes for expressing the GBS *date* type (e.g., **xsd:date**). While for some purposes they may be enough for representing and handling this type, having an ontological description of time may allow more powerful reasoning. Recently, a specific OWL ontology of temporal concepts, OWL-Time [44], has been proposed for describing and handling temporal properties. This might be another solution, in the same spirit as our proposal but more complex and powerful. We reproduce the strategy discussed in Section 3. This option involves adding to the OWL ontology classes 'Day', 'Month', 'Year' and 'Timeline', and object properties 'belongsTo_Day_Month', 'belongsTo_Month_Year' and 'belongsTo_Year_Timeline', as specializations of a general 'belongsTo' relationship, to suitably connect these classes.

4.2.4. Entity Attributes

As usual in databases, attributes in different entities might have the same name but different meaning. Since in OWL each name must identify one element, we disambiguate by merging the attribute name with the entity it belongs to. Therefore, attribute 'attr' of entity 'Ent' will be stored as 'attr_Ent' in the OWL version of the ontology (not ambiguous since underscores are not allowed in entity names).

Attributes of data types *integer*, *real*, *boolean*, *string* and *text* are translated into OWL as datatype properties having the attribute class as the domain and the corresponding primitive OWL datatype as the range (as specified in the previous section).

As shown in the previous section, attributes of types *select* and *tree* are translated into a datatype property having the attribute class as the domain and an Enumerated Type as the range.

In Figure 6, on the left-hand-side, the data properties are shown, all depending from the 'topDataProperty' root. Some correspond to entity attributes. 'buttons_Mouse' is selected, showing its domain class ('Mouse') and the associated datatype ('integer').

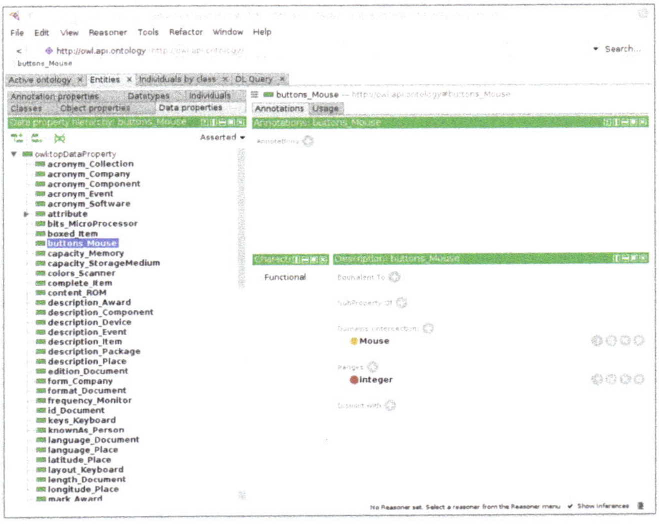

Figure 6. OWL translation of a sample GBS ontology loaded in Protègè: data properties.

Attributes of type *entity*, actually corresponding to a relationship between the instances of the attribute entity and those of the target entity, are translated as object properties having the attribute class as domain and the target class as range. This is compliant with our proposed naming policy, since attribute names start with a lowercase letter just like object property names. Specifically, since the target class individual associated with each domain class instance is unique, we also set this object property in OWL as functional (**owl:FunctionalProperty**).

Finally, according to the what reported in the previous section, attributes of type *date* can be modeled as datatype properties or as object properties.

In Figure 5, some object properties correspond to entity attributes of type 'entity' or 'date', e.g., 'announcementDate_Component_Day' represents the object property expressing the 'announcementDate' attribute (of type 'Date') of entity 'Component' (which is the domain of this object property), linking it to entity 'Day' (acting as the range of this object property).

4.2.5. Relationship Attributes

As previously noted, OWL does not allow expressing attributes (datatype properties) on relationships (object properties). In the ontological practice this is solved by a process of *reification*, by which the object property becomes a class, to which the attributes can be associated, and considering it as the subject of two object properties, linking it respectively to its domain and range. We adopt the same strategy in our translation. After turning the relationship into a class, its attributes are handled as reported in the previous section, e.g., considering again relationship WASIN:

```
<relationship name="wasIn" inverse="hosted">
   <attributes>
      <attribute datatype="string" mandatory="false" name="reason"/>
      <attribute datatype="date" mandatory="false" name="startDate"/>
   </attributes>
</relationship>
```

the OWL classes, datatype properties (for attribute 'reason' and object properties (for attribute 'startDate') generated after reification would be:

```
<owl:Class rdf:about="http://owl.api.ontology#wasIn">
   <rdfs:subClassOf rdf:resource="http://owl.api.ontology#Relationship"/>
</owl:Class>

<owl:DatatypeProperty rdf:about="http://owl.api.ontology#reason_wasIn">
   <rdfs:domain rdf:resource="http://owl.api.ontology#wasIn"/>
   <rdfs:range rdf:resource="http://owl.api.ontology#string"/>
</owl:DatatypeProperty>

<owl:ObjectProperty rdf:about="http://owl.api.ontology#startDate_wasIn_Day">
   <rdfs:subPropertyOf rdf:resource="http://owl.api.ontology#RelationshipProperty"/>
   <rdfs:domain rdf:resource="http://owl.api.ontology#wasIn"/>
   <rdfs:range rdf:resource="http://owl.api.ontology#Day"/>
</owl:ObjectProperty>

<owl:ObjectProperty rdf:about="http://owl.api.ontology#startDate_wasIn_Month">
   <rdfs:subPropertyOf rdf:resource="http://owl.api.ontology#RelationshipProperty"/>
   <rdfs:domain rdf:resource="http://owl.api.ontology#wasIn"/>
   <rdfs:range rdf:resource="http://owl.api.ontology#Month"/>
</owl:ObjectProperty>

<owl:ObjectProperty rdf:about="http://owl.api.ontology#startDate_wasIn_Year">
   <rdfs:subPropertyOf rdf:resource="http://owl.api.ontology#RelationshipProperty"/>
   <rdfs:domain rdf:resource="http://owl.api.ontology#wasIn"/>
   <rdfs:range rdf:resource="http://owl.api.ontology#Year"/>
</owl:ObjectProperty>
```

While this transformation is required only for relationships having attributes, it may not be appropriate to have some relationships translated as object properties (those with no attributes) and others translated as classes. Therefore, we translate all relationships both as classes (possibly with attributes), reproducing their hierarchy under the RELATIONSHIP top-level class, and as object properties.

4.3. Logical Architecture and Workflow

Figure 7 provides a high-level graphical description of the involved components and the flow of information in GraphBRAIN. The GraphBRAIN system is shown as a grey box, including the graph DB that stores the data, the GBS schemas and the API. Shapes denote kinds of information: the schemas (empty shapes) define the allowed information patterns and information (filled shapes) is stored in the DB based on these patterns (the shape of the information blocks is the same as that of the schema they refer to). Some information may belong to different schemas (shown as overlapping shapes in the DB). Note that the schemas are kept apart from the data, that several schemas may be used on the same DB and that the API is independent of the schemas (the same API may be used on all DBs, since the schema to be used are provided as an input during the operations).

All interactions between external entities and the system pass through the API. Applications (e.g., the Web Application described in Section 3) may ask the API to provide information about the patterns in one of the available schemas and use them to inform their data handling requests. When they request to store (insert/update) or retrieve (read) information based on a schema, the API checks that their structure is consistent with the patterns defined in the specified schemas, in which case the request is fulfilled. Requests for information patterns not defined in the scheme (the triangle in the figure) are blocked. Given an existing KG, its ontological part can be imported in a schema; if required, also its instances can be imported into the DB based on the imported schema. Conversely, a schema can be exported to an ontology for a KG and possibly the corresponding data in the DB can be exported as instances to the KG, as well.

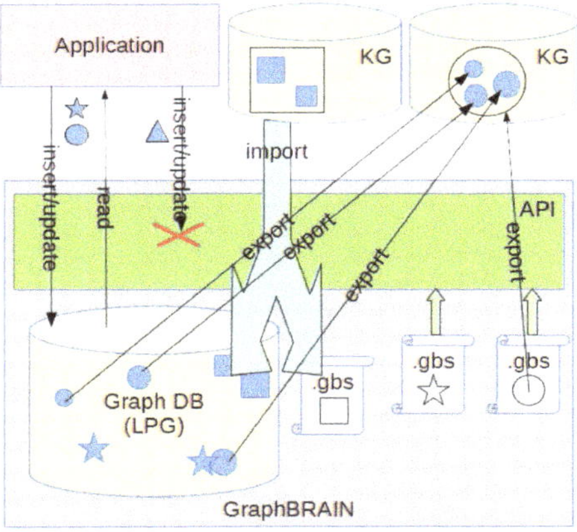

Figure 7. Interplay among components and roles.

5. Conclusions

Formal ontologies, described as RDF graphs, have traditionally been investigated as a means to formalize an application domain so as to carry out automated reasoning on it. The union of the terminological and assertional parts of an ontology is known as a Knowledge Graph. On the other hand, database technology has ever since focused on the optimal organization of data so as to boost efficiency in their storage, management and retrieval. Graph databases, based on the Labeled Property Graphs (LPG) model, are a recent technology specifically focusing on element-driven data browsing rather than on batch processing. Furthermore, graph databases are typically schema-less, preventing uniform interpretation of the data by, and interoperability of, the applications. In spite of the patent and intuitive complementarity and connections between these technologies, the underlying graph models are partially incompatible and little exists to bring them to full integration and cooperation.

Whilst most efforts in the literature are OWL-centric and aimed at mapping RDF ontologies to LPGs, we place more emphasis on the database, so as to benefit from efficient data handling, and aim at enriching it with reasoning capabilites that exploit as much as possible the flexibility of the LPG model. To the best of our knowledge this is a completely novel perspective in the literature.

For this purpose, we proposed to express database schemas in the form of ontologies, so as to clearly describe the database content and to allow users to carry out complex reasoning on it, beyond the queries allowed by the database query language. Specifically, we defined an intermediate format (GBS) that can be easily mapped onto formal ontology standards on one hand and onto the graph database structure on the other. A peculiarity of our approach is that many schemas/ontologies can be applied to the same graph to express different domains or perspectives on its content. These ontologies may share classes and relationships, allowing cross-fertilization of the knowledge from the corresponding domains. The use of ontologies enables multistrategy formal, automated reasoning on the data, that goes much beyond what simple queries can do.

In this paper, for the first time, we provided the full specification for GBS and discussed how its components can be mapped on a most famous graph DB (Neo4j) and on a standard formal ontology (OWL). Operationally, this framework is supported by an API that is meant to act as a wrapper for the DB, ensuring that its content is compliant with a GBS schema, and that can connect the instances in the DB with an ontological reasoner using the same schema as an ontology. Based on this API many different applications may exploit this powerful combinations of databases and ontologies in their functions. Among these applications we developed a tool to build, browse and edit GBS schemas, and a tool to add, edit and consult the DB content according to a pre-specified schema. Such a tool is described in this paper, as well.

The API and tools are continuously under development to be extended and refined, and research is ongoing to further improve the mapping between the GBS and OWL formalisms, so as to fully exploit their respective advantages in both the instance (database) and the schema (ontology) part of the knowledge graph. In particular, we are working at the extension of the schema format with additional tags/features to express information that may improve the effectiveness of reasoning at the ontological level.

Funding: This research received no external funding.

Institutional Review Board Statement: Not applicable.

Informed Consent Statement: Not applicable.

Data Availability Statement: Not applicable.

Acknowledgments: The author would like to thank Domenico Redavid for the useful discussions on the methodology, and Davide Di Pierro for their contribution in the implementation of the schema management section. Grateful thanks go to Artificial Brain S.r.l. for implementing most of the Web Application.

Conflicts of Interest: The author declares no conflict of interest.

References

1. Ehrlinger, L.; Wolfram, W. Towards a definition of knowledge graphs. In Proceedings of the SEMANTICS 2016: Posters and Demos Track, CEUR Workshop Proceedings, Leipzig, Germany, 12–15 September 2016; CEUR-WS.org: Aachen, Germany, 2016; Volume 1695.
2. Schrader, B. *What Is the Difference between an Ontology and a Knowledge Graph? (White Paper)*; Technical Report; Enterprise Knowledge: Arlington, VA, USA, 2021.
3. Available online: https://www.ontotext.com/knowledgehub/fundamentals/what-is-a-knowledge-graph/ (accessed on 8 September 2021).
4. Hogan, A.; Blomqvist, E.; Cochez, M.; d'Amato, C.; Melo, G.D; Gutierrez, C.; Kirrane, S.; Gayo, J.E.L; Navigli, R.; Neumaier, S.; et al. Knowledge Graphs. *ACM Comput. Surv.* **2021**, *54*, 1–37. [CrossRef]
5. Noy, N.; Gao, Y.; Jain, A.; Narayanan, A.; Patterson, A.; Taylor, J. Industry-Scale Knowledge Graphs: Lessons and Challenges. *Commun. ACM* **2019**, *62*, 36–43. [CrossRef]
6. Ferilli, S.; Redavid, D. The GraphBRAIN System for Knowledge Graph Management and Advanced Fruition. In *Foundations of Intelligent Systems*; Springer: Berlin/Heidelberg, Germanny, 2020; Volume 12117, *LNAI*, pp. 308–317.
7. Ferilli, S.; De Carolis, B.; Buono, P.; Di Mauro, N.; Angelastro, S.; Redavid, D. Una piattaforma intelligente per la gestione integrata del settore turistico. In Proceedings of the Primo Convegno Nazionale CINI sull'Intelligenza Artificiale—Workshop on AI for Cultural Heritage, Rome, Italy, 18–19 March 2019; CINI: Rome, Italy; p. 2. (In Italian)
8. Ferilli, S.; Redavid, D. An Ontology and a Collaborative Knowledge Base for History of Computing. In Proceedings of the 1st International Workshop on Open Data and Ontologies for Cultural Heritage (ODOCH-2019), at the 31st International Conference on Advanced Information Systems Engineering (CAiSE 2016), Central Europe (CEUR) Workshop Proceedings, Rome, Italy, 3 June 2019; CEUR-WS.org: Aachen, Germany, 2019; Volume 2375, pp. 49–60.
9. Ferilli, S.; Redavid, D. An Ontology and Knowledge Graph Infrastructure for Digital Library Knowledge Representation. In *Digital Libraries: The Era of Big Data and Data Science*; Communications in Computer and Information Science; Springer: Berlin/Heidelberg, Germany, 2020; Volume 1177, pp. 47–61.
10. Studer, R.; Benjamins, R.; Fensel, D. Knowledge engineering: Principles and methods. *Data Knowl. Eng.* **1998**, *25*, 161–198. [CrossRef]
11. Rudolph, S. Foundations of Description Logics. In *Reasoning Web. Semantic Technologies for the Web of Data: 7th International Summer School 2011, Galway, Ireland, 23–27 August 2011, Tutorial Lectures*; Springer: Berlin/Heidelberg, Germany, 2011; pp. 76–136.
12. Available online: https://www.w3.org/OWL/ (accessed on 23 October 2021).
13. Available online: http://owl.cs.manchester.ac.uk/tools/list-of-reasoners/ (accessed on 23 October 2021).
14. Available online: https://www.w3.org/RDF/ (accessed on 23 October 2021).
15. Rodriguez, M.; Neubauer, P. Constructions from dots and lines. *Bull. Am. Soc. Inf. Sci. Technol.* **2010**, *36*, 35–41. [CrossRef]
16. Available online: https://neo4j.com/blog/rdf-triple-store-vs-labeled-property-graph-difference/ (accessed on 8 September 2021).
17. Shao, B.; Wang, H.; Li, Y. Trinity: A distributed graph engine on a memory cloud. In Proceedings of the 2013 ACM SIGMOD International Conference on Management of Data (SIGMOD'13), New York, NY, USA, 22–27 June 2013; ACM: New York, NY, USA, 2013; pp. 505–516.
18. Robinson, I.; Webber, J.; Eifrem, E. *Graph Databases*, 2nd ed; O'Reilly Media: Sebastopol, CA, USA, 2015.

19. Available online: https://db-engines.com/en/ranking (accessed on 23 October 2021).
20. Available online: https://db-engines.com/en/system/GraphDB%3BNeo4j (accessed on 8 September 2021).
21. Available online: https://neo4j.com/use-cases/ (accessed on 23 October 2021).
22. Krötzsch, M. Ontologies for Knowledge Graphs? In Proceedings of the 30th International Workshop on Description Logics, Montpellier, France, 18–21 July 2017; CEUR Workshop Proceedings; CEUR-WS.org: Aachen, Germany, 2017; Volume 1879.
23. Drakopoulos, G.; Kanavos, A.; Mylonas, P.; Sioutas, S.; Tsolis, D. Towards a framework for tensor ontologies over Neo4j: Representations and operations. In Proceedings of the 8th International Conference on Information, Intelligence, Systems & Applications, IISA 2017, Larnaca, Cyprus, 27–30 August 2017; pp. 1–6.
24. Elbattah, M.; Roushdy, M.; Aref, M.; Salem, A.B.M. Large-scale ontology storage and query using graph database-oriented approach: The case of Freebase. In Proceedings of the 2015 IEEE Seventh International Conference on Intelligent Computing and Information Systems (ICICIS), Cairo, Egypt, 12–14 December 2015; pp. 39–43.
25. Chiba, H.; Yamanaka, R.; Matsumoto, S. G2GML: Graph to Graph Mapping Language for Bridging RDF and Property Graphs. In *The Semantic Web—ISWC 2020*; Springer: Cham, Switzerland, 2020; pp. 160–175.
26. Available online: https://protegeproject.github.io/owl2lpg (accessed on 8 September 2021).
27. Available online: https://github.com/SciGraph/SciGraph/wiki/Neo4jMapping (accessed on 8 September 2021).
28. Available online: http://owlcs.github.io/owlapi (accessed on 23 October 2021)
29. Available online: https://github.com/VirtualFlyBrain/neo4j2owl (accessed on 8 September 2021).
30. Available online: https://github.com/cmungall/owlstar (accessed on 8 September 2021).
31. Hartig, O. Foundations to Query Labeled Property Graphs using SPARQL. In *Proceedings of the CEUR Workshop Proceedings Joint Proceedings of the 1st International Workshop on Semantics for Transport and the 1st International Workshop on Approaches for Making Data Interoperable Co-Located with 15th Semantics Conference (SEMANTiCS 2019)*; CEUR-WS.org: Aachen, Germany, 2019; Volume 2447.
32. Available online: https://neo4j.com/blog/ontologies-in-neo4j-semantics-and-knowledge-graphs/ (accessed on 8 September 2021).
33. Available online: https://neo4j.com/labs/apoc/4.1/export/json/ (accessed on 23 October 2021).
34. Available online: https://www.w3.org/2016/01/json2rdf.html (accessed on 23 October 2021).
35. Available online: https://neo4j.com/docs/labs/nsmntx/current/importing-ontologies/ (accessed on 23 October 2021).
36. Abburu, S.; Babu, G.S. Survey on Ontology Construction Tools. *Int. J. Sci. Eng. Res.* **2013**, *4*, 1748–1752.
37. Knublauch, H. An AI Tool for the Real World: Knowledge Modeling with Protégé. *JavaWorld*, 20 June 2003. Available online: https://www.infoworld.com/article/2073547/an-ai-tool-for-the-real-world.html?page=2 accessed on 23 October 2021).
38. Available online: https://protege.stanford.edu (accessed on 23 October 2021).
39. Rubin, D.; Knublauch, H.; Fergerson, R.; Dameron, O.; Musen, M. Protégé-OWL: Creating Ontology-Driven Reasoning Applications with the Web Ontology Language. In *AMIA Annual Symposium Proceedings*; American Medical Informatics Association: Rockville, MD, USA, 2005; Volume 2005.
40. Knublauch, H.; Fergerson, R.; Noy, N.; Musen, M. The Protégé OWL Plugin: An Open Development Environment for Semantic Web Applications. In *International Semantic Web Conference*; Springer: Berlin/Heidelberg, Germany, 2004; Volume 3298, pp. 229–243.
41. Gherasim, T.; Harzallah, M.; Berio, G.; Kuntz, P. Methods and Tools for Automatic Construction of Ontologies from Textual Resources: A Framework for Comparison and Its Application. In *Advances in Knowledge Discovery and Management—Volume 3 [Best of EGC 2011, Brest, France]*; Studies in Computational Intelligence; Guillet, F., Pinaud, B., Venturini, G., Zighed, D.A., Eds.; Springer: Berlin/Heidelberg, Germany, 2011; Volume 471, pp. 177–201.
42. Bollacker, K.; Evans, C.; Paritosh, P.; Sturge, T.; Taylor, J. Freebase: A collaboratively created graph database for structuring human knowledge. In Proceedings of the 2008 ACM SIGMOD International Conference on Management of Data, Vancouver, BC, Canada, 10–12 June 2008; pp. 1247–1250.
43. Available online: https://mailman.stanford.edu/pipermail/protege-owl/2007-September/003823.html (accessed on 23 October 2021)
44. Available online: https://www.w3.org/TR/owl-time/ (accessed on 23 October 2021).

electronics

MDPI

Article

An Ontology-Based Approach for Knowledge Acquisition: An Example of Sustainable Supplier Selection Domain Corpus

Agnieszka Konys

Faculty of Computer Science and Information Technology, West-Pomeranian University of Technology in Szczecin, Żołnierska 49 Street, 71-210 Szczecin, Poland; akonys@zut.edu.pl

Abstract: Selecting the right supplier is a critical decision in sustainable supply chain management. Sustainable supplier selection plays an important role in achieving a balance between the three pillars of a sustainable supply chain: economic, environmental, and social. One of the most crucial aspects of running a business in this regard is sustainable supplier selection, and, to this end, an accurate and reliable approach is required. Therefore, the main contribution of this paper is to propose and implement an ontology-based approach for knowledge acquisition from the text for a sustainable supplier selection domain. This approach is dedicated to acquiring complex relationships from texts and coding these in the form of rules. The expected outcome is to enrich the existing domain ontology by these rules to obtain higher relational expressiveness, make reasoning, and produce new knowledge.

Keywords: ontology; knowledge base; sustainable supplier selection; ontology population; information extraction; knowledge acquisition from text

Citation: Konys, A. An Ontology-Based Approach for Knowledge Acquisition: An Example of Sustainable Supplier Selection Domain Corpus. *Electronics* **2022**, *11*, 4012. https://doi.org/10.3390/electronics11234012

Academic Editor: Stefano Ferilli

Received: 20 October 2022
Accepted: 2 December 2022
Published: 3 December 2022

1. Introduction

The concept of sustainable development is based on the intersection of three dimensions: economic, environmental, and social. Each of them deals with different aspects, but together they focus on promoting sustainable development. Globalization forces global manufacturers to attach much importance to partnerships between suppliers. In general, a supply chain is a concept that links upstream, midstream, and downstream. The manufacturers' aim is to reduce costs in this process. Moreover, supply chain management (SCM) receives the applicable information from downstream to improve the quality of the goods provided upstream and downstream [1]. Growing customer, non-governmental organization (NGO), and law enforcement concerns about environmental, social, and corporate responsibility have drawn industry academics and practitioners to the concept of sustainable supply chain management [2].

The assessment of sustainable development is an issue of growing importance among scientists and decision-makers. Sustainability assessment offers a large number of opportunities to measure and evaluate the level of its accomplishment. The search for effective methods of assessing sustainable development and its monitoring of development is now becoming one of the key factors determining the development of a sustainable society. The problem of assessing sustainable development applies to almost all areas. The international environmental policy, government, and people have stimulated enterprises to strictly adopt sustainable concepts in the supply chain networking to obtain a reactive, regulatory, proactive strategic, and competitive merit and abrade the non-sustainable challenges and factors against the world's environment [3]. Due to globalization, sustainable supply chains are becoming more and more important. Hence, it is worth paying attention to ensuring sustainable supplier selection in this process. Sustainable supplier selection is a combined multi-dimensional problem that includes considering both qualitative and quantitative factors. The sustainability paradigm has been considered a comprehensive term in supplier selection, which includes a vital presence of three aspects (economic, environmental, and social) [4].

Ensuring sustainable supply chain complexity is one of the most difficult problems in today's global supply chains and is assumed as the key impediment to business performance. It has a significant influence on competitiveness, costs, customer satisfaction, product innovation, and market share. Therefore the decision-makers must know the criteria causing sustainable supply chain efficiency. Proper identification and prioritizing of sustainable supplier criteria are required for effective monitoring and controlling of supply chain management [5]. Moreover, the timeliness of these criteria is also of great importance. The selection of a sustainable supplier depends on many factors. Thus, the crucial question is to find a reasonable approach between comprehensiveness and a manageable multi-dimensional knowledge base as well as up-to-date information exchange.

This paper presents an ontology-based approach to knowledge acquisition from the text. This approach is dedicated to acquiring complex relationships from texts and coding these in the form of rules. The approach begins with elaborating data using VosViewer to plot knowledge domain maps. Next, existing domain knowledge is implemented as OWL ontology and applies NLP tools and text-matching techniques to deduce different atoms, such as classes, properties, and literals, to capture deductive knowledge in the form of new rules. The expected outcome is to enrich the existing domain ontology by these rules to obtain higher relational expressiveness, make reasoning and produce new facts.

Several research gaps are identified through an in-depth review of the literature. Firstly, lack of a comprehensive knowledge base about criteria, sets of criteria are found by various literature studies but cannot effectively estimate sustainable supplier selection criteria [1,6,7]. Secondly, in most cases, there is a subjective evaluation of the performance of sustainable supplier selection [3,8,9].

Moreover, there is a lack of a systematic framework to handle knowledge about sustainable supplier selection criteria [1,6,7,9]. There is also a lack of a complex approach to both selecting and filtering linguistic information about criteria determining sustainable supplier selection and its categorization in the form of a knowledge base [3,5,8,10,11].

These research gaps are transformed into the author's contribution as follows:

- Plotting knowledge domain maps;
- Development of a framework for selecting sustainable supplier criteria;
- Ontology design and implementation;
- Semi-automated ontology population by knowledge extraction from various resources;
- Rule-based reasoning.

Based on this, it is possible to define the following highlights:

- Development of an ontology-based framework to deal with distributed knowledge representation;
- Development of a domain ontology that stores various information about sustainable suppliers to support various aspects of knowledge management by combining dynamic data provided from external sources with predefined information gathered in the ontology;
- Providing examples of using ontology in various scenarios in the domain of sustainable supplier selection;
- Creating a knowledge base with rules and queries using JAPE and reasoners;
- Demonstrating the effectiveness of rule-based reasoning to increase the ability of logical reasoning in the context of selecting sustainable supplier criteria.

The presented approach begins with creating domain knowledge represented as OWL ontology and applies NLP tools and text-matching techniques to deduce different atoms, such as classes, properties, and literals, to capture new knowledge. This research increases the body of knowledge on the ontology for the sustainable supplier domain by providing a systematic keywords map of the subject and grasping the main criteria in the research field. The results demonstrate that the proposed approach can (1) successfully handle the knowledge domain, (2) reduce the time for searching for relevant information, (3) improve the accuracy of search results that suit users' specific needs, and (4) provide quick updates with new knowledge.

The remainder of this paper is organized as follows. Section 2 presents the related works, in particular, taking into account such topics as sustainable supplier selection, information extraction, NLP, and ontologies. In Section 3, Materials and Methods, a new ontology-based approach for extracting knowledge in the form of rules from texts is described in detail. Section 4 presents the working example of the elaborated approach. Section 5 provides the conclusions and directions for further research.

2. Background and Related Works

2.1. Sustainable Supplier Selection

The growing emphasis on supply chain management among manufacturing companies has made the suppliers' role in the value-addition processes to become strategically significant [8]. The problem of assessing sustainable development applies to almost all areas. Supplier selection is a combined multi-dimensional problem that includes considering both qualitative and quantitative factors [9]. Due to globalization, sustainable supply chains are becoming more and more important. The fast globalization of doing business affects business competition, changing the model from "company versus company" to the model "supply chain versus supply chain" [11]. Therefore, choosing a good combination of suppliers to work with is critical to the success of conducting business [1]. Over the years, the importance of selecting suppliers has been appreciated and emphasized. Adding sustainability aspects to the supplier selection process highlights existing trends in environmental, economic, and social issues related to management and business processes. Moreover, the development of sustainable development allows the integration of environmental, economic, and social thinking with conventional supplier selection [12].

From a systematic point of view, the study of the problem of sustainable supplier selection can be divided into two parts, including criteria and methods [13]. The analysis of the literature provides a set of various methods exploiting different aspects and using single or mixed approaches, as well as examples of selection criteria [11,12,14]. Most of the studies on sustainable supplier selection use MCDM or fuzzy MCDM techniques with complex calculations [1]. A wide range of methods was applied to solve the problem of sustainable supplier selection. The literature reviews [12] point out that the main single and combined approaches used to solve this problem are mathematics methods and artificial intelligence approaches, especially including analytic hierarchy process [10,15], linear programming [10], multi-objective programming [16,17], goal programming [6], data envelopment analysis [13], heuristics [18], statistical [19], cluster analysis [7], multiple regression [20], discriminant analysis [21], neural networks [22], software agent [20], case-based reasoning [23], expert system [21], and fuzzy set theory [14] as well as combinations of selected pairs.

As it is a multi-dimensional concept, the selection of sustainable suppliers is not based on a single criterion but on a set of criteria, which are mostly focused on economic, social, and environmental issues. In general, most companies need to focus on their supply chains to enhance sustainability to meet customer demands and comply with environmental legislation. In order to achieve these goals, companies must focus on criteria that include carbon footprint and toxic emissions, energy use and efficiency, waste generation, and worker health and safety [24]. Therefore, to analyze interrelationships among sustainability criteria, it is necessary to identify the most important ones for a given decision problem and then evaluate suppliers according to these criteria. Since the knowledge about criteria is scattered, a set of hybrid information aggregation is required to provide practical evaluation and link this set of information to the proposed knowledge base. The literature analysis provides many multi-criteria methods to support a balanced selection of suppliers and multiple cuttings of criteria sets, often suited for a given area (e.g., food, industry, and others). There are many comparable approaches; Table 1 shows a small piece of them. However, little attention has been paid to building a complex solution that allows gathering the selection criteria for sustainable suppliers, and there is almost no systemic and structured knowledge-based approach that could be used to evaluate the sustainability of suppliers.

Table 1. Examples of multi-criteria methods to support a selection of sustainable suppliers.

The Used MCDA Method	Domain	Source
AHP	urban water reuse, energy landscape	[25,26]
Multi-attribute Value Theory (MAVT)	electricity system, RES, urban regeneration	[27,28]
PROMETHEE	logistics and distribution, agriculture	[29,30]
TOPSIS	air pollution, transportation sector	[31,32]
MULTIMOORA	energy policy	[33]

2.2. *Information Extraction*

The information extraction (IE) process is based on the automatic extraction of certain types of information from natural language text. IE is the process of extracting information from unstructured text sources to enable entities to be searched, classified, and stored in a knowledge base [34]. The general aim is to parse text in natural language and look for instances of a certain class of objects or events and the instances of relationships between them. Another definition describes information extraction as a form of natural language processing in which certain types of information must be recognized and extracted from a text. Extracting information uses various algorithms and methods for finding information [35]. IE deals with the collection of texts in order to transform them into information that can be easily understood and analyzed [36]. Semantically enhanced information extraction (also known as semantic annotation) links these units to their semantic descriptions and connections from the knowledge graph. Because is much information available on the Internet these days, and the amount of it is constantly growing, this results in information overload. However, the real problem is not the sheer amount of information but the inability to filter it properly [34,37]. IE helps in the automatic detection of new, previously unknown information by automatically extracting information from various unstructured resources [38]. Therefore, the key element is linking the extracted information together to formulate new facts or new knowledge. In other words, in IE, the goal is to discover previously unknown information. Figure 1 displays an illustrative example of how information extraction works in practice.

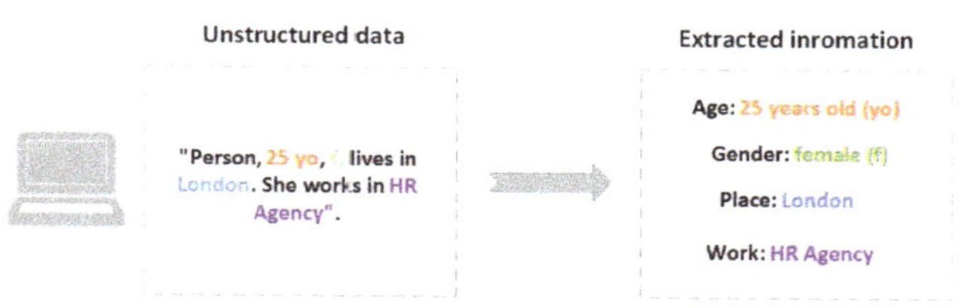

Figure 1. An example of information extraction.

Natural Language Processing (NLP)

NLP aims to analyze, identify and solve problems related to the automatic generation and understanding of human language. NLP aims to perform, decode and understand unstructured information [39]. NLP allows for the following:

- Sorting the data to remove the rubbish from the interesting parts;

- Extracting the relevant pieces of information;
- Linking the extracted information to other sources of information;
- Aggregating the information according to potential new categories;
- Querying the (aggregated) information;
- Visualizing the results of the query.

It is composed of several tasks:

- Text pre-processing—the text is prepared for processing using computational linguistics tools such as tokenization, sentence sharing, morphological analysis, etc.;
- Finding and classifying concepts—the various types of concepts are detected and classified;
- Connecting concepts—this task aims to identify the relationship between the extracted concepts;
- Unify—this task presents the extracted data in a standard form;
- Remove information noise—this task eliminates duplicate data;
- Enriching the knowledge base—the extracted knowledge is processed in the knowledge base for further use.

Overall, the combination of NLP and information extraction extracts new knowledge from the raw data. Finally, unknown information is obtained by automatically extracting information from various unstructured resources.

2.3. Ontology and Ontology Population

2.3.1. Ontology

Recently, the terms ontology and Semantic Web are quite popular and top research areas in computer science. Ontology is a standard recommended by World Wide Web Consortium for representing knowledge in the Semantic Web, and it turns into a fundamental and critical component for developing applications in different real-world scenarios [40]. Ontologies have become an important tool in domain modeling over the years and have been used successfully in several fields. In the artificial intelligence field [41–44], ontologies can also be used to build knowledge databases that will be used in various systems, using the obtained information to perform different tasks [41]. As a result, they help in carrying out real-world representations, establishing axioms, and obtaining conclusions from them [41,45,46].

Ontologies are defined as a set of concepts and relations between them [47]. Concepts can be divided into classes, subclasses, attributes, relationships, and instances. From a technical point of view, ontologies are a formal source of domain-specific knowledge, which is proven to be efficient for search results diversification [48]. In fact, they allow you to express the semantics of a domain in a language that computers can understand, allowing automatic processing of the meaning of the information provided [49]. Ontologies provide a controlled vocabulary of concepts whose semantics are explicitly defined and machine understandable [47]. Ontologies also offer a common understanding of the topics of communication between systems and users and enable the processing of web-based knowledge as well as the sharing and reuse among applications [48]. The most popular definition of ontology was proposed by Gruber, who stated that ontology could be defined as an explicit, formal specification of a shared conceptualization [47]. It contains the following components called concepts, individuals, relations, and attributes. It can be formulated as follows:

$$O = \{I; C; R; A\} \tag{1}$$

where I is the set of individuals, C refers to the set of concepts, R represents the set of relations and the interactions between domain individuals as follows: R is $\subseteq C1 \times C2 \times .$ Cn and A is the set of axioms.

The concepts (classes) correspond to the relevant abstractions of a segment of reality (the domain of the problem). The relations (properties) link the individuals or concepts between them. The individual is defined as a resource that has been placed into the class, but individuals are not classes themselves. The axioms are statements that are asserted to be true in the domain being described [50].

The OWL 2 standard is currently used as a formal language for representing ontologies. The inference process takes place using various ontological reasoners. The main functions of reasoners are ontology consistency checking, class taxonomy building, and ontology querying. Ontology reasoning aims to ensure that the ontology is consistent with its logical semantics. The reasoning is also required to infer new knowledge from ontology. The reasoners enable validation of the ontology, whereas at the end is possible to obtain inferred knowledge against the user's description logic (DL) queries.

2.3.2. Ontology Population

Ontology population is a process for inserting concept and relation instances into an existing ontology [51,52]. The ontology population process has several tasks: the extraction of relation instances and identification values from any information sources and assigning such values to instances. The next task involves extracting instances, or more precisely, identifying values from any information source and assigning them to an instance [51,52]. There are many approaches in the literature related to ontology learning and ontology population. Ontology learning has benefited from the adoption of established techniques such as machine learning, data mining, natural language processing, information retrieval, and knowledge representation [53]. Based on the classification proposed by Alexander Maedche and Steffen Staab [54], ontology learning approaches were distinguished, taking into account the type of input data used for learning. Thus, common classification contains ontology learning from text, dictionary, knowledge base, semi-structured schemata, and relational schemata [53]. Each of them requires multiple research efforts to achieve a common domain conceptualization [55,56].

An automated ontology population is intended to identify concept and relation instances by using a computational tool [52,55,57,58]. Ontology learning techniques apply more complex NLP techniques to the text. Rather than simply extracting terms, they analyze the grammatical structure of sentences to determine how the terms are used. Then they deduce possible IS-A relationships between terms, which will be used to build classification hierarchies.

3. Materials and Methods

This section describes a new ontology-based approach for extracting knowledge in the form of rules from texts. This approach is dedicated to acquiring complex relationships from texts and coding these in the form of rules. The proposed approach is based on different works in the areas of knowledge acquisition, rule-based reasoning, and ontology population. A semi-automated supervised solution has been proposed for extending ontology classes in terms of learning concept attributes, data types, and value ranges. This approach requires two inputs: existing knowledge and free texts. The existing knowledge is OWL ontology. Free texts represent the domain knowledge in unstructured natural language, in this case, English. The selected domain covers sustainable supplier selection criteria.

3.1. Data Preparation and Search Strategy

In this study, we used the following tools: (1) Scopus database for managing bibliographic references [59] and (2) VOSviewer for bibliographic analysis and developing a keywords map [60]. The search strategy encompasses using the Scopus database to retrieve documents related to sustainable supplier selection criteria. In order to support the document filtration process, a formal PRISMA approach [61] was used (Figure 2). However, not all steps from the PRISMA flow diagram were used because the main goal was to search for criteria, and filtering only on the abstract and keywords was insufficient. The list of papers contains 1652 elements. The year of publication of selected documents is between 2003 and 2021. The analysis started in June 2021; hence not all publications from 2021 are included. The analyzed set of papers excluded from the final set of documents the conference reviews, erratum, and review. The query was as follows: (TITLE-ABS-KEY ("sustainable supplier selection"). The extracted documents were exported to Excel spreadsheets as *.csv file. The results can be revised by the author's name, affiliation, document type, source title, or subject area.

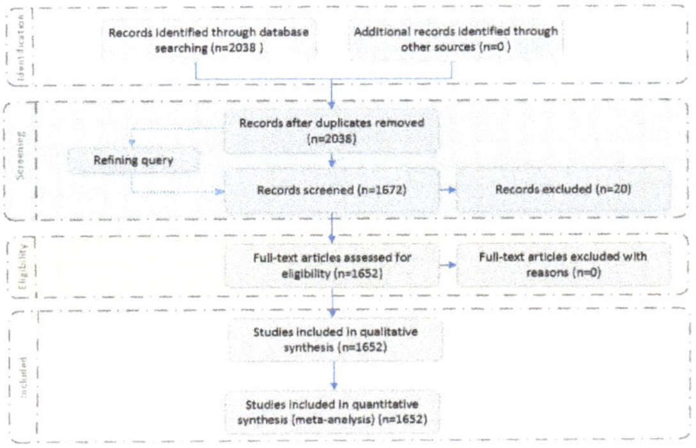

Figure 2. Modified PRISMA procedure [61].

Then the set of papers was manually filtered. The process itself was highly time-consuming but allowed for the identification of an initial set of criteria and sub-criteria. It contained 8261 items. The data set prepared in this way was then subjected to further work using a dedicated tool for plotting knowledge domain maps.

3.2. Plotting Knowledge Domain Maps

The developed data set was used to prepare and plot knowledge domain maps. Previously collected data were processed using the VOS viewer software [62]. VOSviewer enables the user to generate networks from given bibliometric data. VOSViewer allows the user to group criteria and sub-criteria and display the results. The size of a given item displays the density of occurrence of a given criterion (Figure 3).

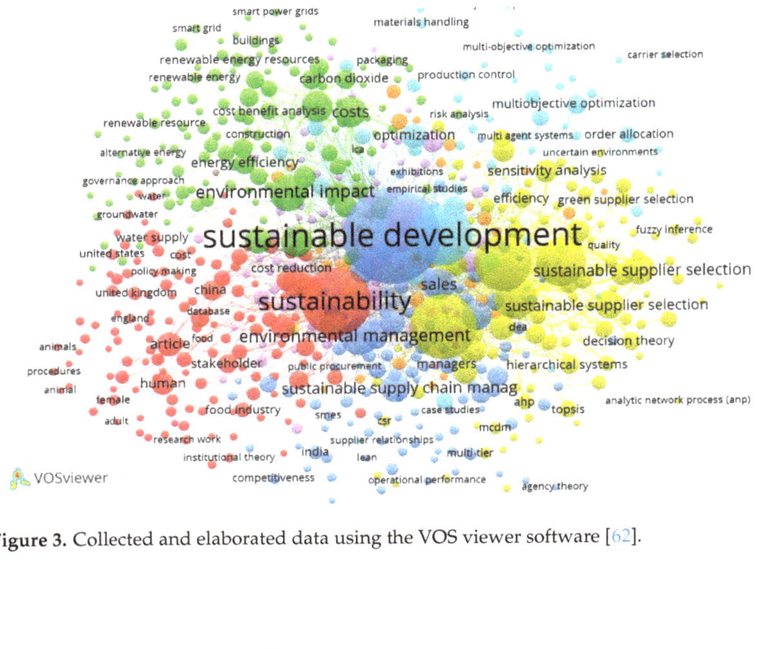

Figure 3. Collected and elaborated data using the VOS viewer software [62].

For the analysis, it was also necessary to clean up the data, so a VOSViewer thesaurus file was created to combine similar criteria names. Due to the relatively large number of criteria, it is not possible to present all changes in this study. Selected limitation rules are defined and shown for example:

1. Merging "Product Quality" and "Quality of Product";
2. Merging "Deliver & Service" and "Delivery and Service";
3. Merging "Technology Capabilities" and "Technology Capability";
4. Merging "Inventory costs" and "Inventory cost";
5. Merging "Service Quality" and "Quality of Service";
6. Merging abbreviations "EMS" and "Environmental management System";
7. Merging synonyms "Green packaging" and "Green packaging ability".

The thesaurus file contains 68 extra items. Ultimately, the set included 8261 criteria as input from 1652 papers. The total number of main clusters is 126. Each cluster contains a set of sub-criteria. The keyword occurrence map was also created. The most common keywords are green, cost, and quality. Table 2 shows the 10 most popular keywords.

Table 2. The top 10 keywords.

Keyword	Occurrences
Green	543
Cost	420
Quality	375
Service	287
Delivery	285
Time	214
Risk	200
Price	154
Technology	152
Waste	140

3.3. *Ontology Representation*

The conducted plotting knowledge domain maps provide the pre-elaborated set of criteria and sub-criteria ready to implement in an OWL ontology. The ontology contains all the identified elements, which are the backbone for taxonomy building/class hierarchy building. This process requires the knowledge engineer's participation. Therefore, the input domain ontology was developed from scratch based on the data set provided. The Protégé OWL-API [63] was selected to work with ontology and to manipulate the different constituents of the ontology (classes, object properties, data type properties, and individuals). It aims to structure knowledge, organize it, and above all, reason about it. The main stages of the development process are inspired by the ontology methodology provided by Noy and McGuiness, as shown in Figure 4.

The first step aims to define the domain and scope of the ontology—in this case, the domain of sustainable suppliers was considered. Since no similar solutions have been found, the second step will be to create an ontology from scratch. Steps 3 through 7 relate directly to ontology construction. In the third step, it is necessary to indicate the most important terms in the ontology. These terms are then detailed. This is the basis for building the class hierarchy in step 4. The class hierarchy represents an "is-a" relation: class X is a subclass of Y if every instance of X is also an instance of Y. It is worth noticing that the whole set contains 126 main classes and 8261 sub-classes. Thus, there are 8378 classes in total. The final set of clusters is attached in supplementary materials (the set of criteria: Sustainable_Supplier_Criteria.xls). Table 3 displays a piece of a class hierarchy.

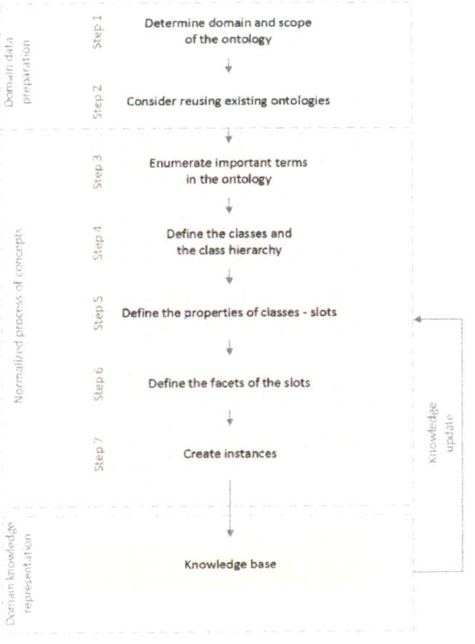

Figure 4. Ontology construction steps.

Table 3. Examples of classes.

Class	Description
Quality	A collection of criteria related to assessing quality (e.g., product quality, quality assurance, QMS)
Green	A collection of criteria related to assessing the level of green (e.g., green competencies, green logistics, green packaging)
Service	A collection of criteria related to assessing the level of offered service (e.g., flexibility of the supplier, payment flexibility)
Delivery	A collection of criteria related to assessing the level of delivery (e.g., delivery lead time, delivery safety, delivery flexibility)
Cost	A collection of criteria related to assessing the level of cost (e.g., cost control, delivery costs, freight costs)
Risks	A collection of criteria related to assessing the level of risks (e.g., economy risk, environmental risk)
Knowledge	A collection of criteria related to assessing the level of knowledge (e.g., sustainable knowledge sharing, IT knowledge)
Supplier's profile	A collection of criteria related to assessing the level of supplier's profile (e.g., supplier's reputation, references)
Logistics	A collection of criteria related to assessing the level of logistics (e.g., reverse logistics, logistics for environment, green logistics)
Pollution	A collection of criteria related to assessing the level of pollution (e.g., energy consumption, pollution control, use of harmful materials)

Therefore, in the 5th step, the constitution of the relations is needed. In Protégé, the slots are also named object properties. Object properties describe the relations between classes or individuals. Another group is datatype property, which aims to describe the relations between individuals and values. Table 4 shows selected object properties and datatype properties with assigned domains and ranges.

Table 4. Examples of object properties and datatype properties.

Type	Property	Domain	Range
Object Property	hasCriterion	Criteria	Sus_Supplier
Object Property	isCriterionOf	Sus_Supplier	Criteria
Object Property	hasFeature	Criteria	Sus_Supplier
Object Property	isFeatureOf	Sus_Supplier	Criteria
Datatype Property	hasValue	Criteria	xsd:double
Datatype Property	hasRating	Sus_Supplier	xsd:int
Datatype Property	hasOpinion	Sus_Supplier	xsd:string
Datatype Property	hasLevel_of_Sustainability	Criteria	xsd:string

In the 6th step, the definitions of facets of the slots take place. The value types, cardinality, range of slots, and other features are determined. The 7th step aims to create instances of the classes in the hierarchy. Defining an individual instance of a class requires (1) selecting the class, (2) creating an individual instance of that class, and (3) filling the slot values [64].

The resulting knowledge base contains 8261 ontological entities such as classes, relations, datatype properties, and individuals. This ontology contains a considerable amount of information representing the sustainable supplier criteria. Moreover, this ontology can be fed with new data from external sources. The ontology is available at: https://webprotege.stanford.edu/#projects/d819c911-a0dc-4208-86a5-3be0df042caa/edit/Classes (accessed on 1 April 2022).

3.4. Ontology Population—Information Extraction and Discovering Specific Concepts from the Text and Semantic Annotation

Ontologies can provide an alternative to storing knowledge at the concept and instance levels. The process of ontology enrichment by adding the names of the concepts and their relationships and instances to populate the ontology is performed by domain experts. However, this process is time-consuming and requires relevant knowledge from domain experts as well as manual skills. Therefore, an ontological population is needed to obtain useful information from texts and includes enrichment with class and relationship instances using an existing ontology as input [52].

The elaborated approach aims to provide a knowledge extraction ontology-based system for texts that helps automatically acquire and formalize this knowledge, limiting the need for expert intervention as much as possible. The proposed approach is based on natural language processing (NLP) and information extraction (IE) techniques. In this work, information extraction techniques are applied as named-entity recognition and co-reference resolution. The process of discovering specific concepts from text requires using a dedicated tool. The approach was developed by using the GATE tool and a pipeline-shaped architecture, i.e., a process should finish for starting the next one. GATE is an architecture, framework, and development environment for language engineering (LE). GATE is a component-based model application that allows for easy coupling and decoupling of the processing resources. GATE includes a core library and a set of reusable LE modules. The framework implements the architecture and provides amenities for processing and visualizing sources, including representation, import, and export of data. The provided reusable modules can perform basic language processing tasks such as POS and semantic tagging [65]. The process is shown in Figure 5.

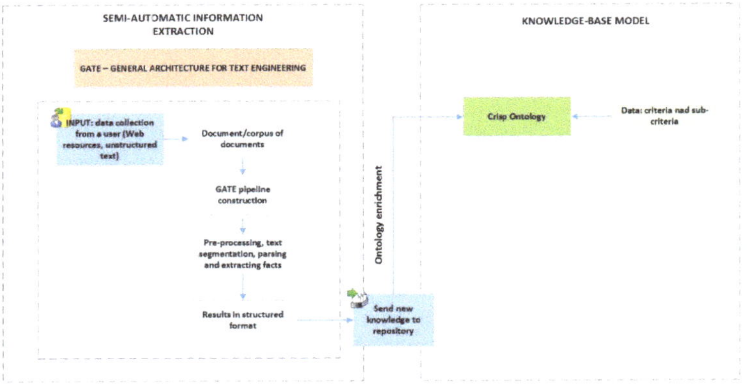

Figure 5. Semi-automatic information extraction and knowledge base model constructions.

The input data are provided by the user in the form of unstructured text or web resources. Therefore, a corpus of documents is created. The corpus consists of a set of various documents related to the sustainable supplier. Apart from scientific papers, it is also possible to use as input various reports and statistics written by specialists. The usage of GATE software enables pipeline construction using various processing resources. Therefore, various steps take place, especially containing:

- Document Reset to remove all previous annotations from the document;
- Tokenizer to split the English text into tokens;
- Gazetteer to find list items in the text and annotate them as "lookup";
- Sentence splitter to split the text into sentences;
- POS Tagger to split the text into parts of speech;
- Transducer NE to identify individuals—e.g., person, location;
- OrthoMatcher to add reference identity relationships between previously annotated entities;
- OntoRoot Gazetter to produce annotations over textual documents, where an ontology is given as input;
- JAPE Transducer to use JAPE rules to transform annotations to property assertions [65].

This semantic annotation using the population of ontology and definition of classes would be impossible without ANNIE (A nearly new information extraction system: Tarnow, Poland). ANNIE is a component of GATE. It is a complete chain dedicated to information extraction. ANNIE is based on the Java Annotation Patterns Engine (JAPE) and includes various annotation modules that are useful for performing various extraction tasks. In selected cases, it is possible to use additional processing resources. Figure 6 shows a simplified procedure related to information extraction and feeding the ontology with new knowledge.

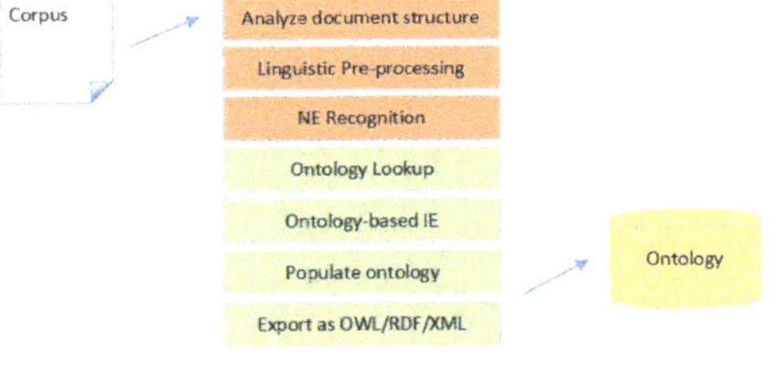

Figure 6. Information extraction procedure and feeding the ontology with new knowledge. Source: Personal elaboration on base of GATE documentation.

3.5. Rule-Based Reasoning

The GATE resource OntoRoot Gazetteer can create annotations over textual documents. It demands implementing an ontology as an input in combination with other generic GATE resources. Another processing resource, the JAPE transducer, applies JAPE rules to transform annotations into property assertions. It allows for defining the rules and recognizing regular expressions in annotations of documents. A single JAPE rule is composed of two parts: LHS and RHS. The LHS contains the patterns to match, whereas the RHS details the annotations to be created. JAPE rules combine to form a specific state. The rules are designed to tag classes, instances, and attribute values. The priority of rules is based on pattern length, rule status, and rule order. The phases combine to create grammar. JAPE rules are used to locate terms in the text that potentially relate to markers, and that will later be used to create new annotations using the JAPE formalism and to identify the body and the head of the produced rules.

Table 5 presents an implemented code of the sample JAPE rule titled "Quality1". In this case, to match a string of text, the "Token" annotation and the "string" feature were used to match text with "Token" annotation quality. The formula combination used in this example is enclosed in parentheses, followed by a colon and label. The sign "->" separates the LHS and the RHS parts, and it begins the RHS part. RHS is responsible for the manipulation of the annotation pattern from LHS, and the label on the RHS must match a label on the LHS. When the LHS part is true, the RHS part should be run [65]. When a rule matches a text sequence, the entire sequence is assigned by the rule to the label. The transducer is informed that the temporary label (quality) will be renamed to "Quality" and the rule that achieves this is "Quality1". Naming a rule is important for the debugging purpose, as when the rule fires, it will be part of the annotation properties that you can see in GATE GUI. In this example, a sample criterion will be annotated as {rule = Quality1}.

Table 5. An implemented code of the sample JAPE rule.

Rule: Quality1
Phase: Quality Input: Token Options: control = appelt Rule: Quality1 Priority:100 ({Token.string == "Quality"}) :quality –> :quality.Quality = {rule = "Quality1"}

The set of syntactic rules was created manually. The categories of developed rules refer to a previously elaborated set of criteria implemented in the OWL ontology. Elaborated rules aim to extract attribute values from any corpus of documents and assign them to a given class. These rules have been implemented in the JAPE language. GATE offers OWLLim as an ontology editor that allows you to add results directly to the ontology. In addition, it is possible to save all extracted information in the XML file. Subsequently, an ontology can be automatically created with all information about classes, attributes, and instances. The XML file may also be used by the Protégé environment as an input file and may be processed and saved in OWL/XML format.

4. Case Study

4.1. Domain Knowledge Acquisition and Cluster Construction

Data were collected from the Scopus database [59]. This data pre-processing and selection process was described in Section 3.1. Manual screening of selected works allows for dividing the data into criteria and sub-criteria. This process enables the initial classification of criteria. The main set of criteria represents keywords specific to a given class. For example, if the criterion "Quality" is analyzed, then the sub-criteria containing this word in the description will belong to that class. Moreover, in many cases, the sub-criteria may belong to other classes l (e.g., the quality of delivery will belong to the quality and delivery classes).

Subsequently, a bibliometric analysis of selected articles takes place in order to obtain and condense a large amount of bibliographic information. The assumptions of this process are described in Section 3.2. The output is a plotted knowledge map containing the criteria of a sustainable supplier. Finally, this process allowed the grouping of a set of clusters with assigned criteria. The input file was modified on the base of a pre-elaborated set of criteria and sub-criteria. As the main purpose is to extract and classify criteria and sub-criteria, other information such as author, publication date, and the title is omitted. Moreover, the analysis of the keywords alone is insufficient, as it does not contain information about the criteria that are crucial for the construction of the knowledge map. Its further elaboration helps in taxonomy construction. Therefore, VOSviewer will be fed data about the items in the network and the links between the items. This process allows for building a map and obtaining a classification of clusters of related items. This map was computed and normalized using the association strength method as the analysis method. This method is used to normalize the strength of connections between items. The association strength method is used for normalizing the strength of the links between items.

Figure 7 depicts the items indicated by a label and, by default, also by a circle. The size of a label and its circle reflects its importance. Overall, the set of 126 various clusters was defined. The items grouped in the cluster represent the criteria that specify the sustainable supplier's selection. Items containing sub-items are arranged in the same cluster and are related to the main criterion. The colors represent the groups of related items. The distance between items tells you how related the items are. The volume of the circle indicates the contribution of the item, while the size of a circle reflects the total number of co-occurrences of the item.

Figure 8 presents the density map, where each point in a map has a color (ranging from blue to green to yellow) that depends on the density of keywords at that point. The color of the point is closer to yellow when there are more items in the neighborhood of the point and the higher weight of these items.

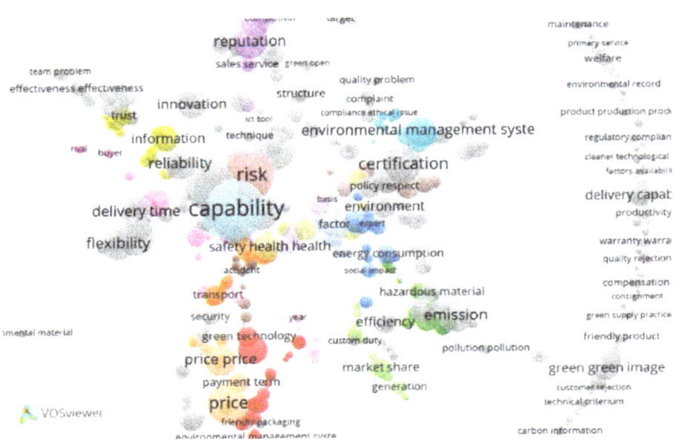

Figure 7. *Cont.*

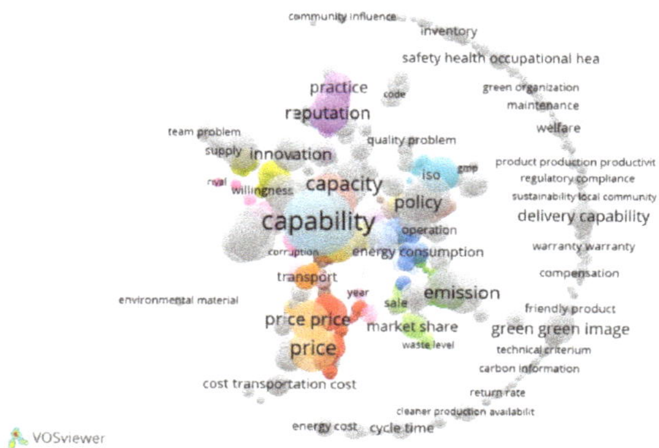

Figure 7. The network visualization of selected items. Source: Personal elaboration using VOSviewer software [62].

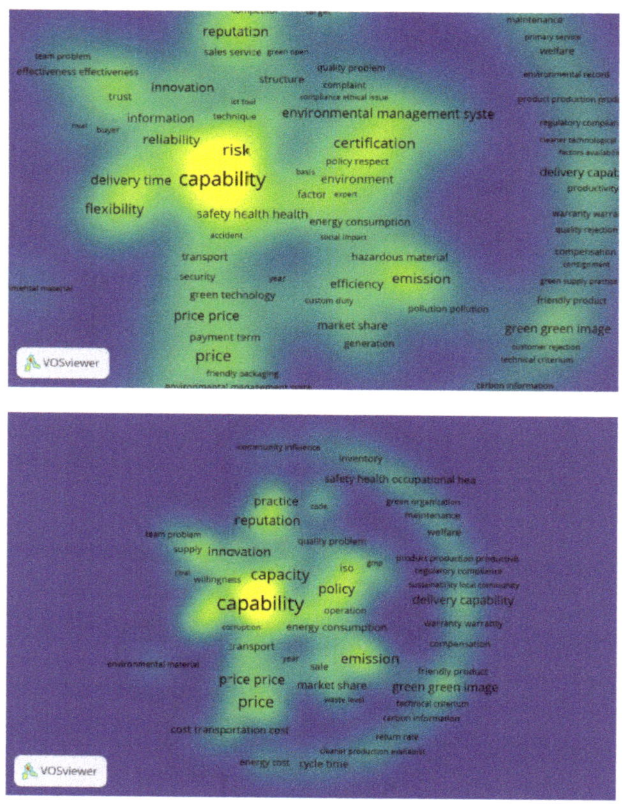

Figure 8. The density visualization of selected items. Source: Personal elaboration using VOSviewer software [62].

In turn, the color of the point is closer to blue when we have a smaller number of items in the neighborhood of the point and the smaller weights of the neighboring items.

As a result, the taxonomic form elaborated on the base of the cluster construction can be implemented in the OWL language. The final set of criteria represents the identified items, and it covers 8261 elements.

4.2. Ontology Construction and Validation

The knowledge acquisition process is described in Section 3.3. The considered domain refers to sustainable supplier criteria. In conclusion, an in-depth analysis of selected articles and the use of bibliometric analysis supports the process of acquiring knowledge and plotting a map of the knowledge domain. This is followed by specification and conceptualization of knowledge, formalization, integration, and implementation in OWL language. Therefore, the knowledge derived from the unstructured data was performed in a structured form. The ontology construction process requires the specification of individuals (concepts), classes, and relations, as well as restrictions, rules, and axioms. The exemplary classes, object properties, and datatype properties were presented in Tables 3 and 4. Figure 9 shows a small piece of a class hierarchy. Each class contains sub-classes. The exemplary class technology is shown in Figure 10 with assigned sub-classes. The ontology also provides information about suppliers' profiles (Figures 11 and 12).

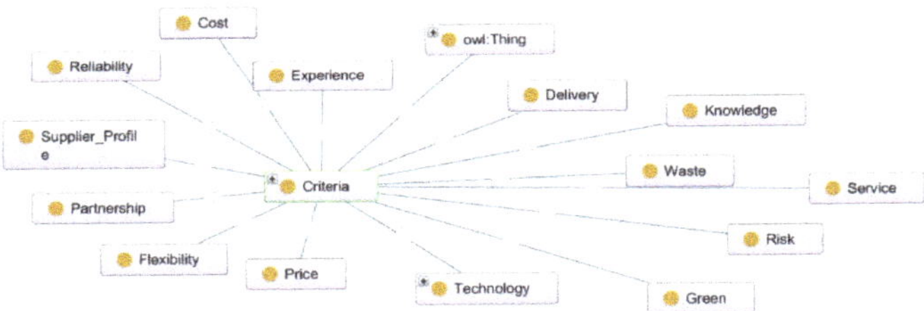

Figure 9. Selected criteria of the constructed ontology. Source: Personal elaboration using Protégé software [63].

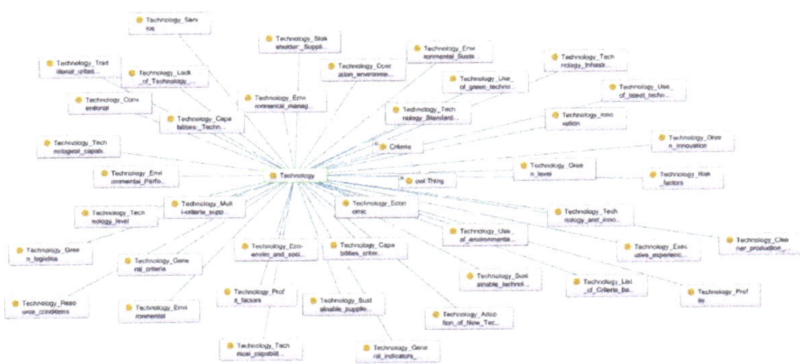

Figure 10. Selected criterion technology with sub-criteria. Source: Personal elaboration using Protégé software [63].

The implementation uses Protégé-OWL API [63] to work with the OWL ontologies and DL query mechanism to manipulate the different constituents of the ontology. The formal description was performed using the description logic (DL) standard. The formal description of the developed knowledge representation using DL allows for machine processing, sharing, reusing, and, finally, populating new knowledge. The evaluation process of the elaborated ontology was performed using the competency questions and implemented using the description logic query mechanism. This process aims to check the coherence and correctness of the constructed ontology using reasoning mechanisms. For a consistent ontology, the output is a result set.

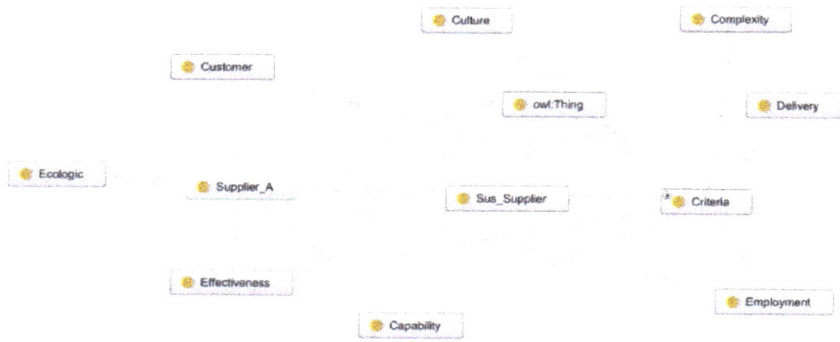

Figure 11. An example of a sustainable supplier profile. Source: Personal elaboration using Protégé software [63].

Figure 12. Description of sustainable supplier profile. Source: Personal elaboration using Protégé software [63].

The first example shows how to ask about sustainable supplier criteria in terms of flexibility, quality, responsiveness, and delivery. A rule-based query is created to find results that meet a defined set of criteria. Query 1 is executed by the code, as shown in Table 6.

Table 6. The working example of the 1st query.

Query 1:

```
<EquivalentClasses>
<Class IRI = "#CQ_1"/>
<ObjectUnionOf>
<ObjectIntersectionOf>
<Class IRI = "#Sus_Supplier"/>
<ObjectSomeValuesFrom>
<ObjectProperty IRI = "#has_Criterion"/>
<Class IRI = "#Flexibility_Technical_capacity"/>
</ObjectSomeValuesFrom>
<ObjectSomeValuesFrom>
<ObjectProperty IRI = "#has_Criterion"/>
<Class IRI = "#Quality_Return_rate"/>
</ObjectSomeValuesFrom>
<ObjectSomeValuesFrom>
<ObjectProperty IRI = "#has_Criterion"/>
<Class IRI = "#Responsiveness"/>
</ObjectSomeValuesFrom>
</ObjectIntersectionOf>
<ObjectSomeValuesFrom>
<ObjectProperty IRI = "#has_Criterion"/>
```

Table 6. *Cont.*

```
<Class IRI = "#Delivery"/>
</ObjectSomeValuesFrom>
<ObjectSomeValuesFrom>
<ObjectProperty IRI = "#has_Criterion"/>
<Class IRI = "#Quality_Discount_rate"/>
</ObjectSomeValuesFrom>
</ObjectUnionOf>
</EquivalentClasses>
```

The second exemplary query aims to demonstrate how to find sustainable supplier criteria in the context of quality, reputation, and delivery. The sub-criteria were predefined, including quality of product, quality ISO 9000, delivery and service, delivery on time, and reputation of the supplier. The query was executed using a reasoner. The code is shown in Table 7.

Table 7. The working example of the 2nd query.

Query 2:

```
<EquivalentClasses>
<Class IRI = "#CQ_2"/>
<ObjectUnionOf>
<ObjectIntersectionOf>
<Class IRI = "#Sus_Supplier"/>
<ObjectSomeValuesFrom>
<ObjectProperty IRI = "#has_Criterion"/>
<Class IRI = "#Quality_ISO_9000"/>
</ObjectSomeValuesFrom>
<ObjectSomeValuesFrom>
<ObjectProperty IRI = "#has_Criterion"/>
<Class IRI = "#Quality_Quality_of_product"/>
</ObjectSomeValuesFrom>
<ObjectSomeValuesFrom>
<ObjectProperty IRI = "#has_Criterion"/>
<Class IRI = "#Delivery_Delivery_Service"/>
</ObjectSomeValuesFrom>
</ObjectIntersectionOf>
<ObjectSomeValuesFrom>
<ObjectProperty IRI = "#has_Criterion"/>
<Class IRI = "#Delivery_On_time_delivery"/>
</ObjectSomeValuesFrom>
<ObjectSomeValuesFrom>
<ObjectProperty IRI = "#has_Criterion"/>
<Class IRI = "#Reputation_Reputation_of_supplier"/>
</ObjectSomeValuesFrom>
</ObjectUnionOf>
</EquivalentClasses>
```

These queries represent only the partial possibilities of using a knowledge base in extracting information. The examples are attached in supplementary materials (see: JAPE examples: JAPE examples.zip). Given the huge number of criteria included in the knowledge base, there are many possibilities to build different combinations of queries. As a result, the user will also be able to indicate the profile of the preferred supplier. It also allows the user to identify the source of the criteria. Combining the knowledge base with additional modules/knowledge bases containing information, for example, on indicators, gives a chance for a comprehensive source of knowledge in the field of sustainable supplies and suppliers.

4.3. Semantic Annotation and Ontology Population

The corpus for tests consists of a set of sustainable supplier reports, papers, and other data gathered from web resources. The use of ANNIE, together with selected processing resources (PR) dedicated to information extraction, enabled the performance of various extraction tasks. (mentioned in detail in Section 3.4). The implementation of these PR begins the process of performing the corpus of documents. The corpus of documents may contain various text documents such as scientific articles, report sheets, plain text, etc., and links to websites. Finally, a set of basic annotations has been provided. In order to extend the built-in set of annotations, the own annotations with specific constraints and rules have been created. The created annotations depend on what a user wants to search for and how to classify it. Figure 13 displays exemplary annotations that aim to find the criteria related to technology, transport, and strategic feature. The criteria found in the document body are highlighted (depending on the color assigned to them). It is also possible to add additional features.

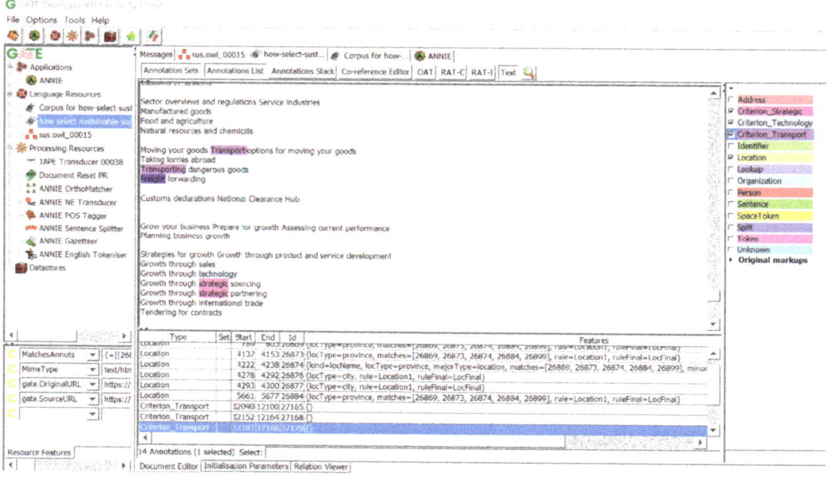

Figure 13. Displaying the exemplary annotations from the text (web resource). Source: Personal elaboration using GATE software [65].

The implementation of the presented approach using semantic annotation and ontology population requires the use of tools included in this environment and, thus, the installation of new plugins for working with ontologies. OWLIM Ontology plugin and GATE Ontology Editor were used to work with ontology (Figure 14). The ontology was created in the Protégé environment [63]; however, to work with GATE and enable semantic annotation and ontology population, available GATE plugins were used in this part of the experiments.

Within the ontology population, it is possible to create specific rules that are designed to find and classify selected concepts. Hence, the next step is to use JAPE Transducer. JAPE Transducer defines the rules and recognizes regular expressions in annotations of documents. Figure 15 displays the partial results of these phases. The working example of the rule named Quality1 demonstrates the applicability of JAPE rules. Many such rules were created to carry out the tests. Of course, the possibilities of creating rules are huge, and it is possible to expand the rules with additional elements. Figure 16 displays the partial results of applied rule Quality1. The execution of the JAPE rule for extracting attribute values for rule Quality1 is shown in Figure 17.

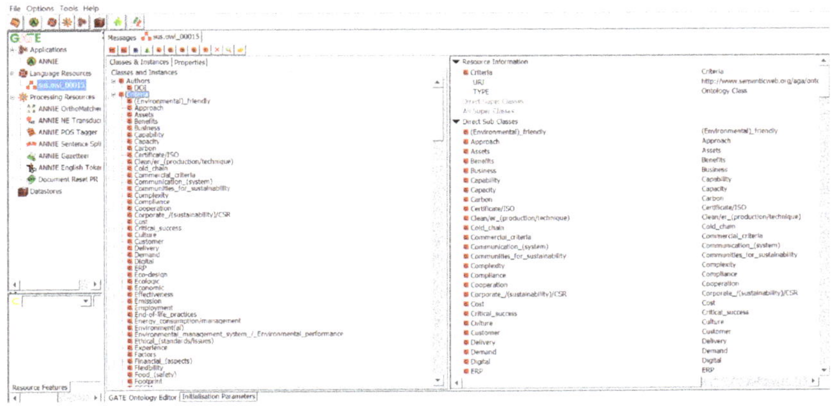

Figure 14. Displaying the ontology using the OWLIM Ontology plugin and GATE Ontology Editor. Source: Personal elaboration using GATE software [65].

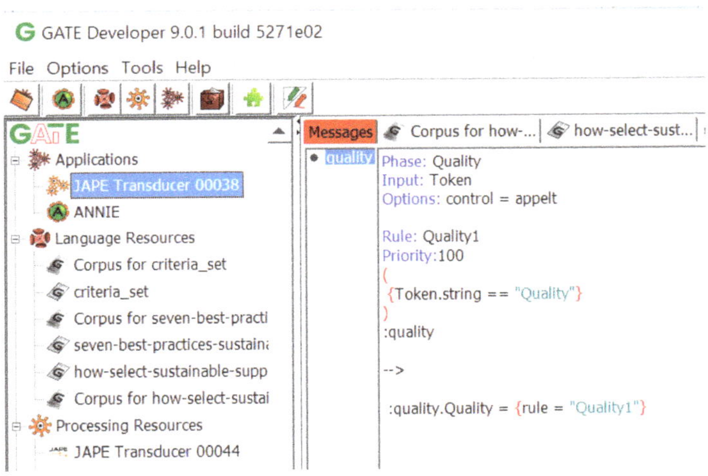

Figure 15. The exemplary JAPE rule "Quality1". Source: Personal elaboration using GATE software [65].

The presented approach offers a semi-automatic, supervised ontology population. By using semantic annotation, it is possible to annotate the relevant word, for example, "Quality of supply" as a criterion related to sustainable suppliers and link it to an ontology instance. As a consequence, new knowledge is added to the ontology. The application of the reasoning mechanism allows classifying the selected word as a criterion of quality. It can therefore be interpreted as follows from the ontology that "Quality of supply" is a criterion associated with a given supplier profile. For implemented ontology, the class feature can be used on the LHS of a JAPE rule. When matching the class value, the ontology is checked for subsumption. If any sub-class on the left side of "==" matches {Lookup.class == Quality}, it will match a lookup annotation with the class feature, whose value is either quality or any subclass of it (Figure 18).

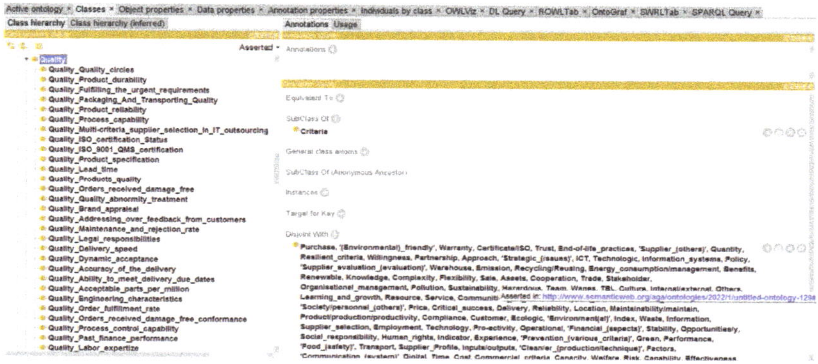

Figure 16. Populated ontology after applying the created rules. Source: Personal elaboration using Protégé software [63].

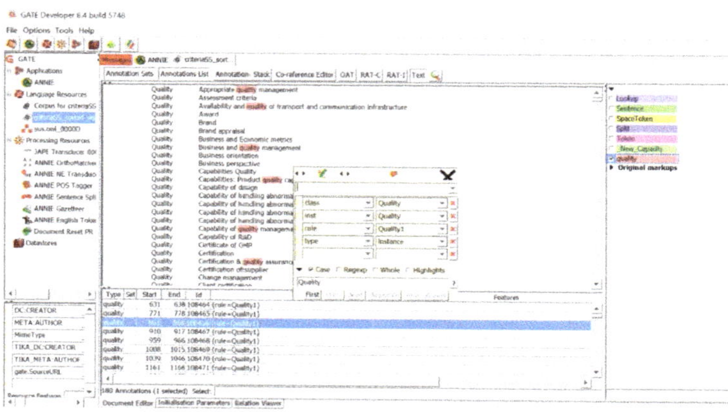

Figure 17. The execution of the JAPE rule for extracting attribute values for rule Quality1. Source: Personal elaboration using GATE software [65].

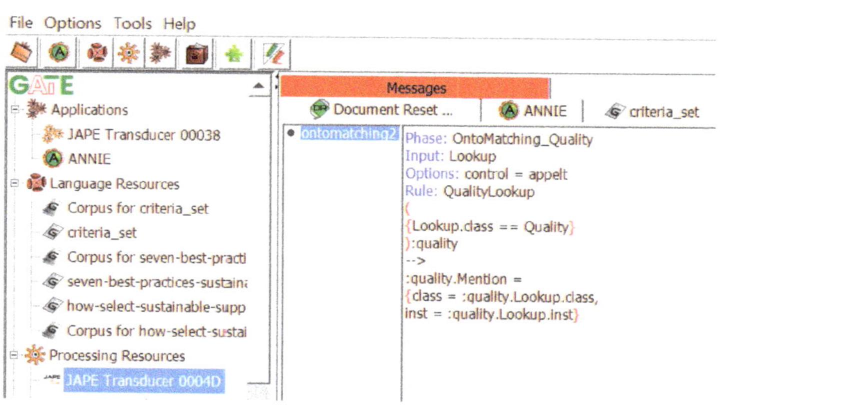

Figure 18. The execution of the LHS JAPE rule for extracting attribute values for rule QualityLookup. Source: Personal elaboration using GATE software [65].

Ontologies are useful for encoding the information found. Applying the created rules for a given corpus of documents makes it possible to extract knowledge using rules and assign this knowledge to classes and instances in the ontology (Figures 16 and 19). The richer NE tagging and application of JAPE rules aim to disambiguate the instances. The modified ontology is then loaded using Protégé software [63]. In this way, the user has control over the development of the ontology and its population and the updating of data. In order to further develop the ontology, rules can be created automatically from a single pattern, with a rule per object property having to be populated.

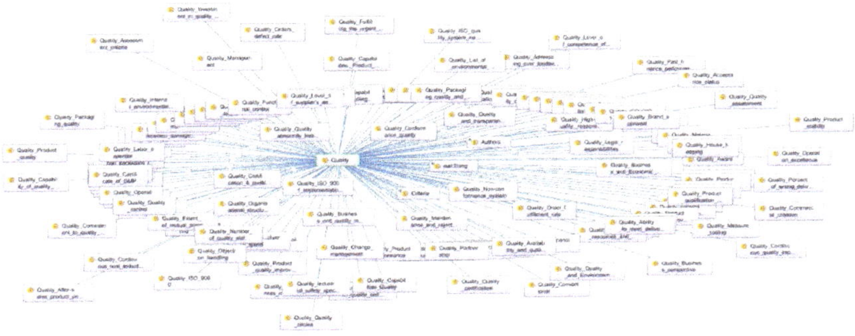

Figure 19. Graphical visualization of the part of populated ontology after applying the created rules. Source: Personal elaboration using Protégé software [63].

4.4. Validation and Evaluation

In order to evaluate and validate the obtained ontology, the application of the reasoning mechanism takes place. Two reasoning mechanisms were applied: HermiT 1.4.3.456 and Pellet. Both of them did not detect the inconsistency of the loaded ontology (Figure 20).

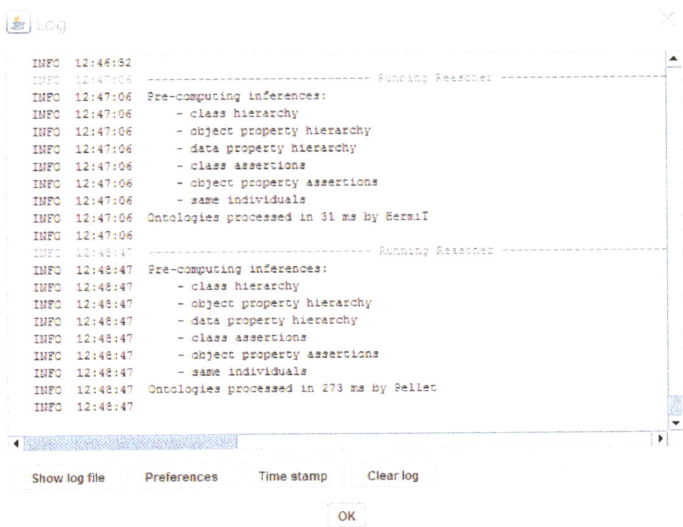

Figure 20. The log results after using HermiT and Pellet reasoners. Source: Personal elaboration using Protégé software [63].

Other ontology assessments and validations require the use of a master ontology. In this case, these measures cannot be used. For example, ontology can be evaluated using metric-based evaluation, including relationship richness, attribute richness, and class richness. However, to evaluate the quality using these metrics, a similar basic ontology is needed. Apart from that, it is possible to evaluate the ontology using dedicated measure balance distance metrics (BDM), but the reference ontology, test set, and training set are also necessary.

5. Conclusions

This paper proposed an ontology-based approach for knowledge acquisition from the text for the sustainable supplier selection domain. The presented solution showed the process of acquiring complex relationships from texts and encoding them in the form of rules. As a result, the enrichment of the existing domain ontology by adding new knowledge and reaching higher relational expression, reasoning, and producing new facts has been successfully implemented and achieved.

This process required the use of various techniques and tools, such as VosViewer for plotting knowledge domain maps, Protégé environment for implementing and managing the OWL ontology, GATE software with NLP tools and text matching techniques and plugins for deducing different atoms, and JAPE rules for capturing deductive knowledge in the form of new rules. The evaluation process was performed using the reasoning mechanisms HermiT 1.4.3.456 and Pellet.

The essential contribution of the work covers the following:

Developing an ontology-based framework to deal with distributed knowledge representation;

Developing a domain ontology that stores various information about sustainable suppliers, which supports various knowledge management aspects, associating dynamic data delivered from external sources with predefined information gathered in the ontology;

Constructing a knowledge base with rules and queries using JAPE;

Checking the consistency and testing the use of the ontology in different scenarios in the domain of sustainable supplier selection and applying rule-based reasoning.

The presented ontology provides independent knowledge about criteria for sustainable supplier selection, which is proved by a scientific literature analysis. The new knowledge can be incorporated into any database, knowledge base, or information system. This form of storing knowledge offers machine-readable access and semantic data handling. Additionally, the proposed approach made it possible to:

Increase the body of knowledge on the ontology for the sustainable supplier domain by providing a systematic keywords map of the subject and grasping the main criteria in the research field;

Handle knowledge domain;

Reduce time for searching for relevant information;

Improve the accuracy of search results that suit user's specific needs;

Provide quick updates with new knowledge.

However, there are still some limitations that need to be addressed in future research. Further refinements to the presented approach include increasing the level of automation of phases that currently require manual work. In particular, a way to automate JAPE rule definitions and prepare patterns is currently under development. The use of the reasoning abilities provided by the ontology to generate new JAPE rules, starting with patterns of manually specified JAPE rules, is also a promising direction and an extension of this work.

Supplementary Materials: The following supporting information can be downloaded at: https://www.mdpi.com/article/10.3390/electronics11234012/s1; the set of criteria: Sustainable_Supplier_Criteria.xls; JAPE examples: JAPE examples.zip.

Funding: This research received no external funding.

Conflicts of Interest: The authors declare no conflict of interest.

References

1. Hoseini, S.A.; Fallahpour, A.; Wong, K.Y.; Mahdiyar, A.; Saberi, M.; Durdyev, S. Sustainable Supplier Selection in Construction Industry through Hybrid Fuzzy-Based Approaches. *Sustainability* **2021**, *13*, 1413. [CrossRef]
2. Govindan, K.; Khodaverdi, R.; Jafarian, A. A Fuzzy Multi Criteria Approach for Measuring Sustainability Performance of a Supplier Based on Triple Bottom Line Approach. *J. Clean. Prod.* **2013**, *47*, 345–354. [CrossRef]
3. Saeed, M.; Kersten, W. Drivers of Sustainable Supply Chain Management: Identification and Classification. *Sustainability* **2019**, *11*, 1137. [CrossRef]
4. Amindoust, A. A Resilient-Sustainable Based Supplier Selection Model Using a Hybrid Intelligent Method. *Comput. Ind. Eng.* **2018**, *126*, 122–135. [CrossRef]
5. Chand, P.; Thakkar, J.J.; Ghosh, K.K. Analysis of Supply Chain Complexity Drivers for Indian Mining Equipment Manufacturing Companies Combining SAP-LAP and AHP. *Resour. Policy* **2018**, *59*, 389–410. [CrossRef]
6. Singh, R.K.; Murty, H.R.; Gupta, S.K.; Dikshit, A.K. An Overview of Sustainability Assessment Methodologies. *Ecol. Indic.* **2012**, *15*, 281–299. [CrossRef]
7. Lee, J.; Jung, K.; Kim, B.H.; Peng, Y.; Cho, H. Semantic Web-Based Supplier Discovery System for Building a Long-Term Supply Chain. *Int. J. Comput. Integr. Manuf.* **2015**, *28*, 155–169. [CrossRef]
8. Kumar, C.V.S.; Routroy, S. Developing the Preferred Supplier Relationships—A Case Study. *Int. J. Intell. Enterp.* **2018**, *5*, 50. [CrossRef]
9. Ware, N.R.; Singh, S.P.; Banwet, D.K. Supplier Selection Problem: A State-of-the-Art Review. *Manag. Sci. Lett.* **2012**, *2*, 1465–1490. [CrossRef]
10. Shaw, K.; Shankar, R.; Yadav, S.S.; Thakur, L.S. Global Supplier Selection Considering Sustainability and Carbon Footprint Issue: AHP Multi-Objective Fuzzy Linear Programming Approach. *Int. J. Oper. Res.* **2013**, *17*, 215. [CrossRef]
11. Awasthi, A.; Govindan, K.; Gold, S. Multi-Tier Sustainable Global Supplier Selection Using a Fuzzy AHP-VIKOR Based Approach. *Int. J. Prod. Econ.* **2018**, *195*, 106–117. [CrossRef]
12. Konys, A. Methods Supporting Supplier Selection Processes–Knowledge-Based Approach. *Procedia Comput. Sci.* **2019**, *159*, 1629–1641. [CrossRef]
13. Ramanathan, R. Supplier Selection Problem: Integrating DEA with the Approaches of Total Cost of Ownership and AHP. *Supply Chain Manag. Int. J.* **2007**, *12*, 258–261. [CrossRef]
14. Arabsheybani, A.; Paydar, M.M.; Safaei, A.S. An Integrated Fuzzy MOORA Method and FMEA Technique for Sustainable Supplier Selection Considering Quantity Discounts and Supplier's Risk. *J. Clean. Prod.* **2018**, *190*, 577–591. [CrossRef]
15. Kahraman, C.; Topcu, Y.I. *Operations Research Applications in Health Care Management*; Springer International Publishing: New York, NY, USA; Heidelberg, Germany; Dordrecht, The Netherlands; London, UK, 2018; ISBN 978-3-319-65455-3.
16. Weber, C.A.; Current, J.R.; Benton, W.C. Vendor Selection Criteria and Methods. *Eur. J. Oper. Res.* **1991**, *50*, 2–18. [CrossRef]
17. Büyüközkan, G.; Feyzioğlu, O.; Havle, C.A. Analysis of Success Factors in Aviation 4.0 Using Integrated Intuitionistic Fuzzy MCDM Methods. In *Intelligent and Fuzzy Techniques in Big Data Analytics and Decision Making*; Kahraman, C., Cebi, S., Cevik Onar, S., Oztaysi, B., Tolga, A.C., Sari, I.U., Eds.; Springer International Publishing: Cham, Switzerland, 2020; Volume 1029, pp. 598–606. ISBN 978-3-030-23755-4.
18. Burki, U.; Ersoy, P.; Dahlstrom, R. Achieving Triple Bottom Line Performance in Manufacturer-Customer Supply Chains: Evidence from an Emerging Economy. *J. Clean. Prod.* **2018**, *197*, 1307–1316. [CrossRef]
19. Sarkis, J.; Dhavale, D.G. Supplier Selection for Sustainable Operations: A Triple-Bottom-Line Approach Using a Bayesian Framework. *Int. J. Prod. Econ.* **2015**, *166*, 177–191. [CrossRef]
20. Yu, C.; Wong, T.N. An Agent-Based Negotiation Model for Supplier Selection of Multiple Products with Synergy Effect. *Expert Syst. Appl.* **2015**, *42*, 223–237. [CrossRef]
21. Kumar, A.; Jain, V.; Kumar, S. A Comprehensive Environment Friendly Approach for Supplier Selection. *Omega* **2014**, *42*, 109–123. [CrossRef]
22. Kuo, R.J.; Wang, Y.C.; Tien, F.C. Integration of Artificial Neural Network and MADA Methods for Green Supplier Selection. *J. Clean. Prod.* **2010**, *18*, 1161–1170. [CrossRef]
23. Zhao, H.; Guo, S. Selecting Green Supplier of Thermal Power Equipment by Using a Hybrid MCDM Method for Sustainability. *Sustainability* **2014**, *6*, 217–235. [CrossRef]
24. Wang, C.-N.; Nguyen, V.T.; Thai, H.T.N.; Tran, N.N.; Tran, T.L.A. Sustainable Supplier Selection Process in Edible Oil Production by a Hybrid Fuzzy Analytical Hierarchy Process and Green Data Envelopment Analysis for the SMEs Food Processing Industry. *Mathematics* **2018**, *6*, 302. [CrossRef]
25. Opher, T.; Friedler, E.; Shapira, A. Comparative Life Cycle Sustainability Assessment of Urban Water Reuse at Various Centralization Scales. *Int. J. Life Cycle Assess.* **2019**, *24*, 1319–1332. [CrossRef]
26. Shaaban, M.; Scheffran, J.; Böhner, J.; Elsobki, M.S. A Dynamic Sustainability Analysis of Energy Landscapes in Egypt: A Spatial Agent-Based Model Combined with Multi-Criteria Decision Analysis. *J. Artif. Soc. Soc. Simul.* **2019**, *22*. [CrossRef]
27. Roinioti, A.; Koroneos, C. Integrated Life Cycle Sustainability Assessment of the Greek Interconnected Electricity System. *Sustain. Energy Technol. Assess.* **2019**, *32*, 29–46. [CrossRef]
28. Bottero, M.; Oppio, A.; Bonardo, M.; Quaglia, G. Hybrid Evaluation Approaches for Urban Regeneration Processes of Landfills and Industrial Sites: The Case of the Kwun Tong Area in Hong Kong. *Land Use Policy* **2019**, *82*, 585–594. [CrossRef]

29. Talukder, B.; Hipel, K.W. The PROMETHEE Framework for Comparing the Sustainability of Agricultural Systems. *Resources* **2018**, *7*, 74. [CrossRef]
30. Melkonyan, A.; Gruchmann, T.; Lohmar, F.; Kamath, V.; Spinler, S. Sustainability Assessment of Last-Mile Logistics and Distribution Strategies: The Case of Local Food Networks. *Int. J. Prod. Econ.* **2020**, *228*, 107746. [CrossRef]
31. Cai, M.; Zhang, W.Y.; Zhang, K. ManuHub: A Semantic Web System for Ontology-Based Service Management in Distributed Manufacturing Environments. *IEEE Trans. Syst. Man Cybern.-Part A Syst. Hum.* **2011**, *41*, 574–582. [CrossRef]
32. Balasbaneh, A.T.; Yeoh, D.; Zainal Abidin, A.R. Life Cycle Sustainability Assessment of Window Renovations in Schools against Noise Pollution in Tropical Climates. *J. Build. Eng.* **2020**, *32*, 101784. [CrossRef]
33. Siksnelyte, I.; Zavadskas, E.K.; Bausys, R.; Streimikiene, D. Implementation of EU Energy Policy Priorities in the Baltic Sea Region Countries: Sustainability Assessment Based on Neutrosophic MULTIMOORA Method. *Energy Policy* **2019**, *125*, 90–102. [CrossRef]
34. Maedche, A. The Text-to-Onto Ontology Extraction and Maintenance System. In Proceedings of the ICDM-Workshop on Integrating Data Mining and Knowledge Management, San Jose, CA, USA, 29 November–2 December 2001.
35. Konys, A. Towards Knowledge Handling in Ontology-Based Information Extraction Systems. *Procedia Comput. Sci.* **2018**, *126*, 2208–2218. [CrossRef]
36. Jain, V.; Singh, M. Ontology Based Information Retrieval in Semantic Web: A Survey. *Int. J. Inf. Technol. Comput. Sci.* **2013**, *5*, 62–69. [CrossRef]
37. Konys, A. A Tool Supporting Mining Based Approach Selection to Automatic Ontology Construction. *IADIS J. Comput. Sci. Inf. Syst.* **2015**, 3–10.
38. Boufrida, A.; Boufaida, Z. Rule Extraction from Scientific Texts: Evaluation in the Specialty of Gynecology. *J. King Saud Univ.-Comput. Inf. Sci.* **2022**, *34*, 1150–1160. [CrossRef]
39. Zhang, Z.; Gentile, A.L.; Ciravegna, F. Recent Advances in Methods of Lexical Semantic Relatedness—A Survey. *Nat. Lang. Eng.* **2013**, *19*, 411–479. [CrossRef]
40. Zhang, F.; Cheng, J.; Ma, Z. A Survey on Fuzzy Ontologies for the Semantic Web. *Knowl. Eng. Rev.* **2016**, *31*, 278–321. [CrossRef]
41. Morente-Molinera, J.A.; Pérez, I.J.; Ureña, M.R.; Herrera-Viedma, E. Building and Managing Fuzzy Ontologies with Heterogeneous Linguistic Information. *Knowl.-Based Syst.* **2015**, *88*, 154–164. [CrossRef]
42. Díaz Rodríguez, N.; Cuéllar, M.P.; Lilius, J.; Delgado Calvo-Flores, M. A Fuzzy Ontology for Semantic Modelling and Recognition of Human Behaviour. *Knowl.-Based Syst.* **2014**, *66*, 46–60. [CrossRef]
43. Poslad, S.; Kesorn, K. A Multi-Modal Incompleteness Ontology Model (MMIO) to Enhance Information Fusion for Image Retrieval. *Inf. Fusion* **2014**, *20*, 225–241. [CrossRef]
44. Pérez, I.J.; Wikström, R.; Mezei, J.; Carlsson, C.; Herrera-Viedma, E. A New Consensus Model for Group Decision Making Using Fuzzy Ontology. *Soft Comput.* **2013**, *17*, 1617–1627. [CrossRef]
45. Fensel, D. *Ontologies*; Springer: Berlin/Heidelberg, Germany, 2001; pp. 11–18.
46. Little, E.G.; Rogova, G.L. Designing Ontologies for Higher Level Fusion. *Inf. Fusion* **2009**, *10*, 70–82. [CrossRef]
47. Gruber, T.R. A Translation Approach to Portable Ontology Specifications. *Knowl. Acquis.* **1993**, *5*, 199–220. [CrossRef]
48. Besbes, G.; Baazaoui-Zghal, H. Fuzzy Ontologies for Search Results Diversification: Application to Medical Data. In Proceedings of the 33rd Annual ACM Symposium on Applied Computing, Pau, France, 9 April 2018; pp. 1968–1975.
49. Shafna, S.; Rajendran, V.V. Fuzzy Ontology Based Recommender System with Diversification Mechanism. In Proceedings of the 2017 International Conference on Intelligent Computing and Control (I2C2), IEEE, Coimbatore, India, 23–24 June 2017; pp. 1–6.
50. Motik, B.; Parsia, B.; Patel-Schneider, P.F. OWL 2 Web Ontology Language XML Serialization. *World Wide Web Consort.* 2009. Available online: https://www.w3.org/TR/2009/WD-owl2-xml-serialization-20090421/all.pdf (accessed on 1 April 2022).
51. Corcoglioniti, F.; Rospocher, M.; Aprosio, A.P. Frame-Based Ontology Population with PIKES. *IEEE Trans. Knowl. Data Eng.* **2016**, *28*, 3261–3275. [CrossRef]
52. Blandón Andrade, J.C.; Zapata Jaramillo, C.M. Gate-Based Rules for Extracting Attribute Values. *Computación y Sistemas* **2021**, *25*, 851–862. [CrossRef]
53. Corcho, O.; Fernández-López, M.; Gómez-Pérez, A.; López-Cima, A. Building Legal Ontologies with METHONTOLOGY and WebODE. In *Law and the Semantic Web*; Benjamins, V.R., Casanovas, P., Breuker, J., Gangemi, A., Eds.; Springer: Berlin/Heidelberg, Germany, 2005; Volume 3369, pp. 142–157. ISBN 978-3-540-25063-0.
54. Maedche, A.; Staab, S. Ontology Learning for the Semantic Web. *IEEE Intell. Syst.* **2001**, *16*, 72–79. [CrossRef]
55. Konys, A. Knowledge Repository of Ontology Learning Tools from Text. *Procedia Comput. Sci.* **2019**, *159*, 1614–1628. [CrossRef]
56. Konys, A. Knowledge Systematization for Ontology Learning Methods. *Procedia Comput. Sci.* **2018**, *126*, 2194–2207. [CrossRef]
57. Cimiano, P. Ontology Learning from Text. In *Ontology Learning and Population from Text*; Springer Science & Business Media: Berlin, Germany, 2006; pp. 19–34. ISBN 978-0-387-30632-2.
58. Ma, C.; Molnár, B. Use of Ontology Learning in Information System Integration: A Literature Survey. In *Intelligent Information and Database Systems*; Sitek, P., Pietranik, M., Krótkiewicz, M., Srinilta, C., Eds.; Springer: Singapore, 2020; Volume 1178, pp. 342–353. ISBN 9789811533792.
59. Available online: www.scopus.com (accessed on 1 April 2022).
60. van Eck, N.J.; Waltman, L. VOS: A New Method for Visualizing Similarities Between Objects. In *Advances in Data Analysis*; Decker, R., Lenz, H.-J., Eds.; Springer: Berlin/Heidelberg, Germany, 2007; pp. 299–306. ISBN 978-3-540-70980-0.

61. Moher, D.; Liberati, A.; Tetzlaff, J.; Altman, D.G. The PRISMA Group Preferred Reporting Items for Systematic Reviews and Meta-Analyses: The PRISMA Statement. *PLoS Med.* **2009**, *6*, e1000097. [CrossRef]
62. van Eck, N.J.; Waltman, L. Software Survey: VOSviewer, a Computer Program for Bibliometric Mapping. *Scientometrics* **2010**, *84*, 523–538. [CrossRef]
63. Musen, M.A. The Protégé Project: A Look Back and a Look Forward. *AI Matters* **2015**, *1*, 4–12. [CrossRef] [PubMed]
64. Noy, N.F.; McGuinness, D.L. *Ontology Development 101: A Guide to Creating Your First Ontology*. 2001. Available online: https://protege.stanford.edu/publications/ontology_development/ontology101.pdf (accessed on 1 April 2022).
65. Ganino, G.; Lembo, D.; Mecella, M.; Scafoglieri, F. Ontology Population for Open-Source Intelligence: A GATE-Based Solution: Ontology Population for OSInt: A GATE-Based Solution. *Softw. Pract. Exp.* **2018**, *48*, 2302–2330. [CrossRef]

electronics

MDPI

Article

Parallel Tiled Code for Computing General Linear Recurrence Equations

Włodzimierz Bielecki †,‡ **and Piotr Błaszyński** *,‡

Faculty of Computer Science and Information Systems, West Pomeranian University of Technology in Szczecin, 70-322 Szczecin, Poland; wbielecki@zut.edu.pl

* Correspondence: pblaszynski@zut.edu.pl

† Current address: Żołnierska 49, 72-210 Szczecin, Poland.

‡ These authors contributed equally to this work.

Abstract: In this article, we present a technique that allows us to generate parallel tiled code to calculate general linear recursion equations (GLRE). That code deals with multidimensional data and it is computing-intensive. We demonstrate that data dependencies available in an original code computing GLREs do not allow us to generate any parallel code because there is only one solution to the time partition constraints built for that program. We show how to transform the original code to another one that exposes dependencies such that there are two linear distinct solutions to the time partition restrictions derived from these dependencies. This allows us to generate parallel 2D tiled code computing GLREs. The wavefront technique is used to achieve parallelism, and the generated code conforms to the OpenMP C/C++ standard. The experiments that we conducted with the resulting parallel 2D tiled code show that this code is much more efficient than the original serial code computing GLREs. Code performance improvement is achieved by allowing parallelism and better locality of the target code.

Keywords: computing-intensive data; dynamic programming; loop nest tiling; parallel code; OpenMP C/C++

Citation: Bielecki, W.; Błaszyński, P. Parallel Tiled Code for Computing General Linear Recurrence Equations. *Electronics* **2021**, *10*, 2050. https://doi.org/10.3390/electronics10172050

Academic Editors: Juan M. Corchado and Xianzhi Wang

Received: 22 June 2021
Accepted: 21 August 2021
Published: 25 August 2021

1. Introduction

The purpose of this document is to show a way to produce a parallel program for computing general linear recurrence equations (GLREs). This code can also be tiled.

A recurrence equation expresses each element of a sequence or multidimensional array of values as a function of the preceding ones. Recurrence equations have a broad spectrum of applications, for example: population dynamics, spatial ecology, analysis of algorithms, binary search, digital signal processing, time series analysis, and theoretical and empirical economics. Such applications deal with multidimensional data and are computing-intensive. To reduce their execution time, those applications should be parallelized and run on modern multicore machines.

Many sequential GLRE solutions have been implemented in a variety of development environments. The main problems of these programs are the long time spent in loops and the low cache performance for a large input data sizes that may make them unapplicable.

Loop parallelization and tiling (blocking) can be used to enhance the efficiency of a sequential program. Blocking is a commonly used technique to improve code performance. It allows us to generate a parallel program with greater granularity of code and data locality that will be executed in a multithreaded environment with both distributed and shared memory.

A classic way to automatically parallelize and tile a loop nest is based on affine transformations and includes the following steps: extracting dependencies available in that nest, forming time partition constraints using the obtained dependencies, finding the maximum number of linear independent solutions to those constraints, and finally generating target code [1,2]. If the number of linear independent solutions to those constraints is two or more, then tiled parallel code can be generated [1,2].

In this paper, we examine the loop nest in the C language presented in Listing 1, which computes GLREs. That code is taken from https://www.netlib.org/benchmark/livermorec (accessed on 24 August 2021).

Listing 1: Original loop nest computing GLREs.

```
for ( l=1 ; l<=loop ; l++ ) {
    for ( i=1 ; i<n ; i++ ) {
        for ( k=0 ; k<i ; k++ ) {
            w[i] += b[k][i] * w[(i-k)-1];
        }
    }
}
```

In Section 2, we demonstrate that there exists a single solution to the time partition constraints for the program in Listing 1; hence, direct parallelization and tiling of the program is not feasible using affine transformations.

Our proposal is to modify the code in Listing 1 to another one whose dependencies allow us to form time partition constraints for which there exist two linear independent solutions that allow us to generate parallel 2D tiled code.

The main contributions of the article are as follows.

- Proposal to convert the code from Listing 1 to another sequential code allowing the generation of parallel 2D tiled code computing GLREs.
- Demonstration of GLRE computation with parallel 2D tiled code.
- Comparison of the target code performance with that of the original one presented in Listing 1.

The rest of this article is organized as follows. In Section 2, we provide background on dependency analysis and parallel code generation. In Section 3, we show how to generate GLREs with the parallel 2D tiled program. In Section 4, we analyze similar work of parallel code generation for comparable cases. In Section 5, we discuss the results of the experiments performed. Conclusions of the work are presented in Section 6.

2. Background

Generating serial tiled code allows for significant improvements in data locality that results in improved program performance. Generating parallel tiled code lets on additional increasing code performance due to running such a code by means of multiple threads on many cores.

To our best knowledge, there is no technique allowing us to generate any tiled code examined in this paper and presented in Listing 1. All known techniques based on affine transformations and/or transitive closure of dependence graphs [1,2] are unable to generate any tiled code (serial and/or parallel) for the code in Listing 1. Those techniques are within the class of reordering transformations. They do not introduce any additional computations to generated target code in comparison with an original code; they only reorder iterations of the original code allowing for tiling and/or parallelism.

The other known techniques, for example, [3,4], that are not within reordering transformations allow us to parallelize and tile code similar to some extent to the code in Listing 1 but not exactly the same. All of those techniques introduce additional computations to generated target code in comparison with those in an original one. That prevents achieving the maximal target code performance.

Thus, there is the need to develop techniques belonging to the class of reordering transformations to tile and/or parallelize the code presented in Listing 1, allowing for generating code that does not include any additional computations in comparison with those of an original one. In this paper, we present an approach to resolving this challenge, allowing for generation of parallel tiled code.

In the loop nest, we can find dependencies between instruction instances in the iteration space of that loop nest—the collection of all statements executed inside that loop nest. A dependence is a situation when two instruction instances access the same location in memory and at least one of those accesses is a write. Each dependence is represented with its source and destination but only if the source is executed before the destination. Most commonly, dependencies are expressed by relations that map dependency sources to dependency destinations. Dependence sources and destinations are represented with iteration vectors. The notation of such a relation is the following.

$R := [PARAMS] \rightarrow \{[input\ tuple] \rightarrow [output\ tuple] \mid constraints\}$,
where $[PARAMS]$ is the list of relation parameters, $[input\ tuple]$ represents dependence sources, $[output\ tuple]$ represents dependence destinations, and $constraints$ are the constraints—a system of affine equalities and inequalities on parameters and tuple variables.

In the case of dependencies when the dimensions of the left and right tuples of the relation are the same, the distance vector is the difference between the iteration vector of a dependence target and the iteration vector of the corresponding dependence source.

A dependence vector is uniform if all its elements are constants.

To extract dependencies present in the loop nest, we use the polyhedral model, which is returned by PET [5], and we use the iscc calculator [6], which performs calculations on polyhedral sets and relations [7]. The iscc is an interactive interface to the PET library and barvinok library that will let you count points in polytopes. Barvinok is available online at https://repo.or.cz/barvinok.git (accessed on 24 August 2021). We also used the iscc calculator to calculate distance vectors and generate target code.

To parallelize and tile the loop nest, a time partitioning constraint should be created [1] that states that if iteration I of statement $S1$ depends on iteration J of statement $S2$, then I must be assigned to a time partition that is executed no earlier than the partition containing J, i.e., schedule(I) $\leq$ schedule(J), where schedule(I) and schedule(J) denote the discrete execution time of iterations I and J, respectively.

Linear independent solutions to time partition constraints are needed to create schedules for each occurence of a single instruction of the loop nest allowing for parallelization and tiling of code. The schedule defines a strict partial order, i.e., an irreflexive and transitive relation on the statement instances that determines the order in which they are or should be executed. Details of use of linear independent schedules for generating parallel tiled code can be found in a number of articles, for example, in article [2].

We should extract as many linear independent solutions to time partition constraints as possible. The degrees of parallelism of the target code and the dimension of the tile are higher when more independent solutions are extracted [1]. When there is a single solution to the time partition constraints, parallelization and tiling of the corresponding loop nest using affine transformations is not possible [1].

3. Methods. Parallel Tiled Code Generation

Using PET and the iscc calculator, we extract dependencies available in the code in Listing 1; they are presented with the following relation.

$R := (loop, n) \rightarrow \{ (l, i, k) \rightarrow (l', i, k') \mid l > 0 \wedge i < n \wedge 0 \leq k < i \wedge l < l' \leq loop \wedge 0 \leq k' < i \} \cup (loop, n) \rightarrow \{ (l, i, k) \rightarrow (l', -1 + i - k, k') \mid l > 0 \wedge i < n \wedge k \geq 0 \wedge l < l' \leq loop \wedge 0 \leq k' \leq -2 + i - k \} \cup (loop, n) \rightarrow \{ (l, i, k) \rightarrow (l', i', -1 - i + i') \mid l > 0 \wedge 0 \leq k < i \wedge l \leq l' \leq loop \wedge i < i' < n \} \cup (loop, n) \rightarrow \{ (l, i, k) \rightarrow (l, i, k') \mid 0 < l \leq loop \wedge i < n \wedge k \geq 0 \wedge k < k' < i \}$,
where R is the relation name; $loop$ and n are parameters; $\cup$ is the union operation of sets (relation R is composed as the set union of simpler relations); the tuple before the sign $\rightarrow$ of each simpler relation is the left tuple of this relation, for example, for the first simpler relation, the left tuple is presented with variables (l, i, k); the tuple after the sign $\rightarrow$ of each simpler relation is the right tuple of this relation, for example, for the first simpler relation,

the right tuple is presented with variables (l', i, k'); the expressions after the sign | are the constraints of each simpler relation, each constraint is represented with the conjunctions of inequalities built on tuple variables and parameters; and $\wedge$ is the logical AND operator.

The left tuple of each simpler relation represents dependence sources, whereas the right represents dependence destinations.

Applying the *deltas* operator of the iscc calculator to relation R, we obtain the three distance vectors presented with set D below.

$D := (loop, n) \rightarrow \{ (l, i, k) \mid 0 \leq l < loop \wedge ((l > 0 \wedge i < 0 \wedge i < k < n + 2i) \vee$
$(i > 0 \wedge -n + 2i < k < i)) \} \cup$
$(loop, n) \rightarrow \{ (0, 0, k) \mid loop > 0 \wedge 0 < k \leq -2 + n \} \cup$
$(loop, n) \rightarrow \{ (l, 0, k) \mid 0 < l < loop \wedge 2 - n \leq k \leq -2 + n \}.$

where the notations used are the same as for relation R above except from the set is represented with a single tuple.

Each conjunct in the set above represents a particular distance vector.

Taking into account the constraints of those distance vectors, we simplify them to the following form.

$D := \{(a_1, a_2, a_3) \mid a_1 \geq 0 \wedge -\infty \leq a_2 \leq \infty \wedge -\infty \leq a_3 \leq \infty;$
$(0, 0, b_3) \mid b_3 > 0;$
$(c_1, 0, c_3) \mid c_1 > 0 \wedge -\infty \leq c_3 \leq \infty\}.$

The time partition constraints created from the resulting distance vectors according to article [1] are as follows.

$$h_1 * a_1 + h_2 * a_2 + h_3 * a_3 \geq 0, \quad (1)$$
$$h_3 * b_3 \geq 0, \quad (2)$$
$$h_1 * c_1 + h_3 * c_3 \geq 0, \quad (3)$$

where h_1, h_2, h_3 are the unknowns.

Taking into consideration that $-\infty \leq a_2 \leq \infty$, $-\infty \leq a_3 \leq \infty$, $-\infty \leq c_3 \leq \infty$, we can suppose that to satisfy all the above constraints, h_2 and h_3 should be 0, i.e., $h_2 = h_3 = 0$. Thus, the above constraints can be rewritten as follows.

$$h_1 * a_1 \geq 0, \quad (4)$$
$$h_1 * c_1 \geq 0. \quad (5)$$

Hence, we may cease that there exists a single solution to constraints (1), (2), and (3), namely $(1, 0, 0)^T$. This means that all the three loops in the code in Listing 1 cannot be parallelized and tiled by means of affine transformations.

Next, we try to parallelize and tile only two inner loops i and k in the loop nest in Listing 1. For this purpose, we make the outermost loop l to be serial and extract dependencies for inner loops i and k described with the relation below.

$R := (n) \rightarrow \{ (i, k) \rightarrow (i, k') \mid i < n \wedge k \geq 0 \wedge k < k' < i \} \cup n \rightarrow \{ (i, k) \rightarrow (i', -1 - i + i') \mid 0 \leq k < i \wedge i < i' < n \}.$

Applying the *deltas* operator of the iscc calculator to relation R, we obtain the two distance vectors presented with set D below.

$D := (n) \rightarrow \{ (i, k) \mid i > 0 \wedge -n + 2i < k < i \} \cup$
$(n) \rightarrow \{ (0, k) \mid 0 < k \leq -2 + n \}.$

Next, we simplify the representation of the distance vectors above to the form.

$D := \{(a_1, a_2) \mid a_1 > 0 \wedge -\infty \leq a_2 \leq \infty;$
$(0, b_2) \mid b_2 > 0\}.$

The time partition constraints formed on the basis of the distance vectors above are the following.

$$h_1 * a_1 + h_2 * a_2 \geq 0, \quad (6)$$
$$h_2 * b_2 \geq 0, \quad (7)$$

where h_1, h_2 are the unknowns.

Taking into consideration that $-\infty \leq a_2 \leq \infty$, we can deduce that to satisfy constraints (6) and (7), h_2 should be 0, i.e., $h_2 = 0$.

So, there exists a single solution to constraints (6) and (7), namely $(1,0)^T$, and we conclude that provided the outermost loop l is serial, the two inner loops i and k cannot be parallelized and tiled by means of affine transformations.

To cope with that problem, we transform the code in Listing 1 to improve dependence properties. With this goal, we apply the following schedule to each iteration of the code in Listing 1:

$(l,i,k)^T \to (l,t=i-k)^T$.

This schedule implies that each iteration of the code in Listing 1, represented with iteration vector $(l,i,k)^T$, is mapped to the two-dimensional time $(l,t=i-k)^T$. It means that iterations of loop l should be executed serially, while for a given value of iterator l, iteration $(i,k)^T$ should be executed at time $t=i-k$. This time guarantees that each iteration $(i,k)^T$ is executed when all its operands are ready. To justify that fact, let us noting that for iteration $(i,k)^T$, operand $b[k][i]$ is input data; hence, its value is ready at time 0, and operand $w[(i-k)-1]$ is ready at time $(i-k)-1$ when an actual value of this operand is already calculated and written in memory. Thus, iteration $(i,k)^T$ can be executed at time $t=i-k$, i.e., at time, which is one more than the time when operand $w[(i-k)-1]$ is ready.

In other words, the schedule above is based on data flow software paradigm [8].

To generate target serial code, we form the following relation, which maps each statement instance within the iteration space of the code in Listing 1 to the two-dimensional schedule below.

$CODE := (loop,n) \to \{\,(l,i,k) \to (l,t=i-k) \mid loop > 0 \wedge 0 < l \leq loop \wedge 0 < i < n \wedge 0 \leq k < i\,\}$,

where the constraints

$loop > 0 \wedge 0 < l \leq loop \wedge 0 < i < n \wedge 0 \leq k < i$

define the iteration space of the code in Listing 1. Applying the *iscc* codegen operator to the relation above, we get the pseudocode shown in Listing 2.

Listing 2: Target serial pseudocode.

```
for (int counter = 1; counter <= loop; counter += 1)
  for (int var1 = 1; var1 < n; var1 += 1)
    for (int var2 = var1; var2 < n; var2 += 1)
      ( counter, var2, -var1 + var2); //pseudostatement
```

We transform the pseudocode code in Listing 2 to C code, taking into account that in that pseudocode, variables $counter, var1$, and $var2$ correspond to variables l, t, and i, respectively, in the tuple of set $CODE$; the second variable $var2$ in the pseudostatement relates to variable i, while the third expression $-var1 + var2$ corresponds to variable k in the tuple of set $CODE$. Thus, we replace the pseudostatement in the code in Listing 2 with the statement

$w[i]+=b[k][i]*w[(i-k)-1]$;

from Listing 1 changing variables i and k with variable $var2$ and the expression $-var1 + var2$, respectively. As a result, we obtain the compilable program fragment presented in Listing 3.

Listing 3: Target sequential compilable program fragment.

```
for (int counter = 1; counter <= loop; counter += 1)
  for (int var1 = 1; var1 < n; var1 += 1)
    for (int var2 = var1; var2 < n; var2 += 1)
      w[var2] += b[-var1 + var2][var2] * w[(var2 - (-var1 +
          var2)) - 1];
```

The target serial code in Listing 3 is in the scope of reordered transformations. It performs the same computations as those performed with the initial code in Listing 1 but in a different order. It is well-known that a reordered transformation of a code is correct if it executes the same computations as those executed with the initial one (1) and respects all the dependencies that appear in that code (2) [1]. The transformed code is correct as it performs the same computations as those executed with the initial one (1) and it respects all the dependencies available in the initial one as explained below (2).

There exist three kinds of dependencies in the code presented in Listing 3: data flow dependencies (some statement instance first generates a result, then that result is used with another statement instance, those instances belong to different time units represented with the value of iterator counter), antidependencies (some statement instance first reads a result, then that result is updated with another statement instance, and output dependencies (two statement instances write their results to the same memory location).

Data flow dependencies are respected due to the fact that in the target code, the execution of a statement instance being the target of each data dependence starts only when all the arguments (data) of this operation are prepared, i.e., the processing of all the instruction instances generating these arguments has already finished, and the operand values are stored in the shared part of memory. This is guaranteed because the source of each data dependence is executed at a time unit defined with the value of iterator counter that is less than the one when the corresponding target is executed.

Anti- and output dependencies are honored due to the lexicographical order of the execution of dependent statement instances within each time partition represented with the value of iterator var1.

We also experimentally confirmed that the both loop nests presented in Listing 1 and Listing 3 generate correct results. The experiments used for input data prepared deterministically and randomly.

Dependencies available in the code in Listing 3 are represented with the following relation.

$R := (loop, n) \to \{\, (counter, var1, var2) \to (counter', 1 + var2, var2') \mid counter > 0 \wedge var1 > 0 \wedge var2 \geq var1 \wedge counter \leq counter' \leq loop \wedge var2 < var2' < n \,\} \cup (loop, n) \to \{\, (counter, var1, var2) \to (counter', var1', var2) \mid counter > 0 \wedge var1 > 0 \wedge var1 \leq var2 < n \wedge counter < counter' \leq loop \wedge 0 < var1' \leq var2 \,\} \cup (loop, n) \to \{\, (counter, var1, var2) \to (counter', var1', -1 + var1) \mid counter > 0 \wedge var1 \leq var2 < n \wedge counter < counter' \leq loop \wedge 0 < var1' < var1 \,\} \cup (loop, n) \to \{\, (counter, var1, var2) \to (counter, var1', var2) \mid 0 < counter \leq loop \wedge var1 > 0 \wedge var2 < n \wedge var1 < var1' \leq var2 \,\},$

where $loop$ and n are parameters.

Applying the *deltas* operator of the iscc calculator to relation R, we obtain the three distance vectors presented below.

$D := (loop, n) \to \{\, (counter, var1, var2) \mid 0 \leq counter < loop \wedge ((var1 > 0 \wedge 0 < var2 < n - var1) \vee$
$(counter > 0 \wedge var1 < 0 \wedge -n - var1 < var2 < 0)) \,\} \cup$
$(loop, n) \to \{\, (0, var1, 0) \mid loop > 0 \wedge 0 < var1 \leq -2 + n \,\} \cup$
$(loop, n) \to \{\, (counter, var1, 0) \mid 0 < counter < loop \wedge 2 - n \leq var1 \leq -2 + n \,\}.$

Taking into account the constraints of those distance vectors, we simplify their representation to the following form.

$D := \{(a_1, a_2, a_3) \mid a_1 \geq 0 \wedge -\infty \leq a_2 \leq \infty \wedge -\infty \leq a_3 \leq \infty;$
$(0, b_2, 0) \mid b_2 > 0;$
$(c_1, c_2, 0) \mid c_1 > 0 \wedge -\infty \leq c_2 \leq \infty\}.$

The time partition constraints constructed according to article [1] are as follows.

$$h_1 * a_1 + h_2 * a_2 + h_3 * a_3 \geq 0, \tag{8}$$
$$h_2 * b_2 \geq 0, \tag{9}$$
$$h_1 * c_1 + h_2 * c_2 \geq 0, \tag{10}$$

where h_1, h_2, h_3 are the unknowns.

Taking into account that $-\infty \leq a_2 \leq \infty$, $-\infty \leq a_3 \leq \infty$, $-\infty \leq c_2 \leq \infty$, we can deduce that h_2 and h_3 should be 0, i.e., $h_2 = h_3 = 0$ for the constraints (8), (9), and (10) to be compatible. Therefore, these constraints can be written with the following formulas.

$$h_1 * a_1 \geq 0, \tag{11}$$

$$h_1 * c_1 \geq 0. \tag{12}$$

Thus, we may conclude that there exists a single solution to constraints (8), (9), and (10), namely $(1,0,0)^T$. This means that all thee loops in the code in Listing 3 cannot be parallelized and tiled by means of affine transformations.

Next, we try to parallelize and tile only two inner loops *var*1 and *var*2 in the loop nest presented in Listing 3. For this purpose, we make the outermost loop *counter* to be serial and extract dependencies for inner loops *var*1 and *var*2. They are expressed with the relation below.

$R := (n) \rightarrow \{\, (var1, var2) \rightarrow (1 + var2, var2') \mid var1 > 0 \wedge var2 \geq var1 \wedge var2 < var2' < n \,\} \cup n \rightarrow \{\, (var1, var2) \rightarrow (var1', var2) \mid var1 > 0 \wedge var2 < n \wedge var1 < var1' \leq var2 \,\}$.

Applying the *deltas* operator of the iscc calculator to relation R, we obtain the two distance vectors presented below.

$D := (n) \rightarrow \{\, (var1, var2) \mid var1 > 0 \wedge 0 < var2 < n - var1 \,\} \cup$
$n \rightarrow \{\, (var1, 0) \mid 0 < var1 \leq -2 + n \,\}$.

After the simplification of the representation of the distance vector above, we obtain the following vectors.

$D := \{(a_1, a_2) \mid a_1 > 0 \wedge a_2 > 0;$
$(b1, 0) \mid b_1 > 0\}$

The time partition constraints created from the distance vectors above are the following:

$$h_1 * a_1 + h_2 * a_2 \geq 0, \tag{13}$$

$$h_1 * b_1 \geq 0, \tag{14}$$

where h_1, h_2 are the unknowns. There are two linear independent solutions to the constraints above: $(1,0)^T$ and $(0,1)^T$. Applying those solutions, we are able to parallelize and tile the two inner loops of the code in Listing 3 using the technique presented in paper [2].

The target parallel tiled code presented by means of the OpenMP C/C++ API is shown in Listing 4. It is generated for the best tile size equal to 24 × 54; choosing the best tile size is explained in Section 5.

Listing 4: Transformed parallel loop nest.

```

#define min(lhs,rhs)     ((lhs) < (rhs) ? (lhs) : (rhs))
#define max(lhs,rhs)     ((lhs) > (rhs) ? (lhs) : (rhs))
#define floord(val,d) (((val)<0) ? -((-(val)+(d)-1)/(d)) : (
    val)/(d))
#define ceild(val,d)   ceil(((double)(val))/((double)(d)))

for(int i0 = 1; i0 <= loop; i0 += 1) {
  for(int w0 = 0; w0 <= floord(26*n-26, 675); w0+=1) {
    #pragma omp parallel for
    for(int h0 = max(0, w0 - (n + 49) / 50 + 1); h0 <= min((
        n - 1) / 54, (25 * w0 + 24) / 52); h0 += 1) {
      for(int i1 = max(1, 54 * h0); i1 <= min(min(n - 1, 50 *
          w0 - 50 * h0 + 49), 54 * h0 + 53); i1 += 1) {
        for(int i2 = max(50 * w0 - 50 * h0, i1); i2 <= min(
            n - 1, 50 * w0 - 50 * h0 + 49); i2 += 1) {
          w[i2] += (b[-i1 + i2][i2] * w[i1 - 1]);
        }
      }
```

```
        }
    }
}
```

In that code, outermost loop $i0$ is serial nontiled, and loops $w0$ and $h0$ enumerate tile identifiers, while loops $i1$ and $i2$ enumerate iterations within each tile. Parallelism is extracted with the wavefront technique [2] and presented with the OpenMP directive *#pragma omp parallel for* inserted before loop $h0$ that means that this loop is parallel.

4. Related Work

Related techniques can be divided into the following two classes: the class of reordering transformations and the one of nonreordering transformations. There are numerous publications concerned with both of the classes. Approaches based on affine transformations [1,2,9–11] and those based on the transitive closure of dependence graphs [12–16] belong to reordering transformations. Reordering techniques are code-independent and are used in optimizing compilers, for example [17–19], which automatically generate optimized target code for source code.

Nonreordering transformations are code-dependent, i.e., for a given code, a transformation is fulfilled manually. The following publications within nonreordering transformations concern the problem similar to that implemented with the code in Listing 1 but not exactly the same problem [3,4,8,20–26].

Both classes allow for generating the target program that is semantically identical to the original one. However, there are the following differences in target code generated using techniques of those classes.

Reordering transformations do not introduce any additional computations to generated target code in comparison with those of original code; they only reorder loop nest iterations of the original code allowing for tiling and/or parallelism. They are code-independent and are aimed at automatic code generation.

Nonreordering transformations allow us to parallelize and tile code similar to some extent to the code in Listing 1 but not exactly the same. All of those techniques introduce additional computations to generated target code in comparison with those in the original code. That increases the computational complexity of the algorithm and prevents achieving the maximal target code performance, and it is the main drawback in comparison with reordering transformations.

Each technique is manually created for the code that should be optimized. Adapting such a technique even to a slightly different problem can require additional work that can be time-consuming and not always possible.

After an extensive analysis of many nonreordering techniques mentioned above, we did not find any one that exactly implements the problem presented with the code in Listing 1. Without extensive research, it is not clear how any of those techniques can be adapted to implement exactly the same problem that implements the code in Listing 1.

In the class of reordering transformations, we examined the PLUTO [17] and TRACO compilers [15]. PLUTO is based on affine transformations and automatically generates tiled and/or parallel code. TRACO uses the transitive closure of dependence graphs to tile and/or parallelize input code. For the code in Listing 1, both PLUTO and TRACO are unable to generate any tiled and/or parallel code. In Section 3, we presented the reason why affine transformations fail to generate any parallel and/or tiled code for the serial code in Listing 1.

Below, we discuss some nonreordering transformations, which allow for generation of code implementing algorithms similar to that realizing with the code in Listing 1 but not exactly the same. Without extensive research, it is not clear how to adapt any of those techniques to generate target code fulfilling the same calculations as those performed with the code in Listing 1.

Karp et al. [20] discussed parallelism in recurrence equations. They proposed a decomposition algorithm that decides if a system of uniform recurrence equations (SURE) is computable or not. If so, multidimensional schedules can be derived and applied to extract parallelism.

Papers [3,4] introduced a recursive doubling strategy to compute recurrence equations in parallel. Recursive doubling envisages the splitting of the computation of a function into two subfunctions whose evaluation can be performed simultaneously in two separate processors. Successive splitting of each of these subfunctions allows for the computation over more processors.

Maleki and Burtscher [21] introduced two phase approach to compute recurrence equations. The first phase iteratively merges pairs of adjacent chunks by correcting the values in the second chunk of each pair. The second phase produces the resulting chunks in a pipelined mode to compute the final solution.

Sung et al. [22,23] proposed the idea to divide the input into blocks and decompose the computation over each block. Interblock parallelism is exploited to enhance code performance.

Nehab et al. [24] also suggested splitting the input data into blocks that are processed in parallel by modern GPU architectures and overlapped the causal, anticausal, row, and column filter processing.

Marongiu and Palazzari [25] addressed the parallelization of a class of iterative algorithms described as the system of affine recurrence equations (SARE). It introduces an affine timing function and an affine allocation function that perform a space-time transformation of the loop nest iteration space. It considers algorithms dealing with only uniform dependence vectors, while the approach presented in this paper deals with nonuniform vectors.

Ben-Asher and Haber [26] defined recurrence equations called "simple indexed recurrences" (SIR). In this type of equation, for extending capabilities, ordinary recurrences are generalized to $X[g(i)] = op_i(X[f(i)], X[g(i)])$, where f and g are affine functions $op_i(x,y)$ is a binary associative operator. In that paper, the authors proposed a parallel solution to the SIR problem. This case of recurrences is simpler than that considered in our paper and any tiled code is not considered.

Summing up, we may conclude that in the class of reordering transformations, there does not exist any technique allowing for parallelizing and/or tiling the code in Listing 1. In the class of nonordering transformations, to our best knowledge, no technique has been published to generate parallel tiled code implementing the problem addressed in this paper.

The main contribution of our paper is presenting a novel technique, which for the first time allows us to parallelize and tile the examined loop nest implementing computing general linear recurrence equations by means of reordering transformations. The novelty consists in adding an additional phase to classical reordering transformations: to source code, we first apply a reordering schedule that respects all data dependencies; then, we apply classical affine transformations to the serial code obtained in the first phase. This increases target code generation time but does not introduce any additional computations to the source code. Generated target code is still within the class of reordering transformations.

5. Results

The primary reason for writing a parallel program is speed. We strive that the parallel program execution should be completed at a shorter time in comparison with that of the serial one. We need to know what is the benefit from tiling and parallelism. For this purpose, we need to compute the parallel program speedup.

The speedup of a parallel program over a corresponding sequential program is the ratio of the compute time for the sequential program to the time for the parallel program. The value of speedup shows how efficient is a parallel program.

Perfect linear speedup occurs when the value of speedup is the same as the number of threads used for running a parallel program. In practice, perfect linear speedup seldom occurs because of parallel program overhead and the fact that all computations of an original program cannot be parallelized.

According to Amdahl's law, the parallel code speedup, S, is limited to $S <= 1/s$, where s is the serial fraction of code, i.e., the fraction of code that cannot be parallelized. For example, if $s = 0.2$, the maximal speedup is 5 regardless of the number of threads used for running a parallel program.

To evaluate the performance of the parallel tiled code presented in Listing 4, we carried out experiments aimed at measuring the execution time of the original program and parallel one (for the different number of threads) and next calculated the speedup of the parallel program.

Below, we present the results of experiments carried out with the codes shown in Listing 1 (serial code) and Listing 4 (parallel code). As we mentioned in the previous section, we cannot find any related parallel code that fulfills exactly the same computations as those executed with the code in Listing 1. Thus, we limited our experiments to the codes mentioned above.

To carry out experiments, we used a processor Intel Xeon X5570, 2.93 GHz, 2 physical units, 8 (2 × 4) cores, 16 hyper-threads, and an 8 MB cache. Executable parallel tiled code was generated by means of the g++ compiler with the -O3 flag of optimization.

Experiments were carried out for ten different lengths of the problem defined with parameter N from 1000 to 5000 for the codes presented in Listing 1 (serial code) and Listing 4 (parallel code).

All of the source code to perform the experiments and the program to run the tested codes can be found at https://github.com/piotrbla/livc (accessed on 24 August 2021).

We carried our experiments to choose the optimal size of a tile. The size of a tile is optimal if (i) all data associated with that tile can be held in cache, (ii) those data occupy almost the entire capacity of cache, and (iii) tiled code execution time is minimal provided that the conditions (i) and (ii) above are satisfied.

For this purpose, we fulfilled three trials whose results are shown in Figure 1. The curve "trial1" represents how the time of tiled program execution depends on the block size along axis $h0$ when the block size along axis $w0$ is fixed equal to 16. After first trial 1, we chose the best size along the $h0$ axis equal to 32.

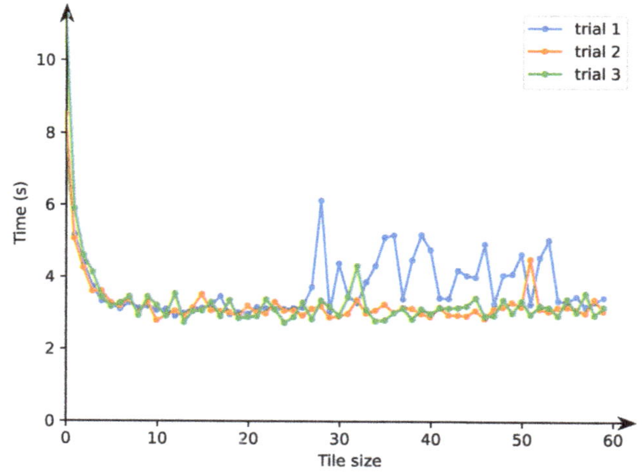

Figure 1. Time for different tile sizes. All phases.

The curve "trial2" demonstrates how the time of tiled program execution depends on the block size along axis $w0$ when the block size along axis $h0$ is equal to 32 (the result of trial 1). After trial 2, we chose the best size along axis $w0$ equal to 24.

The curve "trial3" shows how the time of tiled program execution depends on the block size along axis $h0$ when the block size along axis $w0$ is equal to 24 (the result of trial 2). After trial 3, we chose the best tile size along axis $h0$ equal to 54. Finaly, as the best size of a 2D tile in the parallel tiled code, we chose 24 × 54.

For the best tile size, Table 1 presents execution times and speedup of the serial program in Listing 1 and the parallel tiled one presented in Listing 4 for 32 OpenMP threads used. Figure 2 depicts the data presented in Table 1 in a graphical way. As presented, the execution time of parallel tiled program grows practically in a linear manner exposing considerable speedup (the ratio of the serial program execution time to that of the corresponding parallel one) presented in Figure 3.

Table 1. Time in seconds and speedup for Intel Xeon X5570 and 32 OpenMP threads.

N	Serial	Parallel	Speedup
1000	0.27	0.13	2.08
1500	0.75	0.21	3.57
2000	1.46	0.32	4.56
2500	2.38	0.49	4.86
3000	4.14	0.70	5.91
3500	6.22	0.89	6.99
4000	8.22	1.12	7.34
4500	10.6	1.33	7.97
5000	12.0	1.52	7.89

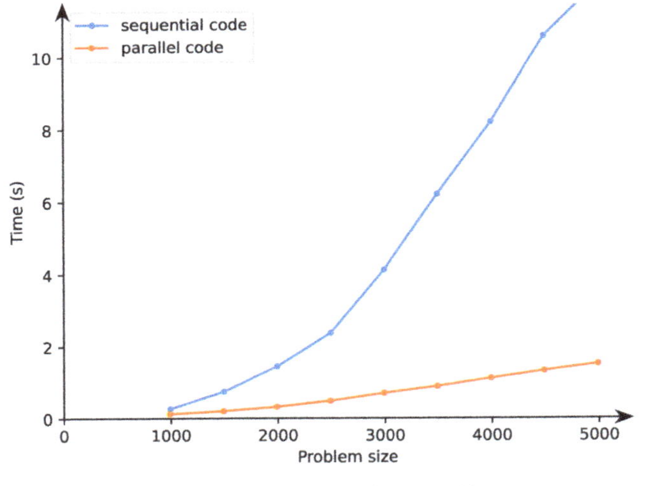

Figure 2. Time for Intel Xeon X5570 and different problem sizes.

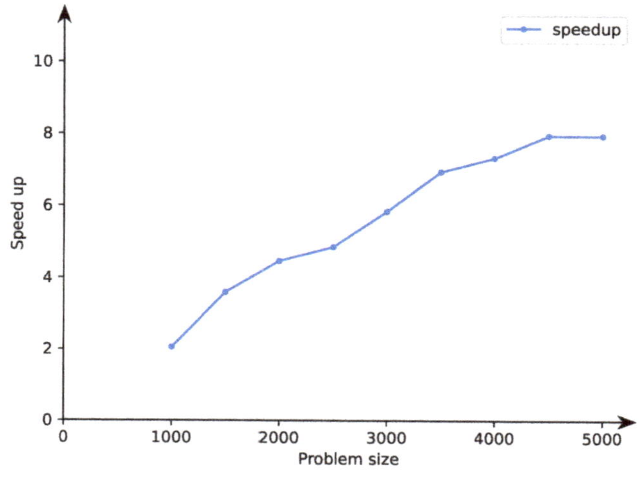

Figure 3. Speedup for Intel Xeon X5570 v3 and 32 OpenMP threads.

Figure 4 presents how parallel tiled code speedup depends on the thread number for the maximal problem size used for experiments, i.e., for $N = 5000$. The parallel tiled code speedup grows practically linear for the number of threads 1 to 12. Linear speedup for the number of threads ≥ 12 is prevented with the serial fraction of code (Amdahl's law)—parallel loop initialization fulfilled with a single thread and serial input–output operations. Speedup is also limited with parallel program overhead—there is thread synchronization in the examined parallel code, after each wavefront, barrier synchronization is inserted because the following wavefront can be executed after completing the calculations of the previous wavefront.

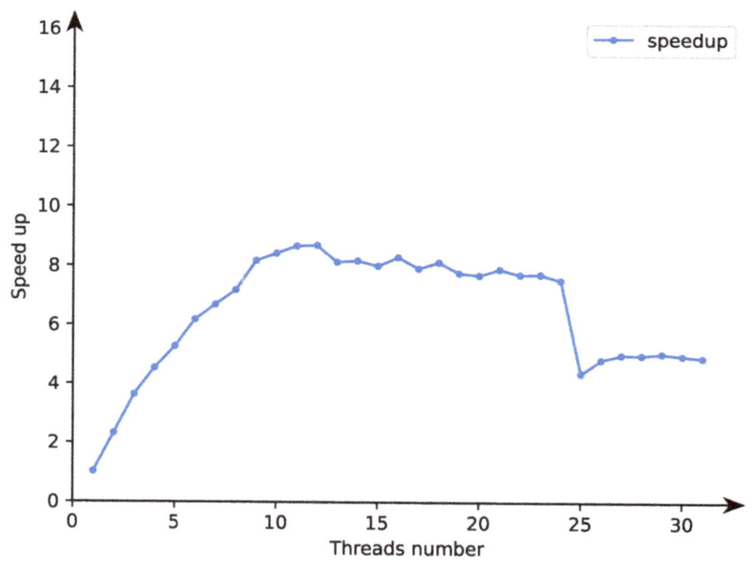

Figure 4. Speedup for Intel Xeon X5570 for different threads number.

We may conclude that the generated parallel tiled code implementing computing-intensive general linear recurrence equations and presented in Listing 4 can be successfully run on modern multicore machines with a large number of cores.

6. Conclusions

We presented an approach to generate parallel tiled code for computing general linear recurrence equations (GLREs) presented in Listing 1. That code is computing-intensive and must be run on modern multicore computers to reduce execution time. We demonstrated how to transform that code to obtain the modified code shown in Listing 3, which exposes dependencies such that there exist two linear independent solutions to the time partition constraints formed on the basis of those dependencies. This allows us to apply the affine transformation framework and generate parallel 2D tiled code computing GLREs presented in Listing 4. The parallelism is achieved using the wavefront technique and presented with the code that conforms to the OpenMP standard. To our best knowledge, the target parallel tiled code generated by us and presented in Listing 4 is the first to allow for enumerating 2D tiles and the first that does not require any additional computations in comparison with those of the original serial code. This code is derived by means of tiling the loop nest iteration space. Our experiments with the resulting parallel tiled code show that the code significantly outperforms the original GLREs computing serial code. The code performance improvement is achieved due to the parallelism and better locality of the target code.

Author Contributions: Conceptualization and methodology, W.B. and P.B.; software, P.B.; validation, W.B. and P.B.; data curation, P.B.; original draft preparation, W.B.; writing—review and editing, W.B. and P.B.; visualization, P.B. All authors have read and agreed to the published version of the manuscript.

Funding: This research received no external funding.

Institutional Review Board Statement: Not applicable.

Informed Consent Statement: Not applicable.

Data Availability Statement: Source code to reproduce all the results described in this paper can be found at: https://github.com/piotrbla/livc (accessed on 24 August 2021).

Conflicts of Interest: The authors declare no conflict of interest.

Abbreviations

The following abbreviations are used in this manuscript:

GLRE	General Linear Recurrence Equations
TRACO	Compiler based on the TRAnsitive ClOsure of dependence graphs

References

1. Lim, A.W.; Cheong, G.I.; Lam, M.S. An affine partitioning algorithm to maximize parallelism and minimize communication. In Proceedings of the 13th international conference on Supercomputing, Rhodes, Greece, 20–25 June 1999; pp. 228–237.
2. Bondhugula, U.; Hartono, A.; Ramanujam, J.; Sadayappan, P. A practical automatic polyhedral parallelizer and locality optimizer. In Proceedings of the 29th ACM SIGPLAN Conference on Programming Language Design and Implementation, Tucson, AZ, USA, 7–13 June 2008; pp. 101–113.
3. Stone, H.S. An efficient parallel algorithm for the solution of a tridiagonal linear system of equations. *J. ACM (JACM)* **1973**, *20*, 27–38. [CrossRef]
4. Kogge, P.M.; Stone, H.S. A parallel algorithm for the efficient solution of a general class of recurrence equations. *IEEE Trans. Comput.* **1973**, *100*, 786–793. [CrossRef]
5. Verdoolaege, S.; Grosser, T. Polyhedral extraction tool. In Proceedings of the Second International Workshop on Polyhedral Compilation Techniques (IMPACT'12), Paris, France, 23 January 2012; pp. 1–16.
6. Verdoolaege, S. Counting affine calculator and applications. In Proceedings of the First International Workshop on Polyhedral Compilation Techniques (IMPACT'11), Chamonix, France, 3 April 2011
7. Verdoolaege, S. isl: An integer set library for the polyhedral model. In *International Congress on Mathematical Software*; Springer: Berlin/Heidelberg, Germany, 2010; pp. 299–302.

8. Stephens, R. A survey of stream processing. *Acta Inform.* **1997**, *34*, 491–541. [CrossRef]
9. Wolf, M.E.; Lam, M.S. A loop transformation theory and an algorithm to maximize parallelism. *IEEE Trans. Parallel Distrib. Syst.* **1991**, *2*, 452–471. [CrossRef]
10. Benabderrahmane, M.W.; Pouchet, L.N.; Cohen, A.; Bastoul, C. The polyhedral model is more widely applicable than you think. In Proceedings of the 19th Joint European conference on Theory and Practice of Software, International Conference on Compiler Construction, Paphos, Cyprus, 20–28 March 2010, pp. 283–303.
11. Irigoin, F.; Triolet, R. Supernode partitioning. In Proceedings of the 15th ACM SIGPLAN-SIGACT Symposium on Principles of Programming Languages, San Diego, CA, USA, 10–13 January 1988; pp. 319–329.
12. Kelly, W.; Pugh, W.; Rosser, E.; Shpeisman, T. Transitive closure of infinite graphs and its applications. *Int. J. Parallel Program.* **1996**, *24*, 579–598. [CrossRef]
13. Pugh, W.; Rosser, E. Iteration Space Slicing for Locality. In Proceedings of the Languages and Compilers for Parallel Computing, La Jolla, CA, USA, 4–6 August 1999; pp. 164–184.
14. Bielecki, W.; Palkowski, M. Tiling arbitrarily nested loops by means of the transitive closure of dependence graphs. *Int. J. Appl. Math. Comput. Sci. (AMCS)* **2016**, *26*, 919–939. [CrossRef]
15. Palkowski, M.; Bielecki, W. TRACO: Source-to-Source Parallelizing Compiler. *Comput. Inform.* **2016**, *35*, 1277–1306.
16. Palkowski, M.; Bielecki, W. Tuning iteration space slicing based tiled multi-core code implementing Nussinov's RNA folding. *BMC Bioinform.* **2018**, *19*, 12. [CrossRef] [PubMed]
17. Bondhugula, U.K. Effective Automatic Parallelization and Locality Optimization Using the Polyhedral Model. Ph.D. Thesis, The Ohio State University, Columbus, OH, USA, 2008.
18. Verdoolaege, S.; Carlos Juega, J.; Cohen, A.; Ignacio Gomez, J.; Tenllado, C.; Catthoor, F. Polyhedral parallel code generation for CUDA. *ACM Trans. Archit. Code Optim. (TACO)* **2013**, *9*, 54. [CrossRef]
19. Dave, C.; Bae, H.; Min, S.J.; Lee, S.; Eigenmann, R.; Midkiff, S. Cetus: A Source-to-Source Compiler Infrastructure for Multicores. *Computer* **2009**, *42*, 36–42. [CrossRef]
20. Karp, R.M.; Miller, R.E.; Winograd, S. The organization of computations for uniform recurrence equations. *J. ACM (JACM)* **1967**, *14*, 563–590. [CrossRef]
21. Maleki, S.; Burtscher, M. Automatic hierarchical parallelization of linear recurrences. *ACM SIGPLAN Not.* **2018**, *53*, 128–138. [CrossRef]
22. Sung, W.; Mitra, S. Efficient multi-processor implementation of recursive digital filters. In Proceedings of the IEEE International Conference on Acoustics, Speech, and Signal Processing, ICASSP'86, Tokyo, Japan, 7–11 April 1986; Volume 11, pp. 257–260.
23. Sung, W.; Mitra, S.K.; Jeren, B. Multiprocessor implementation of digital filtering algorithms using a parallel block processing method. *IEEE Comput. Archit. Lett.* **1992**, *3*, 110–120. [CrossRef]
24. Nehab, D.; Maximo, A.; Lima, R.S.; Hoppe, H. GPU-efficient recursive filtering and summed-area tables. *ACM Trans. Graph. (TOG)* **2011**, *30*, 1–12. [CrossRef]
25. Marongiu, A.; Palazzari, P. Automatic mapping of system of N-dimensional affine recurrence equations (SARE) onto distributed memory parallel systems. *IEEE Trans. Softw. Eng.* **2000**, *26*, 262–275. [CrossRef]
26. Ben-Asher, Y.; Haber, G. Parallel solutions of simple indexed recurrence equations. *IEEE Trans. Parallel Distrib. Syst.* **2001**, *12*, 22–37. [CrossRef]

electronics

MDPI

Article

Design of Automatic Correction System for UAV's Smoke Trajectory Angle Based on KNN Algorithm

Pao-Yuan Chao *, Wei-Chih Hsu and Wei-You Chen

Department of Computer and Communication Engineering, National Kaohsiung University of Science and Technology (NKUST), Kaohsiung 807618, Taiwan
* Correspondence: i107109103@nkust.edu.tw

Abstract: Unmanned aerial vehicles (UAVs) have evolved with the progress of science and technology in recent years. They combine high-tech, such as information and communications technology, mechanical power, remote control, and electric power storage. In the past, drones could be flown only via remote control, and the mounted cameras captured images from the air. Now, UAVs integrate new technologies such as 5G, AI, and IoT in Taiwan. They have a great application value in a high-altitude data acquisition, entertainment performances (such as night light shows and UAV shows with smoke), agriculture, and 3D modeling. UAVs are susceptible to the natural wind when spraying smoke into the air, which leads to a smoke track offset. This study developed an autocorrect system for UAV smoke tracing. An AI model was used to calculate smoke tube angle corrections so that smoke tube angles could be immediately corrected when smoke is sprayed. This led to smoke tracks being consistent with flight tracks.

Keywords: unmanned aerial vehicle; machine learning; UAV smoke show; mobile networks; artificial intelligence

Citation: Chao, P.-Y.; Hsu, W.-C.; Chen, W.-Y. Design of Automatic Correction System for UAV's Smoke Trajectory Angle Based on KNN Algorithm. *Electronics* **2022**, *11*, 3587. https://doi.org/10.3390/electronics11213587

Academic Editor: Juan M. Corchado

Received: 23 September 2022
Accepted: 31 October 2022
Published: 3 November 2022

1. Introduction

Flexible, safe, stable, high-speed, and low-cost UAVs have been developed in the past few years. This was achieved due to the continuous development of the modules, including the process materials, electric power storage, and sensors [1–12]. So far, UAVs have been widely used in civilian, commercial, and government units. Industrial UAVs can be applied in environmental monitoring, infrastructure inspection, disaster or accident rescue, agriculture, forestry, fishery, animal husbandry management, spatial information measurement, land and guard patrol, media communication, telecommunications services, home delivery logistics, and the military. Most application fields can be further subdivided. For example, environmental monitoring can be divided into the monitoring and investigating of air pollution, oil pollution, nuclear pollution, marine pollution, and river pollution, and even includes the study of weather changes. Many types of infrastructure are subjects for inspection, including roads, railways, transmission towers, and oil fields. Regarding the rescue, drones can be used for video recording, a real-time image transmission, and material delivery. Regarding the environmental conditions, there are waters, mountainous areas, or buildings. Agriculture, forestry, fishery, and animal husbandry management includes pesticide or fertilizer spraying and the observation of crops, trees, pastures, and fish farms. The work in spatial information includes aerial mapping, a terrain attribute classification and survey, a national land survey, urban planning, a land survey and development, water control and flood control planning, and 3D real scene modeling. Guard patrol includes a coastal patrol, criminal chasing, and general security work. Regarding media communication, in addition to providing real-time news about disaster areas and war zones, they can be applied to business and tourism marketing. There are diverse applications in the military, and they can be used as reconnaissance aircraft, target aircrafts, and bombers. Therefore, industrial UAVs have unlimited business opportunities. In Ghana, there have been about 275,000 UAVs flying and delivering medical kits containing vaccines [13]. The edge computing technology is used in UAVs for the power transmission line inspection [14].

On many major holidays and celebrations in Taiwan, people can see the colorful smoke from fighters flying by the military in the sky. In addition to the purchase of fighters, such activities often cost a lot of money (military aircraft maintenance, personnel training, and aviation gasoline), which consumes gasoline and imposes a burden on the environment. Therefore, the use of UAVs carrying smoke tubes for aerial smoke spraying has grown in recent years. Today, most UAVs are mainly powered by electricity and do not emit exhaust gas like gasoline engines. Moreover, UAVs cost less than traditional fighters in smoke spraying. For example, in Taiwan, according to the Regulations of Drone of the Civil Aeronautics Administration [15], if pilots hold G2 licenses (flying more than 400 feet above the ground or water, operating beyond the range of visibility, dropping, or spraying objects) and UAVs are registered in the UAV system of the Civil Aeronautics Administration and apply for airspace in advance, regulatory restrictions can be exempted and UAV shows with smoke can be performed.

UAVs cannot carry large smoke tubes and smoke tubes installed at different positions are susceptible to the natural wind and lead to a smoke track offset. In this study, detectors and smoke tube correctors were installed above the UAV. An AI model was used for training to correct the offset tracks. In this way, small low-altitude UAVs could achieve the same visual effects as traditional high-altitude fighters in UAV shows with smoke, in a way which is cheap and environmentally friendly.

2. Materials and Methods

2.1. Unmanned Aerial Vehicle

The UAV used in this study has the following basic flying parts: a flight controller, motor, electronic transmission, frame, and GPS positioning module. In addition, a servo motor, smoke tube, Raspberry Pi, 4G communication module, and lithium battery are installed. The overall weight is about 2.5 kg. Considering the motor load in the flight process and flight time, a UAV with a wheelbase of 450 mm and an EDU450 carbon fiber frame was selected [16], as shown in Figure 1.

Figure 1. EDU450 carbon fiber frame.

A Pixhawk2 CUBE consisting of two processors was used as the flight controller of the UAV. In the Pixhawk2 CUBE, the main processor was STM32F427 V3, and the coprocessor was STM32F1. The built-in sensor included a tri-axis accelerometer (L3GD20), an accelerometer and magnetometer (LS303D), a gyroscope (MPU9250), and a barometer (MS5611). With a weight of 73 g, this light and efficient flight controller supported open-source flight control software PX4 and Ardupilot. Three–eight axis multi-rotor models and multiple interfaces, including the Mavlink interface, I2C interface, and PWM signal output system, are used in this study. Raspberry Pi 3B+ has a USB interface which can be used for Internet access apart from the built-in WiFi card. If the 4G network card fails, it can switch to the built-in WiFi [17], a backup network, as shown in Figure 2.

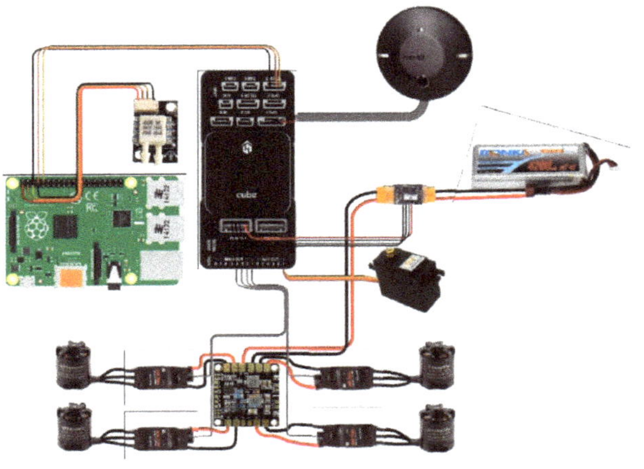

Figure 2. System configuration.

2.2. Smoke Tube Position

In this study, a four-axis multi-rotor UAV was used for the experimentation. The smoke from the smoke tube is vulnerable to the natural wind and tube position, leading to a smoke track offset. In that case, the audience is unable to enjoy the performance. As shown in Figure 3, when the smoke tube was placed directly below or above the UAV, the smoke from the smoke tube was affected by the downdraft generated by the propeller. Regardless of the rotation of the servo motor and adjustment of the smoke spraying direction, the smoke would be affected by the airflow and sprayed downward. In this study, self-made 3D material parts were attached to the UAV. An additional extension area was built to install the servo motor and smoke tube for adjusting the spraying angle. This is to prevent the influence of downdraft generated by the propeller as much as possible and to make the smoke from the smoke tube be sprayed backward, as shown in Figure 4a. To avoid the pendulum effect and the excessive output energy of the rear motor, the length of the extension area is adjusted to 29 cm after multiple outdoor flight tests. The aerial testing shows that the smoke track is significantly improved, as shown in Figure 4b. After improving the smoke emission direction, the track angle correction was discussed below, so that the audience could enjoy the best effect of the smoke track.

(**a**) (**b**)

Figure 3. Spraying effects at different smoke tube positions, should be listed as: (**a**) below the UAV; (**b**) above the UAV.

(a) (b)

Figure 4. Spraying effects at the redesigned smoke tube position, should be listed as: (**a**) the smoke tube extension frame is at the rear; (**b**) effect after modification.

2.3. AI Model Selection and Design

K-nearest neighbors (KNN) are one of the most popular machine learning algorithms. It has been widely used in HPC applications, such as image/video retrieval, big data analysis, machine learning, and computer vision [18,19]. It is a nonparametric statistical method for regression and classification. The K-nearest training samples in the feature space ware input [20,21] and the k-value were used to determine which classification group the data were nearest to. The classification criteria were decided by majority voting, and the Euclidean distance was used to calculate the distance, as shown in Equation (1).

$$P = \sqrt{(x_2 - x_1)^2 + (y_2 - y_1)^2} \tag{1}$$

In the KNN classifier, the output was the group classification, and its neighbors determined the category corresponding to the input object by a majority voting. KNN adopted the vector space modal for a classification, and objects in the same category were highly similar. The similarity could be calculated by the known category cases to evaluate the possible categories of the input objects. The training samples were multi-dimensional eigenspace vectors, in which each training sample had a classification label. The algorithm included eigenvector access and training sample labels at the training stage.

The KNN classifier assigns a weight of 1/k to the k-nearest neighbors and zero to all other neighbors. This can be applied to the weighted nearest neighbor classifier. The weight w_{ni} is given to the nearest neighbor i. In (2), a similar result holds for the strong consistency of the weighted nearest neighbor classifier [21].

$$\sum_{i=i}^{n} w_{ni} = 1 \tag{2}$$

Let C_n^{wnn} denote the weighted nearest classifier with the weight $\{w_{ni}\}_{i=1}^{n}$. According to the regularity condition of the category distribution, the excess risk has (3) an asymptotic expansion.

$$R_R(C_n^{wnn}) - R_R\left(C^{Bayes}\right) = \left(B_1 s_n^2 + B_2 t_n^2\right)\{1 + o(1)\} \tag{3}$$

$$s_n^2 = \sum_{i=i}^{n} w_{ni}^2 \tag{4}$$

$$t_n = n^{-\frac{2}{d}} \sum_{i=i}^{n} w_{ni} \left\{ i^{1+\frac{2}{d}} - (i-1)^{1+\frac{2}{d}} \right\} \tag{5}$$

The optimal weighting method $\{ w_{ni}^{*} \}_{i=1}^{n}$ is used to balance the two items above. Let $k^* = \lfloor B_n^{\frac{4}{d+4}} \rfloor$, (6) correspond to i = 1, 2, . . . , k^*, and $w_{ni}^{*} = 0$ correspond to i = k^* + 1, . . . , n.

After using the optimum weight, the dominant term in the asymptotic expansion of excess risk is $\mathcal{O}\left(n^{-\frac{4}{d+4}}\right)$.

$$w^*_{ni} = \frac{1}{k^*}\left[1 + \frac{d}{2} - \frac{d}{2k^{*2/d}}\left\{i^{1+2/d} - (i-1)^{1+2/d}\right\}\right] \quad (6)$$

At the classification stage, k is a user-defined constant. A vector without a category label (query or test point) will be classified into the most frequently used category among the k sample points nearest to the point.

When flying in the air, the UAV is affected by the crosswind w_{wind} and deviates from its route. At this point, to correct the route, the flight controller will give a Roll value to adjust the pitch angle θ_1 of the UAV, as shown in Figure 5. When the airframe is corrected, an angle adjustment θ_2 is given to the smoke tube to make the smoke tube turn to the windward face to face the direction the wind comes from, as shown in Figure 6. According to the angle adjustment θ_1 of the flight controller and the angle θ_2 of the smoke tube, the direction and magnitude of the wind the UAV is exposed to in the air can be known, as shown in Equation (7).

$$w_{wind} \rightarrow \theta_1 + \theta_2 \quad (7)$$

Figure 5. Roll angle correction θ_1.

Figure 6. Smoke tube angle correction θ_2.

Based on the above conclusion, the direction and magnitude of the wind are related to the value of θ_1. The operator can adjust θ_2 according to the value of θ_1 when flying the UAV. θ_1 and θ_2 will be trained by the machine learning-based KNN classification method, and their relationship is shown in Equation (8).

$$w \propto \theta_1 \rightarrow \theta_2 \quad (8)$$

3. Experimental Results and System Validation

During its flight, the UAV is affected by the natural wind and deviated from its course. At this point, the flight controller corrects the pitch angle in real-time to make the UAV return to its course. Its flight direction was mainly changed by correcting the pitch, yaw, and roll parameters. In addition to the airspeed meter sensor installed above the UAV, the three sensing values of the pitch, yaw, and roll parameters of the UAV can be used to learn the changes in the wind fields in the air.

Five angles, namely −60°, −30°, 0°, 30°, and 60° were designed in this study. These angles are the output y to be estimated by the KNN model. The input features include the pitch, yaw, roll, and airspeed meter values read from the flight controller. A KNN model was built for training.

Finally, the trained model was stored in the Joblib package and then put into the Raspberry PI in the UAV. Later, the designed system was used to read the model so that the real-time wind speed data read could be directly put into the AI model in the Raspberry PI for calculation. The results could be transmitted to the flight controller via the system to adjust the servo motor that controlled the smoke tube direction. It helped adjust the smoke tube to the optimal angle. The correction flow chart is shown in Figure 7.

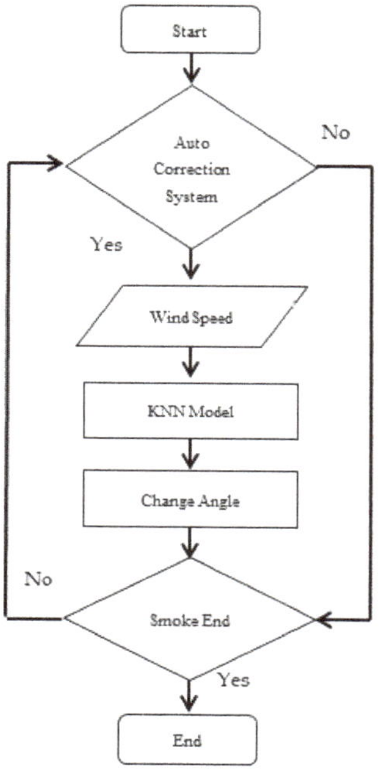

Figure 7. Autocorrection flow chart for smoke trailing.

As the UAV which was selected could not be fitted with a larger smoke tube, each spraying took about 30 to 40 s. The operator could collect about five pieces of data on each flight, which is not much. During the training, the pitch, yaw, roll, and airspeed meter data were put into the KNN model, and the accuracy was 50%. To improve the accuracy, more data is required. Therefore, the pitch and yaw were discarded, and the roll value was kept. The UAV was designed to fly back and forth in a straight line automatically. In the case of the deviation caused by a crosswind, the UAV could return to its route mainly by correcting the roll value. A roll value of 0 indicates that the UAV flew horizontally without any roll. A positive roll value indicates that the wind blew from the left side of the UAV. A higher value represents a higher wind speed. On the contrary, a negative roll indicates that the wind blew from the right side of the UAV. A smaller value reflects a higher wind speed. In this study, the roll value was collected to determine the speed and direction of the wind. This is to make up for the limitation of the airspeed meter under the breeze. The roll value was sensitive and thus could detect detailed data. In this study, 64 data items from the database were put into the AI training model, and the accuracy was 71%, as shown in Figure 8.

```
In [15]:
        def knn_predict_rev():

            accuracy = []
            open_dataset = "datasets/real_data_roll.csv"
            dataset = pd.read_csv(open_dataset)
            array = dataset.values


            X = array[:,0:1]

            Y = array[:,1]



            knn = KNeighborsClassifier(n_neighbors=3)
            knn.fit(X, Y)
            print(knn.predict(np.array([[-0.191497]])))
            y_pred = knn.predict(X)
            accuracy.append(metrics.accuracy_score(Y, y_pred))
            joblib.dump(knn,'./model')

            print(accuracy)

In [16]: knn_predict_rev()

        [60.]
        [0.71875]
```

Figure 8. KNN training accuracy.

The roll data read by the UAV was recorded and put into the KNN model for testing. The test data and predicted angles are shown in Table 1. The table shows that the rolls and angle corrections were as expected. The smoke tube should have been shifted to the right when the roll was positive and left when the roll was negative.

Table 1. KNN model test data and results.

Roll Input	Predicted Angle
0.00364547478966	0° (Middle)
0.00348061858676	0° (Middle)
0.00342658907175	0° (Middle)
0.00615163240582	0° (Middle)
0.0677677094936	−60° (Right)
0.0777317807078	−60° (Right)
0.08781837672	−30° (Right)
0.0836124494672	−30° (Right)
−0.0313124507666	30° (Left)
−0.157921299338	60° (Left)
−0.180348366499	60° (Left)
−0.105139121413	30° (Left)

Images taken behind the smoke tube show that the smoke tube could automatically change the direction and angle according to the direction and speed of the wind. Figure 9 shows that the UAV deviated to the left due to the wind from the right side. The roll of the angle correction to the right given by the flight controller was 0.061363. The smoke tube should be adjusted 30° to the right according to the calculation by the AI model. Figure 10 shows that the UAV deviated to the right due to the wind from the left side. The smoke tube should be adjusted 30° to the left, according to the calculation by the AI model.

Figure 9. The smoke tube shifted to the right.

Figure 10. The smoke tube shifted to the left.

The smoke from the smoke tube would be adjusted to the direction of the windward face. Based on the wind speed, the smoke tube would be adjusted to 30° or 60°. In this case, when the windward face of the UAV was in the front and rear directions, the smoke tube angle would not be adjusted.

Based on the observation of the actual flight, an accuracy of 71% indicated a significant improvement. Figure 11 illustrates the case without correction by the AI model. From the audience's angle, it could be clearly seen that the track (yellow arrow) of the smoke from the smoke tube was offset due to the natural wind. Figure 12 shows the case with a correction by the AI model. From the audience's angle, the smoke from the smoke tube could be observed (yellow arrow). Despite the influence of the wind field in the air, the smoke track could be manipulated to be almost the same as the flight route.

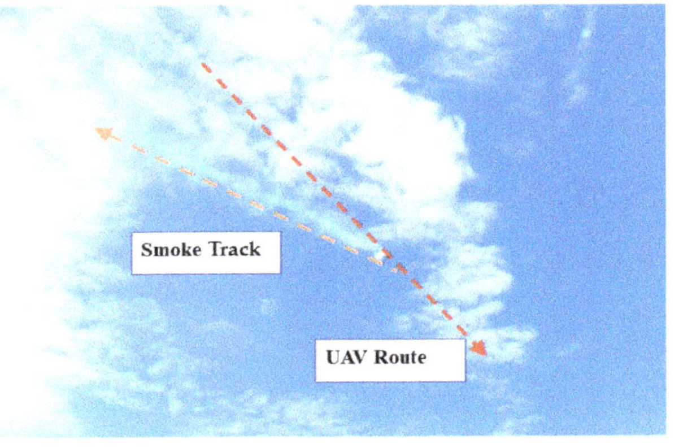

Figure 11. Smoke track correction without the AI model.

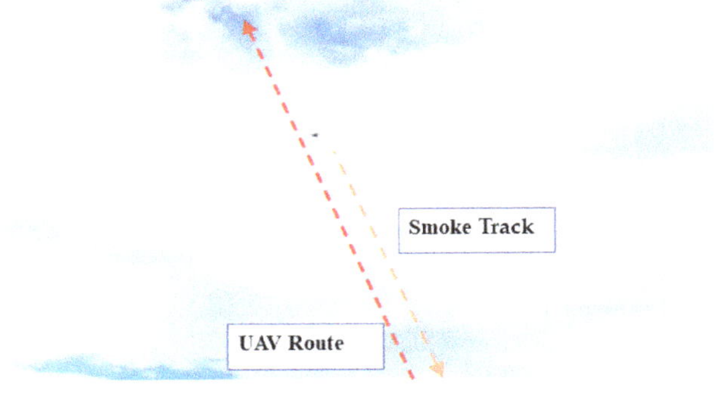

Figure 12. Smoke track correction with the AI model.

4. Conclusions and Future Work

In this study, Raspberry PI was used as the microcomputer for transmissions with the server. The flight data were received to control the smoke tube and sensor of a quad-axis UAV. To avoid the influence of the airflow, 3D-printed parts were used to refit the UAV. Its frame was extended to install a smoke tube and an electronic igniter so that the smoke tube could be lit for shows with smoke. As the smoke from the smoke tube was susceptible to the natural wind and became offset, a servo motor was installed to adjust the direction of the smoke from the smoke tube. A manned aircraft flew to record the angle adjustment, wind direction, and wind speed. The KNN was used to train a modified AI model. After applying the AI model to the Raspberry PI, the UAV emitted smoke in the air. The Raspberry PI and the flight controller could directly read the wind field data, and the angle could be immediately calculated and then sent back to the flight controller if it needed to be corrected. In this way, the spraying angle of the smoke tube could be adjusted immediately to make the smoke track the same as the flight route as much as possible. According to the results, the correction accuracy was 71%, which can demonstrate the difference between before and after the correction.

In the past, fighters sprayed smoke in the air to celebrate major festivals, which was expensive and polluted the environment. Our design is expected to make UAV shows with smoke possible on small occasions so that such events can be enjoyed on many occasions other than major festivals. The UAV designed by us is powered by electricity. Compared with a fuel-powered aircraft, it causes less environmental pollution and is cheaper.

The architecture in this study was designed for single UAVs. If the information of multiple UAVs can be displayed simultaneously on the web page and the crowd control can be carried out through function buttons on the web page, multiple shows with smoke can be performed simultaneously to spray. As for the collection of the wind speed data, this study collected various data, such as the roll, pitch, yaw, and anemometer values. Due to insufficient sample data, only the roll data were used for the machine learning. If more data can be collected in the future and all data collected can be imported for machine learning, the AI model will be more effective, and the overall correction effect will be perfect.

Author Contributions: Conceptualization, P.-Y.C. and W.-Y.C.; methodology, W.-C.H. and W.-Y.C.; resources, P.-Y.C. and W.-C.H.; writing—original draft preparation, P.-Y.C. and W.-Y.C.; writing—review and editing, W.-C.H. and W.-Y.C.; visualization, W.-Y.C.; supervision, P.-Y.C. and W.-C.H. All authors have read and agreed to the published version of the manuscript.

Funding: This research received no external funding.

Data Availability Statement: Not applicable.

Conflicts of Interest: The authors declare no conflict of interest.

References

1. Valavanis, K.V. *Advances in Unmanned Aerial Vehicles—State of Art and the Road to Autonomy*; Springer: Berlin/Heidelberg, Germany, 2007.
2. Papa, U. *Embedded Platforms for UAS Landing Path and Obstacle Detection: Integration and Development of Unmanned Aircraft Systems*; Springer: Berlin/Heidelberg, Germany, 2018; Volume 136.
3. Lillian, B. FAA Predicts Future UAS Growth. 2019. Available online: https://unmanned-aerial.com/faa-predicts-future-uas-growth (accessed on 15 May 2020).
4. González-Jorge, H.; Martínez-Sánchez, J.; Bueno, M.; Arias, P. Unmanned Aerial Systems for Civil Applications: A Review. *Drones* **2017**, *1*, 2. [CrossRef]
5. Sigala, A.; Langhals, B. Applications of Unmanned Aerial Systems (UAS): A Delphi Study Projecting Future UAS Missions and Relevant Challenges. *Drones* **2020**, *4*, 8. [CrossRef]
6. Department of Transportation. *Unmanned Aircraft Systems (UAS) Service Demand 2015–2035: Literature Review & Projections of Future Usage*; Technical Report, Version 0.1; UASF Aerospace Management Systems Division, Air Traffic Systems Branch (AFLCMC/HBAG): Bedford, MA, USA, 2013; 151p.
7. Ollero, A. UAV Applications. In *Handbook of Unmanned Aerial Vehicles*; Valavanis, K.P., Vachtsevanos, G.J., Eds.; Springer: Berlin/Heidelberg, Germany, 2015; pp. 2638–2860.
8. Mahmoud Zadeh, S.; Powers, D.M.W.; Zadeh, R.B. *Autonomy and Unmanned Vehicles—Augmented Reactive Mission and Motion Planning Architecture*; 66C-PRT; Springer Nature Singapore Pte Ltd.: Singapore, 2019; Volume I, pp. 562–566.
9. Mustapha, B.; Zayegh, A.; Begg, R.K. Multiple sensors based obstacle detection system. In Proceedings of the 4th International Conference on Intelligent and Advanced Systems (ICIAS2012), Kuala Lumpur, Malaysia, 12–14 June 2012.
10. Gageik, N.; Benz, P.; Montenegro, S. Obstacle Detection and Collision Avoidance for a UAV with Complementary Low-Cost Sensors. *IEEE Access* **2015**, *3*, 599–609. [CrossRef]
11. Engel, J.; Sturm, J.; Cremers, D. Camera-based navigation of a low-cost quadrocopter. In Proceedings of the 2012 IEEE/RSJ International Conference on Intelligent Robots and Systems, Vilamoura, Portugal, 7–12 October 2012; pp. 2815–2821.
12. Shen, H.; Jiang, Y.; Deng, F.; Shan, Y. Task Unloading Strategy of Multi UAV for Transmission Line Inspection Based on Deep Reinforcement Learning. *Electronics* **2022**, *11*, 2188. [CrossRef]
13. Aswini, N.; Uma, S.V. Obstacle Detection in Drones Using Computer Vision Algorithm. In *Advances in Signal Processing and Intelligent Recognition Systems, SIRS 2018*; Communications in Computer and Information, Science; Thampi, S., Marques, O., Krishnan, S., Li, K.C., Ciuonzo, D., Kolekar, M., Eds.; Springer: Singapore, 2019; Volume 968.
14. Zipline. Zipline Delivers 1 Million COVID-19 Vaccines in Ghana. Available online: https://flyzipline.com/press/zipline-delivers-1-million-covid-19-vaccines-in-ghana (accessed on 4 July 2022).
15. Hellaoui, H.; Bekkouche, O.; Bagaa, M.; Taleb, T. Aerial control system for spectrum efficiency in UAV-to-cellular communications. *IEEE Commun. Mag.* **2018**, *56*, 108–113. [CrossRef]
16. Hexsoon EDU450—Description and Technical Data. 2019. Available online: https://ardupilot.org/copter/docs/reference-frames-hexsoon-edu450.html (accessed on 12 August 2021).
17. Motlagh, N.H.; Bagaa, M.; Taleb, T. UAV-based IoT platform: A crowd surveillance use case. *IEEE Commun. Mag.* **2017**, *55*, 128–134. [CrossRef]
18. Feng, H.; Eyers, D.; Mills, S.; Wu, Y.; Huang, Z. Principal component analysis based filtering for scalable high precision k-nn search. *IEEE Trans. Comput.* **2018**, *67*, 252–267. [CrossRef]

19. Wu, Z.; Ke, Q.; Isard, M.; Sun, J. Bundling features for large scale partial-duplicate web image search. In Proceedings of the 2009 IEEE Conference on Computer Vision and Pattern Recognition, Miami, FL, USA, 20–25 June 2009; pp. 25–32.
20. Altman, N.S. An Introduction to Kernel and Nearest-Neighbor Nonparametric Regression. *Am. Stat.* **1992**, *46*, 175–185. [CrossRef]
21. Stone, C.J. Consistent Nonparametric Regression. *Ann. Stat.* **1977**, *5*, 595–620. Available online: https://www.jstor.org/stable/2958783 (accessed on 21 July 2022). [CrossRef]

electronics

Article

Prediction of Offshore Wave at East Coast of Malaysia—A Comparative Study

Mohammad Azad [1,*] and Md. Alhaz Uddin [2]

[1] Department of Computer Science, College of Computer and Information Sciences, Jouf University, Sakaka 72441, Saudi Arabia
[2] Department of Civil Engineering, College of Engineering, Jouf University, Sakaka 72441, Saudi Arabia
* Correspondence: mmazad@ju.edu.sa

Abstract: Exploration of oil and gas in the offshore regions is increasing due to global energy demand. The weather in offshore areas is truly unpredictable due to the sparsity and unreliability of metocean data. Offshore structures may be affected by critical marine environments (severe storms, cyclones, etc.) during oil and gas exploration. In the interest of public safety, fast decisions must be made about whether to proceed or cancel oil and gas exploration, based on offshore wave estimates and anticipated wind speed provided by the Meteorological Department. In this paper, using the metocean data, the offshore wave height and period are predicted from the wind speed by three state-of-the-art machine learning algorithms (Artificial Neural Network, Support Vector Machine, and Random Forest). Such data has been acquired from satellite altimetry and calibrated and corrected by Fugro OCEANOR. The performance of the considered algorithms is compared by various metrics such as mean squared error, root mean squared error, mean absolute error, and coefficient of determination. The experimental results show that the Random Forest algorithm performs best for the prediction of wave period and the Artificial Neural Network algorithm performs best for the prediction of wave height.

Keywords: prediction; artificial neural network; support vector machine; random forest; regression; offshore wave; wind speed

Citation: Azad, M.; Uddin, M.A. Prediction of Offshore Wave at East Coast of Malaysia—A Comparative Study. *Electronics* **2022**, *11*, 2527. https://doi.org/10.3390/electronics11162527

Academic Editor: Giuseppe Ciaburro

Received: 25 April 2022
Accepted: 23 July 2022
Published: 12 August 2022

1. Introduction

Human activities in the offshore region are continually increasing due to oil and gas exploration. Adverse conditions can often occur due to environmental disasters during offshore operations. Predicting wave characteristics is a crucial prerequisite for offshore oil and gas development (Figure 1), considering the safety of lives and the avoidance of economic damage. Wave height and period are typically significantly increased by the wind associated with storms passing across the ocean's surface. Weather forecasting departments usually predict wind forces rather than wave periods and heights. There are several empirical approaches for estimating wave height and wave period from wind force [1–5]. Estimating wave height and period is inherently inaccurate and random, making it difficult to simulate using deterministic equations [6]. The numerical approach for calculating wave height and period is a difficult and complex procedure that, despite substantial breakthroughs in computational tools, produces solutions that are neither dependable nor consistently applicable. Machine learning methods are perfect for modeling inputs with corresponding outputs since they do not necessitate an understanding of the underlying physical mechanism [7].

Several studies have been performed using artificial neural networks (ANN) to measure important wave heights and mean-zero-up-crossing wave period history for different locations in the seas. These parameters were predicted three, six, twelve, and twenty-four hours in advance using two different neural network methods [8–10]. The time series of these wave parameters has been investigated using simulations in Portugal's western coast area. Time series with wave height have been disintegrated into multi-resolution time series using wavelet transformation hybridized with ANN and wavelet transformation [11,12]. As the input of the ANN, the multi-resolution time series has been used to predict the important wave height at an unlike multi-step lead time near Mangalore, India's west coast. Mandal et al. [13] expected wave heights from observed ocean waves off the west coast of India, Marmugao. To predict wave height, recurrent neural networks with a resilient propagation (Rprop) update algorithm have been implemented.

Figure 1. Offshore oil and gas development

The time series of wave height and mean-zero-up crossing wave period were simulated using data from the local wind. The application of various types of machine learning models [14–17] was performed to improve the accuracy of the prediction. The authors [18] used ANN to estimate wave height from wind force at 10 chosen locations in the Baltic Sea. The WAM4 wave model was used to figure out the time series for the waves that had been forecasted in the past. There were two different techniques, Feed-forward Back-propagation (FFBP) and Radial Basis Function (RBF), used to develop a machine learning system for predicting wave heights at a particular coastal point in deeper offshore areas [19–21]. Data on wave height, average wave period, and wind speed were obtained from remotely sensed satellite data on India's west coast. Tsai et al. [22] employed the ANN with a back-propagation algorithm to estimate wave height and cycle from wind input based on the wind-wave relationship [23]. Time records of waves at either station may be predicted based on data from the neighboring station. Several deterministic neural network models have been developed by Deo et al. [24] to predict wave height and wave periods from generated wind speed. However, the model can offer adequate results in deep water and open areas, and the prediction periods are extensive.

This research uses wind force/metocean data to correctly predict wave height and wave period. Critical comparisons are performed for the current investigation to provide more accurate findings in predicting the wave parameters using three machine learning algorithms: ANN (Artificial Neural Network), SVM (Support Vector Machine), and RF (Random Forest).

The main contributions of the paper are (i) performing the experiments on the data sets obtained on the east coast of Malaysia and (ii) understanding the best predictive models for future prediction.

We arrange the remaining parts of the manuscript as follows. First, we explain the data collection procedures, the three regressor models, and performance metrics used for the comparison among the models in Section 2. Then, we show the experimental results and discuss the findings in Section 3. Finally, Section 4 contains a concise conclusion.

2. Materials and Methods

In this section, first, the data collection procedure is described, followed by the regression methods, and finally the evaluation metrics are described.

2.1. Data Collection

On a 2° × 2° grid in South China, environmental data are collected along Malaysia's east coast, including the basins of Sabah (longitude 114.39° E, latitude 5.83° N) and Sarawak (longitude 111.82° E, latitude 5.15° N) (Figure 2). The Malaysian Meteorological Service and Fugro OCEANOR provided the data that was acquired from satellite altimetry [25,26] using oceanographic SEAWATCH meteorological (metocean) buoys and sensors, calibrated and corrected by Fugro OCEANOR [27].

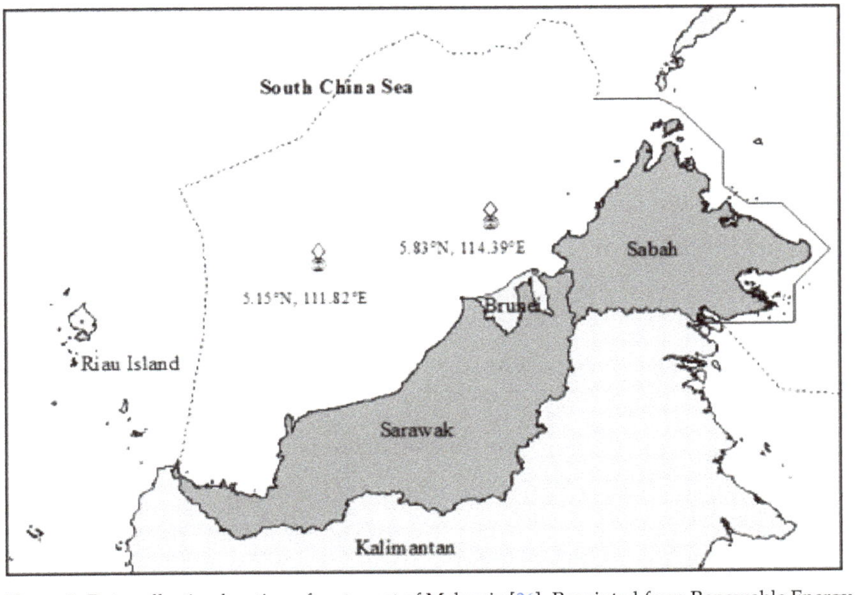

Figure 2. Data collection location of east coast of Malaysia [26]. Reprinted from Renewable Energy, 88, Omar Yaakob,Farah Ellyza Hashim,Kamaludin Mohd Omar,Ami Hassan Md Din,Kho King Koh, Satellite-based wave data and wave energy resource assessment for South China Sea, 359-371, 2016, with permission from Elsevier.

The summary of common statistical values of the collected data is given in Table 1.

Table 1. Summary of basic statistics of the collected data.

Statistics	Wind Speed	Wave Height	Wave Period
Number of Samples	1460	1460	1460
Minimum	0.00	0.17	4.05
Average	4.90	1.29	6.67
Maximum	15.41	5.13	13.68
Standard Deviation	3.34	0.75	1.24

There are 1460 data samples and three features: wind speed (m/s), wave height (m), and wave period (s).

2.2. Methods

There are a lot of algorithms [28–33] that can be used for regression analysis. In this study, three well-known and commonly used regression algorithms (ANN, SVM, and RF) were chosen to predict wave height and period from wind force. The wind force is employed as an input for training the network, while the wave height and period are used as outputs. Before discussing the details of each method, Table 2 shows the advantages and disadvantages of each method [34]:

Table 2. Advantages and disadvantages of different regressors.

Name	Advantages	Disadvantages
ANN	state-of-the-art algorithm; complex relations can be modeled	blackbox and hard to understand; many hyperparameters need to be tuned
SVM	provide non-linear solutions minimum generalization error	require knowledge of kernels weak regressor for large dataset
RF	most commonly used with good performance can handle missing values	interpretability of ensembles are questionable can lead to overfitting

The ANN algorithm is the most frequently utilized algorithm in such problems. The difference between the state-of-the-art ANN algorithm and two commonly used algorithms, SVM and RF, is then determined. Note that ANN is a parametric regression algorithm while SVM and RF are nonparametric regression algorithms. However, because the data set is small enough, it is not possible to employ advanced deep learning techniques (such as convolutional neural networks or recurrent neural networks).

2.2.1. Artificial Neural Network (ANN)

Inspiring by biological neurons, the researchers created artificial neurons [35]. It is possible to create a network of artificial neurons to predict the desired functions (i.e., the target variables).

The basic idea behind a single artificial neuron is that it takes an input function I, which is a product of weight vectors with the sample input vector, and feeds it into an activation function g to produce an output (Figure 3). Generally, the usual practice is to create a network of neurons based on many layers: input layers, hidden layers, and output layers. The input layers basically consist of one or more neurons based on the supplied inputs (for the present study, only one input is used), then there can be one or more hidden layers (only one hidden layer is used), and finally, there can be one output layer consisting of the expected outputs (Figure 4). The general trend is to use a back-propagation algorithm to update the weights involved in each connection between neurons so that it is possible to minimize the errors of regression between true output and the predicted output.

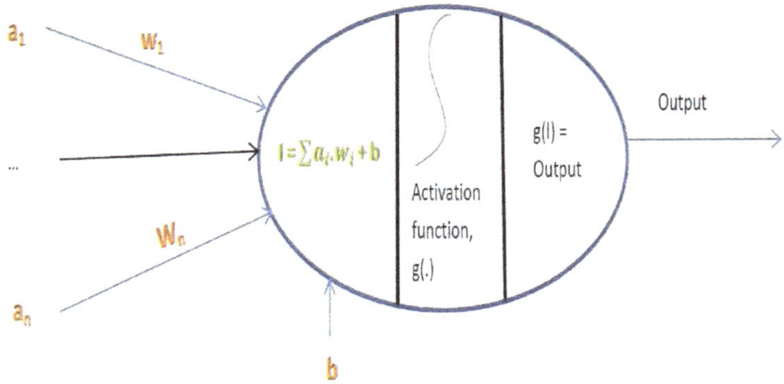

Figure 3. The architecture of a single artificial neuron.

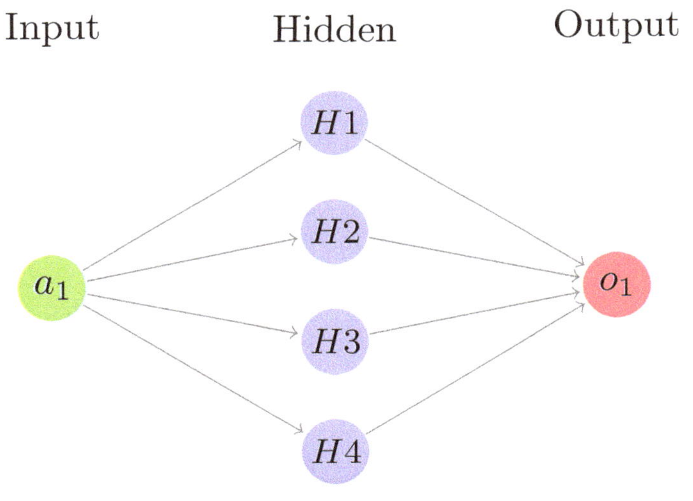

Figure 4. The architecture of ANN.

The hyperbolic tangent function, $\tanh(x) = \frac{e^x - e^{-x}}{e^x + e^{-x}}$ is used as the activation function. The steps of the Python implementation are described in Algorithm 1.

Algorithm 1 The Python implementation of the ANN algorithm

Input: Data set containing features and target
Output: Prediction of ANN for the given data set
1. Read the data set using the Python read_excel function;
2. Extract features and target variables;
3. Scale the features and target using the MinMaxScaler function;
4. Split the data set into training and test parts using the train_test_split function;
5. Choose the best hyperparameter using the GridSearchCV function;
6. Derive the MLPRegressor using the above results;
7. Predict using the obtained regressor and calculate the performance metrics on the test data set;

2.2.2. Support Vector Machine (SVM)

The basic idea behind regression using SVM (i.e., also popularly known as Support Vector Regression) is that given input examples $\{(i_1, o_1), \ldots, (i_n, o_n)\} \subset X \times R$, where X

denotes the example space (e.g., $X = R$ for our problem), a function $f(x)$ is obtained that has at most ϵ deviation from the actual outputs oi for all the input examples (Figure 5) [36].

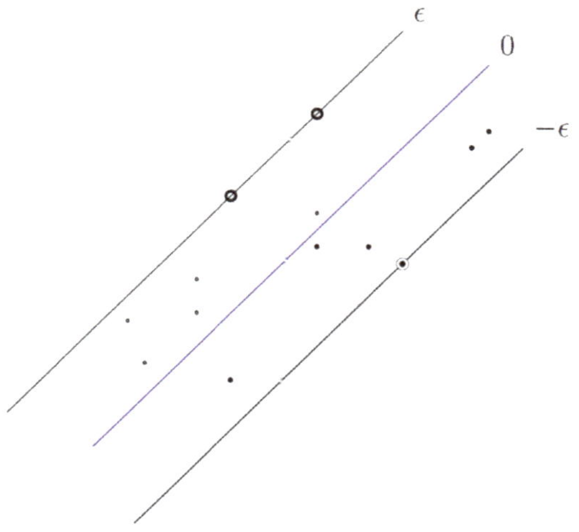

Figure 5. The soft margin for the support vector regression for the linear case

Generally, it is easy to describe the case of linear functions [37] f,
$f(x) =< w, x > +b$, with $w \in X, b \in R$
where $<\ ,\ >$ denotes the dot product in X. Furthermore, it is possible to rewrite this problem as a convex optimization problem:
minimize $\frac{1}{2}||w^2||$ subject to

$$y_i - < w, x > -b \leq \epsilon$$

$$< w, x > +b - y_i \leq \epsilon$$

The steps of the Python implementation are described in Algorithm 2.

Algorithm 2 The python implementation of the SVR algorithm

Input: Data set containing features and target
Output: Prediction of SVR for the given data set
1. Read the data set using the Python read_excel function;
2. Extract features and target variables;
3. Scale the features and target using the StandardScaler function;
4. Split the data set into training and test parts using the train_test_split function;
5. Choose the best hyperparameter using the GridSearchCV function;
6. Derive the SVR using the above results;
7. Predict using the obtained regressor and calculate the performance metrics on the test data set;

2.2.3. Random Forest (RF)

A decision tree has been widely used from the very beginning of machine learning research to find the best hypothesis as the classifier and regressor. A decision tree is a tree-like structure where a sequence of actions is taken from the root to the leaf nodes based on the values of each decision node in the concerned path. A sample decision tree is depicted in Figure 6, where the course of action of playing outside is taken based on the value of whether it is training outside or not.

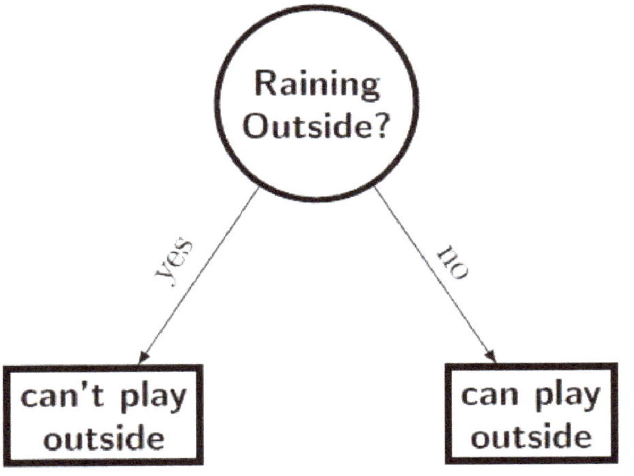

Figure 6. A simple decision tree model.

There are a number of different variants of decision tree algorithms, from ID3 [38], C4.5 [39], and CART [40] to ensembles of decision trees (e.g., Random Forest [28])) that have been proposed to improve the accuracy of the classifiers and regressors. For a single decision tree construction, the most widely used splitting criteria, "Gini index" or "Entropy", can be mathematically stated as follows [32,38–41]:

$$p_t = \frac{N_t(T)}{N(T)}$$

- Entropy $ent(T) = -\sum_{t \in D(T)} p_t \log_2(p_t)$;
- Gini index $gini(T) = 1 - \sum_{t \in D(T)} p_t^2$.

where T is the data set and $D(T)$ is the set of labels in the data set T. In addition, $N(T)$ represents the number of samples and $N_t(T)$ represents the number of samples with the label t.

A Random Forest (RF) is an ensemble of decision tree models such that each tree in the ensemble is built based on the bootstrap samples of the training data set. Furthermore, during the construction of the decision trees, the random forest selects only a subset of the features at each split point. In this way, the constructed decision trees are more different from each other to reduce the correlation among them and to have a better prediction. In contrast to the single decision tree (CART), random forests do not use any pruning of the tree.

Like any ensemble algorithm, the random forest also takes the average among all decision tree regressors to produce the final predictions (Figure 7).

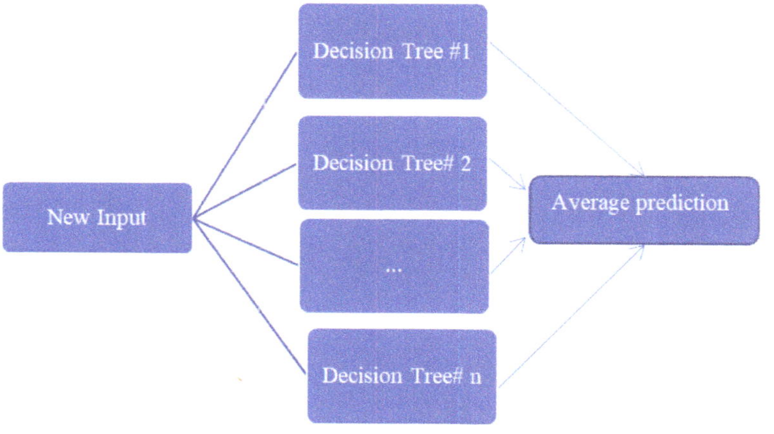

Figure 7. The average prediction of n decision trees in a random forest prediction model

The steps of the Python implementation are described in Algorithm 3.

Algorithm 3 The Python implementation of the RF algorithm

Input: Data set containing features and target
Output: Prediction of RF for the given data set
1. Read the data set using the Python read_excel function;
2. Extract features and target variables;
3. Scale the features and target using the MinMaxScaler function;
4. Split the data set into training and test parts using the train_test_split function;
5. Choose the best hyperparameter using the GridSearchCV function;
6. Derive the RandomForestRegressor using the above results;
7. Predict using the obtained regressor and calculate the performance metrics on the test data set;

2.3. Performance Metrics

Four performance metrics [42] have been used for the comparison of the results.

1. Mean squared error (MSE)
2. Root mean squared error ($RMSE$)
3. Mean absolute error (MAE)
4. Coefficient of determination (R^2)

For each sample, the error (e_i) is the difference between actual output (o_i) and predicted output ($\hat{o}_i$). The average of the squares of the error (e_i) is the MSE. The average is taken by dividing the summation of the square of all the errors by the total number of samples (n).

$$MSE = \frac{\sum_{i=1}^{n}(o_i - \hat{o}_i)^2}{n} \tag{1}$$

Root mean squared error ($RMSE$) is the square root of the average difference in squared error between actual output (o_i) and predicted output ($\hat{o}_i$). It is the most commonly used metric.

$$RMSE = \sqrt{\frac{\sum_{i=1}^{n}(o_i - \hat{o}_i)^2}{n}} \tag{2}$$

Mean absolute error (MAE) is the average absolute difference of error between actual output (o_i) and predicted output ($\hat{o}_i$). It is the most commonly used metric. The advantage of MAE is that it is softer to outliers. The reason is that it does not have any square term associated with the equation, therefore the penalty for the outlier points is not that much compared to MSE and $RMSE$ where there are heavy penalties imposed by the square term.

$$MAE = \frac{\sum_{i=1}^{n} |(o_i - \hat{o}_i)|}{n} \quad (3)$$

The coefficient of determination or R^2 explains the degree to which the independent variables explain the variation of the output variable. It is a measure of how well new samples can be predicted by the model through the proportion of explained variance. If o_i is the true value of the i-th sample and $\hat{o}_i$ is the corresponding output or predicted value, the estimated R^2 can be calculated as:

$$R^2 = 1 - \frac{\sum_{i=1}^{n} (o_i - \hat{o}_i)^2}{\sum_{i=1}^{n} (o_i - \bar{o})^2} \quad (4)$$

where $\bar{o} = \frac{\sum_{i=1}^{n} o_i}{n}$.

3. Results and Discussion

The experiments have been executed based on the available field data from the east coast of Malaysia, which consists of 1460 samples, one input variable, wind speed (m/s), and one target variable, either wave height (m) or wave period (s). The regression results are compared among the three regressors (ANN, SVM, and RF). The Python programming environment (version 3.6) with scikit-learn (version 0.24.1) is used for the implementation.

Initially, the ANN regressor is trained using the training data and then the accuracy is validated using the testing data. The network is chosen with one hidden layer of 4 neurons, and each neuron uses the hyperbolic tangent function. The weight of the ANN is optimized using "lbfgs" in the family of quasi-Newton methods. Furthermore, the SVM regressor is trained using the same training data and validated using the same testing data as mentioned above. For SVM, the standard radial basis function (RBF) is used as the kernel function. Finally, the RF regressor is trained and validated in a similar fashion. For RF, 400 trees are used for the number of trees, 3 is used for the maximum depth of the tree, and $1460 \times 0.05 = 73$ samples are used for the training of each decision tree in the ensembles. Each regressor is compared by the state-of-the-art performance metrics.

3.1. Cross Validation Results

It is a common practice to keep a part of the available data for testing and the remaining part for training. However, the model evaluation is particularly dependent on specific pairs of (training and testing) fractions, and the result can be overfitting. To overcome this problem, cross validation [43] is used to test the model's fitness to predict new data that was not used to train the model and how it can generalize to unknown data.

There are many variants of cross-validation. The standard method is to use k-fold cross-validation. In general, the procedure is to use k rounds in cross-validation. It splits the data into k subsets. In a single round, it completes the analysis on one subset (the training set), and then validates the performance on the remaining $k-1$ subsets (the validation set). In the next round, another subset is chosen for training and the remaining is chosen for validation. To reduce variability, these steps are repeated k times for k subsets so that each subset is selected for training. Finally, an average among the k validation results is reported as the fitness of the model's predictive performance.

Nevertheless, it is common practice to repeat the k-fold cross-validation process multiple times and report the average performance among all folds and all repeats. This approach is called repeated k-fold cross-validation.

The average and standard deviation of the regression performance of wave height were determined using 10-fold cross validation that was done for three methods as presented in Table 3. The four-performance metrics are reported in the same way as in the preceding section.

Table 3. Regression results for 10-fold cross validation with three times repeated for wave height (m).

Regressor	*MAE*		*MSE*		*RMSE*		R^2	
	Avg	**Std**	**Avg**	**Std**	**Avg**	**Std**	**Avg**	**Std**
SVM	0.4920	0.0385	0.4881	0.0920	0.6956	0.0651	0.5015	0.0871
ANN	0.0765	0.0049	0.0107	0.0016	0.1034	0.0079	0.5164	0.0872
RF	0.0775	0.0042	0.0110	0.0016	0.1046	0.0075	0.5104	0.0725

It is clearly evident that the method of ANN has a minimum average value of mean absolute error, mean square error, and root mean square error. Besides, it has the maximum average value of the coefficient of determination. As a result, this approach performs best for predicting wave height.

Similar findings for wave period have been presented in Table 4 using 10-fold cross validation that has been done for the consecutive three methods.

Table 4. Regression results for 10-fold cross validation with three times repeated for wave period (s).

Regressor	*MAE*		*MSE*		*RMSE*		R^2	
	Avg	**Std**	**Avg**	**Std**	**Avg**	**Std**	**Avg**	**Std**
SVM	0.7136	0.0495	0.8317	0.1406	0.9089	0.0753	0.1629	0.0743
ANN	0.0935	0.0056	0.0139	0.0019	0.1176	0.0079	0.1552	0.0614
RF	0.0930	0.0054	0.0136	0.0018	0.1165	0.0076	0.1717	0.0652

It is apparent that the RF approach has the lowest mean absolute error, mean square error, and root mean square error. Furthermore, it has the maximum value of the coefficient of correlation. As a result, this approach is the most accurate for predicting wave periods.

R^2 is close to 1, which indicates that regression predictions perfectly fit the data. However, in our case, the experimental results show the lower value of R^2 due to the inconsistencies in the collected data set.

3.2. Graphical Representation of a Sample Training and Testing Results

To illustrate the model's performance graphically, one sample is chosen randomly as the pair of (training, testing), where the training is 80% and testing is 20% of the data. In Table 5, the results for both training and testing data are shown in the case of the wave height regression problem. It is evident that the mean squared error, mean absolute error, and root mean squared errors are the smallest for RF compared to others in the training data. However, these metrics are the smallest for ANN compared to others in the testing data. Furthermore, the coefficient of correlation (R^2) is largest for RF compared to others in the training data, but it is largest for ANN compared to others in the testing data. As a result, the ANN approach is superior at predicting wave height in the future.

Table 5. A sample regression results for wave height (m).

Regressor	**Training**				**Testing**			
	MAE	MSE	$RMSE$	R^2	MAE	MSE	$RMSE$	R^2
SVM	0.4843	0.4739	0.6884	0.5196	0.5025	0.5009	0.7077	0.5249
ANN	0.0752	0.0105	0.1026	0.5367	0.0792	0.0111	0.1052	0.5449
RF	0.0751	0.0105	0.1024	0.5386	0.0814	0.0115	0.1074	0.5258

The regression results of ANN, SVM, and RF for wave height are depicted in Figure 8. Figure 8a shows the ANN regression findings for wave height. The black dots are actual testing data, and the green dots are the predicted values. It is clear that the predicted values follow the pattern of the actual testing data. Nevertheless, the regression results for the SVM regressor are shown in Figure 8b. The black dots are actual testing data, and the blue dots are the predicted values. It is clear that the predicted values follow the pattern of the actual testing data.

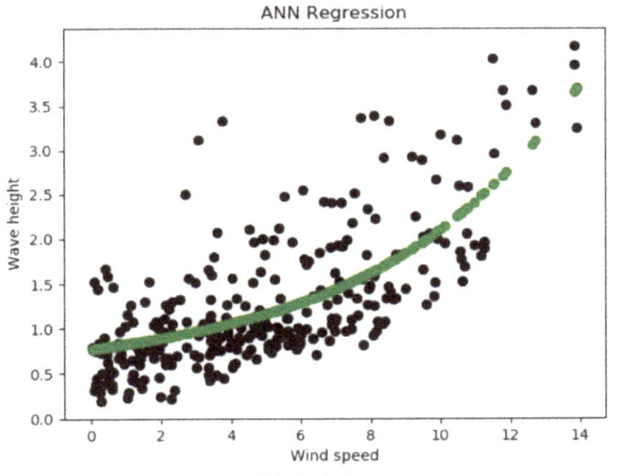

(**a**) ANN

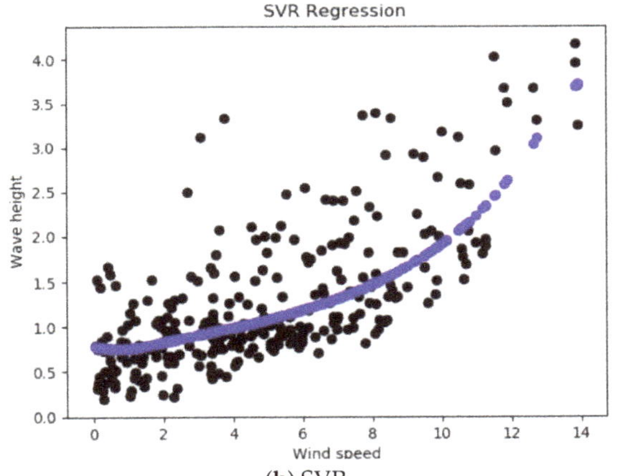

(**b**) SVR

Figure 8. *Cont.*

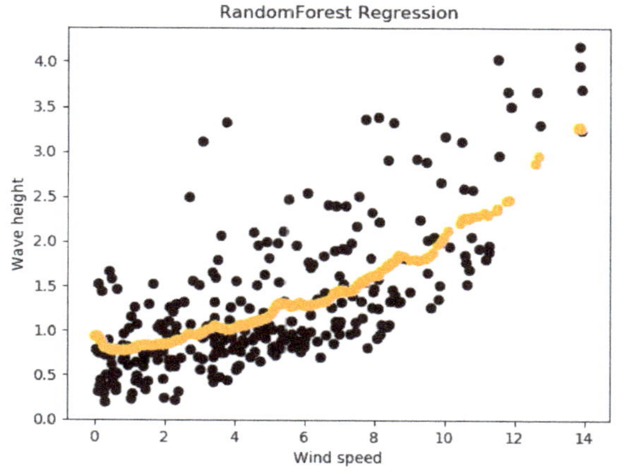

(c) RF

Figure 8. The regression result for wave height (m); (**a**) shows results using ANN, (**b**) shows results using SVR, and (**c**) shows results using RF. The predicted values follow the pattern of the actual testing data in all cases. For RF, the pattern is not as smooth as in ANN or SVM, because RF is not a single decision tree method; rather it is an ensemble method that works by taking the average votes. ANN prediction is smoother than SVR.

The regression results of wave height based on the RF regressor are presented in Figure 8c. The black dots are actual testing data, and the orange dots are the predicted values. It is clear that the predicted values follow the pattern of the actual testing data. The pattern is not as smooth as SVM or ANN because RF is not a single decision tree method; rather it is an ensemble method that works by taking the average of different decision trees' predictions.

In addition, in the case of wave period, Table 6 shows the findings for both training and testing data. It is evident that the mean squared error, mean absolute error, and root mean squared error is the smallest for RF compared to others in both training and testing data. Nevertheless, the coefficient of determination (R^2) is the largest for RF compared to others in both training and testing data. Therefore, for future prediction of wave period, the RF method is best.

Table 6. A sample regression results for wave period (s).

Regressor	Training				Testing			
	MAE	*MSE*	*RMSE*	R^2	*MAE*	*MSE*	*RMSE*	R^2
SVM	0.694	0.7980	0.8933	0.1846	0.7654	0.9366	0.9678	0.1353
ANN	0.0926	0.0137	0.1169	0.1574	0.1006	0.0157	0.1251	0.1284
RF	0.0899	0.0128	0.1129	0.2140	0.099	0.0153	0.1235	0.1507

The regression results of ANN, SVM, and RF for wave period are depicted in Figure 9. For ANN, in Figure 9a, the actual testing data is shown in black dots and the predicted values are shown in green dots. It is clear that the predicted values follow the pattern of the actual testing data for ANN. For SVM, in Figure 9b, the actual testing data are represented by black dots, while the actual predicted values are represented by blue dots. It is clear that the predicted values follow the pattern of the actual testing data for SVM.

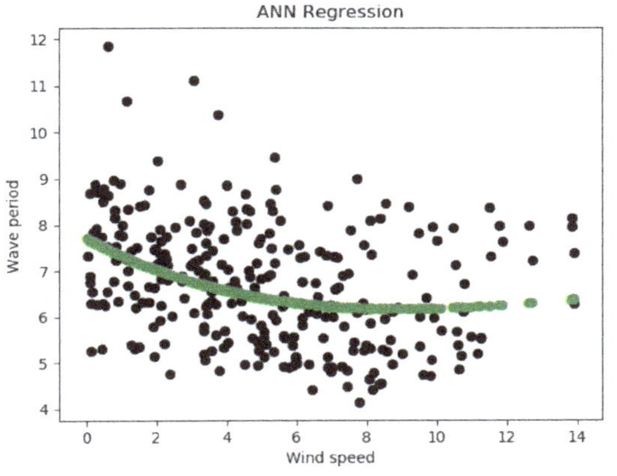

(a) ANN

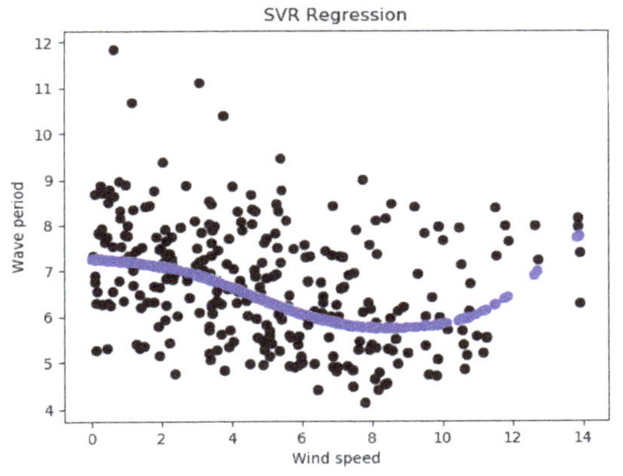

(b) SVR

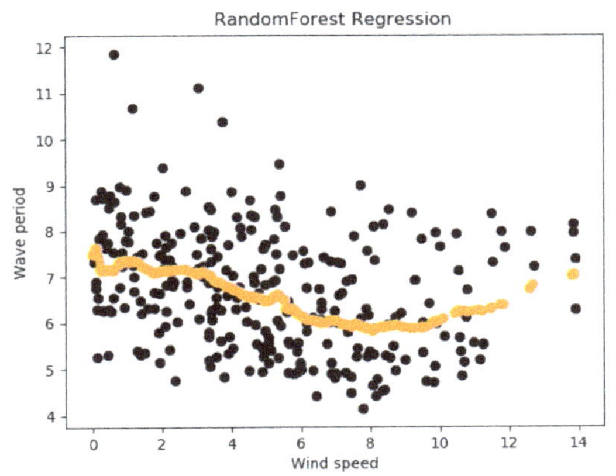

(c) RF

Figure 9. The regression result for wave period (s); (**a**) shows results using ANN, (**b**) shows results using SVR, and (c) shows results using RF. The pattern is not directed positively as like wave height in Figure 8 and rather going horizontally which indicates lower values of R^2 compared to wave height.

Finally, for RF, the real testing data is represented in black dots, whereas the actual predicted values are shown in orange dots in Figure 9c. It is clear that the predicted values follow the pattern of the actual testing data for RF. The pattern is not as smooth as SVM or ANN because RF is not a single decision tree method; rather it is an ensemble method that works by taking the average of different decision trees' predictions.

3.3. Comparison with Standard Non-Parametric Kernel Regression

The non-parametric method of kernel regression (KR) in statistics is used to calculate the conditional expectation of a random variable. The goal is to find a non-linear relationship between two random variables, I and O [44]. The problem under consideration can be modeled using a standard non-parametric regression problem. As an example, I can be wind speed and O can be wave height. In Table 7, the previous results are compared with the results of KR for the same sample.

Table 7. A sample comparison with kernel regression for predicting wave height.

Methods	R^2 for Predicting Wave Height
SVM	0.5249
ANN	0.5449
RF	0.5258
KR	0.5389

It is clear that KR results are not far from those of the SVM, ANN, or RF. In fact, ANN produces the best results in the case of wave height prediction.

3.4. Overall Discussion

The goal of this study is to understand the best predictive model among ANN, SVM, and RF for the prediction of wave height and period from the wind speed. A detailed experiments are performed in terms of cross-validation and sample training and testing results. The summary of this result is given in the Table 8.

Table 8. Summary of prediction results.

Methods	Prediction of Wave Height	Prediction of Wave Period	Final Comment
SVM	worst in MAE, MSE, RMSE and R^2	worst in MAE, MSE, RMSE, but second best in R^2	The results are not the best, however, not very far from the best one
ANN	best in MAE, MSE, RMSE, and R^2	second best in MAE, MSE, RMSE but worst in R^2	The results are promising for wave height and reasonable for wave period
RF	second best in MAE, MSE, RMSE, and R^2	best in MAE, MSE, RMSE, and R^2	The results are really promising for wave period and also for wave height

It is evident from Table 8 that the Random Forest (RF) method is truly performing well across two prediction problems. It is the second best for the prediction of wave height and the best for predicting wave period. Nevertheless, it has the advantage of being a non-parametric method. Moreover, the underlying tree structure has the advantages of interpretability and usage. However, to be specific, RF performs best for the prediction of wave period while ANN performs best for the prediction of wave height.

4. Conclusions

This study carried out three different and well-known machine learning algorithms for the prediction of offshore waves. These approaches are used to predict the wave height and wave period from the given wind forces. Multiple accuracy ranges are obtained in terms of the mean absolute error, mean square error, root mean square error, and coefficient of determination. Overall, these performance measures show average behavior. However, it is possible to compare the three employed methods and analyze the results. The regression analysis in the random forest performs best for the prediction of wave period, and the artificial neural network performs best for the prediction of wave height. Furthermore, it was compared with the standard non-parametric kernel regression and found to have a similar result.

In situations when a traditional analysis would be challenging, these machine learning techniques can produce very quick and reasonable predictions. These studies can benefit the community as a measurement of safety and precaution in the critical marine environment.

In this regard, there are numerous potential future research directions. One disadvantage of using such metocean data is the existence of discrepancies. Future work should incorporate advanced algorithms in addition to the aforementioned models to address such inconsistencies. Additionally, more machine learning techniques should be used to find the ideal answer for this specific prediction problem. The time dependence of the metocean data, which can be examined using time series analysis tools, is another interesting subject of research.

Author Contributions: Conceptualization, all authors; methodology, all authors; software, M.A.; validation, M.A.; formal analysis, all authors; investigation, all authors; resources, all authors; data curation, M.A.U.; writing, all authors; visualization, all authors; supervision, all authors.; project administration, M.A.U.; funding acquisition, M.A. All authors have read and agreed to the published version of the manuscript.

Funding: This research received no external funding.

Institutional Review Board Statement: Not applicable.

Informed Consent Statement: Not applicable.

Data Availability Statement: It is available with a justifiable and reasonable request.

Acknowledgments: The authors are grateful to Jouf University for their assistance with this research. Moreover, the authors would like to express their gratitude to all of the volunteers and anonymous reviewers for their suggestions.

Conflicts of Interest: The authors declare no conflict of interest.

References

1. Krogstad, H.; Barstow, S. Satellite wave measurements for coastal engineering applications. *Coast. Eng.* **1999**, *37*, 283–307. [CrossRef]
2. Tolman, H.; Alves, J.; Chao, Y. Operational Forecasting of Wind-Generated Waves by Hurricane Isabel at NCEP*. *Weather Forecast.* **2005**, *20*, 544–557. [CrossRef]
3. Tolman, H.; Balasubramaniyan, B.; Burroughs, L.; Chalikov, D.; Chao, Y.; Chen, H.; Gerald, V.M. Development and Implementation of Wind-Generated Ocean Surface Wave Modelsat NCEP. *Weather Forecast.* **2002**, *17*, 311–333. [CrossRef]
4. Günther, H.; Rosenthal, W.; Stawarz, M.; Carretero, J.; Gomez, M.; Lozano, I.; Serrano, O.; Reistad, M. The wave climate of the Northeast Atlantic over the period 1955-1994 : The WASA wave hindcast. *Glob. Atmos. Ocean Syst.* **1998**, *6*, 121–163.
5. Muzathik, A.M.; Nik, W.B.W.; Samo, K.; Ibrahim, M.Z. Ocean Wave Measurement and Wave Climate Prediction of Peninsular Malaysia. *J. Phys. Sci.* **2011**, *22*, 79–94.
6. Setiawan, I.; Yuni, S.M.; Miftahuddin, M.; Ilhamsyah, Y. Prediction of the height and period of sea waves in the coastal waters of Meulaboh, Aceh Province, Indonesia. *J. Physics: Conf. Ser.* **2021**, *1882*, 012013. [CrossRef]
7. Wang, J.; Wang, Y.; Yang, J. Forecasting of Significant Wave Height Based on Gated Recurrent Unit Network in the Taiwan Strait and Its Adjacent Waters. *Water* **2021**, *13*, 86. [CrossRef]
8. Makarynskyy, O.; Makarynska, D.; Kuhn, M.; Featherstone, W. Predicting sea level variations with artificial neural networks at Hillarys Boat Harbour, Western Australia. *Estuar. Coast. Shelf Sci.* **2004**, *61*, 351–360. [CrossRef]
9. Makarynskyy, O.; Pires-Silva, A.; Makarynska, D.; Ventura-Soares, C. Artificial neural networks in wave predictions at the west coast of Portugal. *Comput. Geosci.* **2005**, *31*, 415–424. [CrossRef]
10. Kambekar, A.; Deo, M. Real Time Wave Forecasting Using Wind Time History and Genetic Programming. *Int. J. Ocean Clim. Syst.* **2014**, *5*, 249–260. [CrossRef]

11. Deka, P.C.; Prahlada, R. Discrete wavelet neural network approach in significant wave height forecasting for multistep lead time. *Ocean Eng.* **2012**, *43*, 32–42. [CrossRef]
12. Wu, M.; Stefanakos, C.; Gao, Z.; Haver, S. Prediction of short-term wind and wave conditions for marine operations using a multi-step-ahead decomposition-ANFIS model and quantification of its uncertainty. *Ocean Eng.* **2019**, *188*, 106300. [CrossRef]
13. Mandal, S.; Prabaharan, N. Ocean wave forecasting using recurrent neural networks. *Ocean Eng.* **2006**, *33*, 1401–1410. [CrossRef]
14. Wei, C.C. Wind Features Extracted from Weather Simulations for Wind-Wave Prediction Using High-Resolution Neural Networks. *J. Mar. Sci. Eng.* **2021**, *9*, 1257. [CrossRef]
15. Hu, H.; Westhuysen, A.; Chu, P.; Fujisaki-Manome, A. Predicting Lake Erie wave heights using XGBoost and LSTM. *Ocean Model.* **2021**, *164*, 101832. [CrossRef]
16. Wei, C.C.; Cheng, J.Y. Nearshore two-step typhoon wind-wave prediction using deep recurrent neural networks. *J. Hydroinform.* **2019**, *22*, 346–367. [CrossRef]
17. Juliani, V.; Adytia, D.; Adiwijaya. Wave Height Prediction based on Wind Information by using General Regression Neural Network, study case in Jakarta Bay. In Proceedings of the 2020 8th International Conference on Information and Communication Technology (ICoICT), Yogyakarta, Indonesia, 24–26 June 2020; pp. 1–5. [CrossRef]
18. Paplińska-Swerpel, B. Application of Neural Networks to the Prediction of Significant Wave Height at Selected Locations on the Baltic Sea. *Arch. Hydroeng. Environ. Mech.* **2006**, *53*, 183–201.
19. Kalra, R.; Deo, M.; Kumar, R.; Agarwal, V. Artificial neural network to translate offshore satellite wave data to coastal locations. *Ocean Eng.* **2005**, *32*, 1917–1932. [CrossRef]
20. Uddin, M.A.; Jameel, M.; Abdul Razak, H.; Islam, A.B.M. Response Prediction of Offshore Floating Structure Using Artificial Neural Network. *Adv. Sci. Lett.* **2012**, *14*, 186–189. [CrossRef]
21. Ellenson, A.; Özkan Haller, H. Predicting Large Ocean Wave Events Characterized by Bi-Modal Energy Spectra in the Presence of a Low-Level SoutherlyWind Feature. *Weather Forecast.* **2018**, *33*. [CrossRef]
22. Tsai, C.P.; Lin, C.; Shen, J.N. Neural network for wave forecasting among multi-stations. *Ocean Eng.* **2002**, *29*, 1683–1695. [CrossRef]
23. Londhe, S.; Panchang, V. One-Day Wave Forecasts Based on Artificial Neural Networks. *J. Atmos. Ocean. Technol.* **2006**, *23*, 1593–1603. [CrossRef]
24. Deo, M.; Jha, A.; Chaphekar, A.; Ravikant, K. Neural networks for wave forecasting. *Ocean Eng.* **2001**, *28*, 889–898. [CrossRef]
25. Yaakob, O.; Zainudin, N.; Samian, Y.; Malik, A.M.A.; Palaraman, R.A. Developing Malaysian Ocean Wave Database Using Satellite. In Proceedings of the 25th Asian Conference on Remote Sensing, Chiang Mai, Thailand, 22–26 November 2004.
26. Yaakob, O.; Hashim, F.; Omar, K.; Md Din, A.H.; Koh, K. Satellite-based wave data and wave energy resource assessment for South China Sea. *Renew. Energy* **2015**, *88*, 359–371. [CrossRef]
27. FUGRO. Fugro Metocean Services. 2022. Available online: https://www.fugro.com/media-centre/news/fulldetails/2009/10/05/20-years-and-100-countries---worldwaves-a-success-story (accessed on 5 August 2022).
28. Breiman, L. Random Forests. *Mach. Learn.* **2001**, *45*, 5–32.:1010950718922. [CrossRef]
29. Azad, M. Knowledge Representation Using Decision Trees Constructed Based on Binary Splits. *KSII Trans. Internet Inf. Syst.* **2020**, *14*, 4007–4024. [CrossRef]
30. Azad, M.; Chikalov, I.; Moshkov, M. Decision Trees for Knowledge Representation. In Proceedings of the 28th International Workshop on Concurrency, Specification and Programming, CS&P 2019, Olsztyn, Poland, 24–26 September 2019; Ropiak, K., Polkowski, L., Artiemjew, P., Eds.; CEUR-WS.org; 2019; Volume 2571, CEUR Workshop Proceedings.
31. Azad, M.; Chikalov, I.; Hussain, S.; Moshkov, M. Multi-Pruning of Decision Trees for Knowledge Representation and Classification. In Proceedings of the 3rd IAPR Asian Conference on Pattern Recognition, ACPR 2015, Kuala Lumpur, Malaysia, 3–6 November 2015; pp. 604–608.
32. Alsolami, F.; Azad, M.; Chikalov, I.; Moshkov, M. *Decision and Inhibitory Trees and Rules for Decision Tables with Many-Valued Decisions; Intelligent Systems Reference Library*; Springer: Berlin/Heidelberg, Germany, 2020; Volume 156.
33. Azad, M.; Moshkov, M. A Bi-criteria Optimization Model for Adjusting the Decision Tree Parameters. *Kuwait J. Sci.* **2022**, *49*, 1–14. [CrossRef]
34. Juárez-Orozco, L.; Martinez Manzanera, O.; Nesterov, S.; Kajander, S.; Knuuti, J. The machine learning horizon in cardiac hybrid imaging. *Eur. J. Hybrid Imaging* **2018**, *2*, 1–15. [CrossRef]
35. McCulloch, W.S.; Pitts, W., A Logical Calculus of the Ideas Immanent in Nervous Activity. In *Neurocomputing: Foundations of Research*; MIT Press: Cambridge, MA, USA, 1988; pp. 15–27.
36. Boser, B.; Guyon, I.; Vapnik, V. A Training Algorithm for Optimal Margin Classifier. In Proceedings of the Fifth Annual Workshop on Computational Learning Theory, Pittsburgh, PA, USA, 27–29 July 1992; Volume 5. [CrossRef]
37. Smola, A.; Schölkopf, B. A tutorial on support vector regression. *Stat. Comput.* **2004**, *14*, 199–222. [CrossRef]
38. Quinlan, J.R. Induction of decision trees. *Mach. Learn.* **1986**, *1*, 81–106. [CrossRef]
39. Quinlan, J.R. *C4. 5: Programs for Machine Learning*; Morgan Kaufmann: Burlington, MA, USA, 1992.
40. Breiman, L.; Friedman, J.H.; Olshen, R.A.; Stone, C.J. *Classification and Regression Trees*; Wadsworth and Brooks: Monterey, CA, USA, 1984.
41. Azad, M.; Chikalov, I.; Hussain, S.; Moshkov, M.; Zielosko, B. *Decision Trees with Hypotheses (To Appear)*; Synthesis Lectures on Intelligent Technologies; Springer: Berlin/Heidelberg, Germany, 2022.

42. Botchkarev, A. A New Typology Design of Performance Metrics to Measure Errors in Machine Learning Regression Algorithms. *Interdiscip. J. Inf. Knowl. Manag.* **2019**, *14*, 045–076. [CrossRef]
43. Hastie, T.; Tibshirani, R.; Friedman, J. *The Elements of Statistical Learning*; Springer Series in Statistics; Springer Inc.: New York, NY, USA, 2001.
44. Wikipedia Contributors. Kernel Regression—Wikipedia, the Free Encyclopedia. 2022. Available online: https://en.wikipedia.org/w/index.php?title=Kernel_regression&oldid=1099742212 (accessed on 7 July 2022).

electronics

MDPI

Article

Classification of Task Types in Software Development Projects

Włodzimierz Wysocki [1,*], Ireneusz Miciuła [2] and Marcin Mastalerz [3]

1 Department of Software Engineering and Cybersecurity, Faculty of Computer Science and Information Technology, West Pomeranian University of Technology, 70-310 Szczecin, Poland
2 Department of Sustainable Finance and Capital Markets, Institute of Economics and Finance, University of Szczecin, 70-453 Szczecin, Poland
3 Department of Computer Science in Management, Institute of Management, University of Szczecin, 70-453 Szczecin, Poland
* Correspondence: wwysocki@zut.edu.pl

Abstract: Managing software development processes is still a serious challenge and offers the possibility of introducing improvements that will reduce the resources needed to successfully complete projects. The article presents the original concept of classification of types of project tasks, which will allow for more beneficial use of the collected data in management support systems in the IT industry. The currently used agile management methods—described in the article—and the fact that changes during the course of projects are inevitable, were the inspiration for creating sets of tasks that occur in software development. Thanks to statistics for generating tasks and aggregating results in an iterative and incremental way, the analysis is more accurate and allows planning the further course of work in the project, selecting the optimal number of employees in task teams, and identifying bottlenecks that may decide on faster completion of the project with success. The use of data from actual software projects in the IT industry made it possible to classify the types of tasks and the necessary values for further work planning, depending on the nature of the planned software development project.

Keywords: software; knowledge management; reasoning; information extraction; rule mining; knowledge acquisition; engineering

Citation: Wysocki, W.; Miciuła, I.; Mastalerz, M. Classification of Task Types in Software Development Projects. *Electronics* **2022**, *11*, 3827. https://doi.org/10.3390/electronics11223827

Academic Editor: Maria Liz Crespo

Received: 3 November 2022
Accepted: 19 November 2022
Published: 21 November 2022

1. Introduction

The contemporary intensity of changes in the surrounding reality due to technological progress makes management by economic units extremely demanding and difficult. Therefore, it is necessary to react quickly and effectively to changes that create new conditions for business activity. In the era of knowledge-based economy, information, information systems co-created by it, and related information technologies are extremely valuable and inextricably linked with knowledge [1]. The usefulness of IT systems is enormous and directly influences the increase of the possibilities of management units by reflecting their innovativeness and technological potential, which is of great importance when making strategic business decisions.

Globalization, the era of e-economy, and computerization are the topicality of modern economic activity [2]. That is why the production of an IT product (IT product) is so fundamental. Managing the production of IT products is a new scientific and technological discipline that has emerged at the interface between computer science and management engineering. All economic entities that base their activities on IT solutions are interested in the practical results of research in this discipline. In addition, it should be stated that, indirectly from all kinds of improvements in software development and management, any activity that uses any IT software will benefit. The digitization of the economy results in the need for reliability and security of emerging applications, IT systems, or transaction services. A lack of reliability of IT software can cause enormous losses. That is why it is so important to develop and create IT software that is as reliable as possible. Therefore, the work efficiency of teams producing IT software—characterized by a cyclical nature—is so important. Moreover, the literature highlights the dependence of the implementation effectiveness of subsequent software development stages on current project stages and the maturity level of the IT product.

The software development process is constantly evolving; there are technological changes and ways of organizing the work of project teams. In the knowledge economy era, most economic activities require systems and all kinds of IT products [3]. This forces the appropriate acceleration of software development, while at the same time maintaining the required quality. That is why it is so important to have the right team to implement often complex and innovative IT solutions. On the other hand, numerous project teams cause problems in effective communication in the group and cause the need to optimize work organization. To ensure this optimization, tools and IT products designed for managing software development processes are often used.

Traditional approaches to IT project management have become insufficient due to the high variability of requirements and the need to change the work organization of project teams [4]. It is particularly noticeable in innovative software development projects for the e-economy, where flexibility in the production process and originality of solutions are required. Therefore, despite the support of software development processes with increasingly better tools and systems, the pursuit of optimal use of project resources—including human teams—is still a considerable challenge. Since the beginning of the 21st century, the agile approach has been increasingly used in the practice of software development. The literature on the subject shows a significant improvement in the number of successful projects and the optimization of the use of resources necessary in software development through the use of continuously developed agile methods. This is particularly important due to the constant changes taking place in the requirements for the software being developed [5]. Therefore, traditional methodologies cannot be successfully applied to the numerous changes that are natural in innovation projects because they are extremely static and, apart from the first phase, consist of a fixed number of tasks. The next stage of the methodology development is based on iterations, which assumes that we learn about new requirements in the development process; requirements change and the existing ones are detailed. However, in the current agile methodologies, the requirements are incomplete and we do not know the details; this is also due to the nature of the innovative projects [6]. We have a set of requirements in the form of epics and user stories. Then, in the course of the software development process, creating new requirements becomes the cause of new tasks. Another source of new implementation tasks is the tests of current versions that detect defects or reveal new implementation possibilities. Automation of this process by the task generator in agile methodologies allows for the creation of appropriate increments of tasks; the number and the nature of which depends on the size and type of the project. On the other hand, the iterative measurement of tasks allows for subsequent iteration planning, with the assumption of the invariability of certain aspects for the best estimation. Therefore, planning methods based on summary work are nowadays of great practical importance for the management of manufacturing processes.

Project management systems for software development collect a lot of data about tasks and information related to them necessary to perform the work and the necessary exchange of information between the project team [7]. However, there is still room for improvement when it comes to providing information that will optimize IT project management. Because the data and information collected in the systems are semantically poor, it is necessary to analyze the works that are performed during the implementation of tasks to correctly classify them, which will also allow the estimation of their size, thus increasing the quality of planning. The isolation and classification of the characteristic types of tasks, and the work performed within them, will allow for the automatic execution of analyses and estimations to create insightful statistics and reports necessary for optimal forecasting decisions. In addition, taking into account the knowledge about the current technological and business components will give an appropriate insight into the current state of work in the project and will allow for more optimal planning of the software development process.

This work aims to identify the impact of the type of tasks on efficiency and to indicate those factors that positively affect the effectiveness of software development teams. The model of task types was built based on the data analysis of real software projects from the financial sector, managed by the Jira system. Connecting the model with the Jira system enables easy data acquisition for analysis and increases its commercial potential. A separate abstract layer of the model, in combination with a dedicated database, supports the possibility of creating interfaces to other IT project management systems. The article attempts to classify project tasks, thus determining the necessary expenditure on tasks of various types. In conjunction with the results of research on tasks created during the production process and cyclical works, it allows for planning projects. Linking task types with project team roles allows the simulation of project work and supports project management by identifying bottlenecks in the manufacturing process and avoiding over-employment.

The article discusses the basic concepts of the context in which software development projects are implemented and the development of methodology for the optimal management of these processes. At the same time, the principles and practices, as well as the lifecycle of software development projects, are characterized, which is important for the possibility of introducing improvements in this process. The concept of the programming process model is inspired by the agile approach to managing software as it is developed. Extending the original approach to the roles of team members and task types allows more precise work planning in the project and management of the project team, including detecting bottlenecks or unused resources. The article discusses the programming process model that is oriented toward the manual recognition of task types, thus enabling effective support for planning tasks carried out in innovative IT projects.

2. Literature Review

Software development in IT companies is mainly carried out through appropriate organization of the work of project teams. Unfortunately, in projects involving the search and creation of new solutions where the variability and unusual nature of ideas and requirements are natural, there is a constant need to improve the management of this process. Due to the innovative nature of IT projects, project teams are burdened with high risk. This forces the search for optimal project management methods and techniques that will allow for better control and use of the company's resources. Nowadays, analytical methods and tools embedded in IT systems are a great help in supporting project management.

Traditional software development project management methodologies are currently being replaced by, or supplemented with, agile methodologies. The development of project management methods in the IT industry has resulted from dissatisfaction with the small number of successful projects. A big problem was discovered in the management of innovative projects, where rigid and strongly formalized traditional software development methodologies did not work and even made it difficult to introduce changes and the proper functioning of projects [8]. Agile methods are used primarily due to the desire to reduce the number of errors, to shorten the time needed to create finished products, or to reduce the total production costs [9]. However, as noted by J. Shore and S. Warden, when making decisions regarding the implementation of agile practices, it is difficult to unequivocally state greater successes in the implementation of IT projects [10]. This is due to the variety of interpretations of the concept of "IT project success" and the fact that the use of the same methodologies in different companies from the same industry results in different results [11]. There is a common view in the literature that the key to success in project management is learning about appropriate techniques and tools [12]. However, the instrumental layer of an appropriate project management methodology alone will not ensure its success if not properly applied by its members, because the most important part of software development projects are the people [13]. Therefore, an appropriate balance should be found between the hard (budget, schedule, and implementation time) and soft (communication, changes, motivation, and competencies) elements of the project. Therefore, regardless of the methodology used, project management is closely related to people management, and practice shows that most of the problems affecting project success result from the omission of the "purely human aspects" of team management [14]. Accordingly, this article aims to identify the main problems related to people management in the success of an IT project by enabling effective task planning carried out in innovative, and thus variable, software development processes.

Since the beginning of the 21st century, we have witnessed a dynamic development in agile methodologies [15]. This is due to the adaptation of the management process to projects with a high degree of innovation. In these cases, hard methodologies such as PRINCE2 and PMI PMBoK do not work due to the detailed and long-term planning stage that is not able to take into account future changes [16,17]. With these characteristics of the projects, the too-high level of standardization of activities does not work, because there are often unsuccessful attempts at establishing the project lifecycle and detailed requirements already at the project initiation stage. Even though this approach gives a lot of comfort to the implementers, in the case of innovative projects, it does not meet the real needs that will be known only at further stages of implementation. The progress of work on the innovative project reveals the necessity to repeatedly verify many assumptions, actions, and plans, because, only as the implementation progresses, knowledge and actual visualization of the developed solutions in the shape of the final product are acquired. Additionally, certain paths for reaching specific results turn out to be ineffective only after the verification and testing phase. External changes should also be mentioned, i.e., changes that are beyond the control of project teams, e.g., changes in regulations and legal norms or the current market situation. Often, consistent plan implementation leads to the creation of functionalities that will not be adapted to reality or that will be useful only after costly modifications. Therefore, this article attempts to classify the characteristic types of tasks in software development projects that will allow for more effective planning and adaptation of the required work to dynamically changing reality.

All factors that make it difficult, or even impossible, to precisely define and describe the results of the project at the stage of its definition result from the following elements that occur when creating software (especially innovative software) [18,19]:

- Customers and users are not sure what result they expect and have difficulties with formulating requirements;
- Many details and solutions will only be revealed during the project implementation;
- Details of implementing innovative projects at the planning stage are not feasible;
- The way of thinking changes during software development;
- External forces (such as competitors' products or services) lead to changes or expansion of requirements.

In addition, the implementation of innovative IT projects is associated with the risk of other irregularities. Kruchten (2004) lists several main problems that can affect any project, which are [20]:

- Ad hoc requirements' management;
- Ambiguous and imprecise communication;
- Fragile system architecture;
- Overwhelming complexity and undetected inconsistency in requirements, designs, and implementation;
- Insufficient testing;
- Subjective assessment of the project status;
- Uncontrolled introduction of changes;
- Insufficient automation.

The above factors and elements of project implementation management, and the inability to solve them in accordance with the traditional approach, contributed directly to the development of methods for making traditional methodologies more flexible and, as a result, prepared the ground for the Manifesto for Agile Software Development that announced in 2001 the principles for agile software development methodologies [21,22]:

- Individual people and their interactions (rather than processes and tools);
- Working products, i.e., software (more than comprehensive documentation);
- Cooperation with the client (more than negotiating a contract);
- Reacting to changes (rather than sticking to the plan);
- Customer satisfaction is the highest priority and should be achieved through early and continuous delivery of valuable software;
- Variability of requirements applies to both new and changing requirements during the project implementation; the adaptive software development process is able to keep up with changes in advance;
- Frequent delivery of working software in periods from a few to several weeks, with shorter time frames being preferred;
- Communication that should be realized directly;
- Working software, because this is the most important measure of progress in the implementation of works in the project;
- Adaptability of software development consists in the ability to maintain an appropriate pace of work during the implementation of the project by all members of the project team and to adapt the product (software) to frequently changing requirements.

The principles of an adaptive approach to software development project management indicate the direction for project teams, while specific practice is necessary for the actual implementation of works [23]. The process structure and specific practices create a minimal flexible framework for self-organizing teams. IT tools are essential to accelerate software development and to reduce costs. Contracts are crucial for the development of the customer–supplier relationship. Documentation supports communication [24]. However, the key issue is to provide the project team with feedback to answer the question of where the team currently is in the software development process [25]. This is possible thanks to iteration and incremental software lifecycles. The essence of the iterative process is the frequent delivery of working pieces of software (successive increments) that implement selected sets of functions that together make up the usefulness of the final product. The iterative software development cycle leads to a management style where long-term plans are fluid, while a stable plan can be created for a short period of time. Iterative and incremental software development leads to completely new relationships with the business client and different principles of the project team's functioning.

The concept of the iterative and incremental programming development cycle as a remedy for dilemmas of the e-economy era has resulted in the creation of new methodologies for IT project management [26]. These methodologies, called agile, do not cut off completely from document-oriented formalized traditional methodologies, but have specific features (adapted to the requirements of modern software development projects) [27,28]:

- Adaptive, not predictive; traditional methodologies do not cope with frequent changes in requirements, while agile ones accept them on principle;
- People-oriented, not process-oriented;
- Creative style of work.

Organizations are increasingly implementing digital transformation plans, due to the threat of disruptions, in order to keep pace with the growing pace of business. Agile software development plays a huge role in this process. Many of the digital workflows in use today are based on agile principles. Thanks to a flexible scalable IT infrastructure, cloud computing is evolving in line with the needs of agile software development. The DevOps concept removes the traditional distinction between software development and operations. The software is used as a tool in SRE–DevOps implementation and systems management and automation of operational duties [29]. CI/CD methodologies confirm that software will change frequently and provide tools that help developers deliver new code faster [30]. Agile methodologies come in a variety of forms to meet the needs of any project [31]. Even though agile approaches are different, all of them are based on the key ideas contained in the agile approach. For this reason, any framework or behavior that complies with these principles is termed agile. Regardless of the specific agile approaches that a team decides to apply, the benefits of an agile methodology can only be fully realized through the collaboration of all parties involved [32]. In recent years, a large number of agile software development project management methodologies have emerged. The most popular are [33–36]:

- Kanban. The phrase "Kanban" (which comes from Japanese) is translated as "visual board or signboard" and is associated with the idea of "just in time". The Kanban concept gradually found its way into agile development teams. This approach develops project management using visual methods. The Kanban board—divided into columns to illustrate the flow of the software development process—is used to supervise projects. Teams benefit from greater visibility as they can track their progress through each stage of development and can plan upcoming tasks to deliver the product on schedule. To ensure that team members always have a smooth workflow and are aware of the appropriate stage of development, this method requires comprehensive communication and transparency.
- Scrum. The agile scrum development approach, which is represented by multiple development cycles, is one of the best-known examples of an agile methodology. Scrum breaks down development processes into units known as "sprints", much like Kanban. By maximizing and devoting time to developing each sprint, only one sprint is managed at a time. Consistent results, emphasized by scrum and agile techniques, allow designers to modify priorities in such a way that any incomplete or delayed sprint attracts more attention. The daily scrum is where activities are coordinated to develop the best sprint strategy; the scrum team has exclusive design roles such as scrum master and product owner.
- XP (extreme programming). The extreme programming (XP) methodology places great emphasis on collaboration, dialogue, and feedback. It emphasizes the constant improvement and happiness of the client. This approach uses sprints, or short development cycles, similar to scrum. It is created by the team to create a highly effective and productive atmosphere. The extreme programming technique is very helpful in a situation where the customer's needs are continuous and changing. It encourages developers to accept changes to customer requirements, even if these requirements appear at an advanced stage in the development process. In extreme programming, the design is evaluated from the outset by gathering input data that increase system performance. Additionally, it offers a quick way to meet any customer requirements.
- Crystal Clear family, a concept developed by Alistair Cockburn, an expert in object-oriented design. Each project class may have a different methodology (Crystal Clear Method). Crystal is a collection of smaller agile programming approaches that include Crystal Yellow, Crystal Clear, Crystal Red, Crystal Orange, and more. It was first introduced by Mr. Alistair Cockburn, one of the key figures in creating the Agile Manifesto for Software Development. Each one has a unique structure that distinguishes it from the others based on variables such as system criticality, team size, and project priorities. The type of Crystal agile approach is selected depending on the criticality of the project or system. Crystal strives for on-time product delivery, regularity, reduced administration with high user interaction, and customer satisfaction, similar to other agile approaches. The Crystal family, which has earned the title of Lightest Ways of Agile Methodology, promotes the idea that each system or design is unique and requires different practices, processes, and principles to be applied to obtain the best results.
- Adaptative software development, an extensive adaptive methodology developed by Jim Highsmith.
- Dynamic system development. This method of dynamic systems development was created to meet the demand for a unified industry charter for fast software delivery. The software development process can be planned, run, managed, and scaled using the comprehensive structure provided by DSDM, which maintains that quality and on-

time delivery can never be compromised and that design modifications are always to be expected. This belief is based on eight principles and a business-based methodology.

- Lean development, a "lean" software development. The basic idea of the approach is the elimination of losses understood as elements that do not add any value to the product. The aim of such action is to deliver the finished product to the customer as soon as possible. Lean software development is based on the values and principles of adaptive project management. The term "lean software development" is considered by Mary Poppendieck and Tom Poppendieck, who, in their book *Lean Software Development: An Agile Toolkit,* presented, among others, the seven main principles of lean management and a set of 22 techniques supporting the approach.

Agile management methodologies are a group of methodologies that are characterized by an adaptive and variable approach to managing software development projects. Additionally, agile methodologies were developed much earlier than the agile manifesto itself. The first works on adaptive project management methods date back to the 1980s (an example is the rapid application development methodology) and the concept of agile methodologies was introduced in the mid-1990s [37]. The idea of a time frame is a well-defined process dedicated to software development control at the lowest level in an iterative cycle with several review points. Reviews help ensure the quality and efficiency of software development. By delivering the software on time at the lowest level, the timely production at the highest level (i.e., the project level) is ensured [38]. The basic principle of the project plan is to prepare a schedule of planned increments and, within them, the planned time frames, which will create a complete project schedule capable of changes with emerging new requirements. The use of the time frame technique, together with the MoSCoW prioritization technique, ensures no delays in project implementation and the delivery of ready-made software that will meet business goals within a given time [39].

Projects with a high degree of innovation are very difficult to include in a complete schedule and scope of work. Therefore, adaptive methodologies describe functionalities (i.e., independent elements of the subsystem), which in subsequent releases can be quickly changed and handed over for implementation. Agile methodologies, as opposed to traditional methodologies, rule out the validity of long-term planning [40,41]. Therefore, the plans are speculative and not deterministic. This allows you to adapt to all types of changes that appear during the implementation of the project. Additionally, the distinguishing factor of agile methodology is a strong emphasis on the cooperation and integration of the project team, because only in this case is the smooth flow of information and effective communication ensured. The general scheme of the project lifecycle in the case of agile methodologies is based on five phases, indicated by J. Highsmith [18,42]:

1. Creating a vision of the project by defining its scope and principles of cooperation within the project team;
2. Planned speculation by specifying functional elements for the product, creating time-limited iteration plans and major milestones of the project itself;
3. Exploration, i.e., a quick start by providing the user with functional elements and implementing production methods that minimize the costs of changes, including the creation of an adaptable and collaborative project team;
4. Adaptation, i.e., the assessment of the product, process, project, and project team, and then the correction of existing plans and design practices;
5. Closing the project by creating a database of experiences for the next project.

Agile project management methodologies require an appropriate level of project maturity for organizations and project teams [43]. In addition, agile methods continue to be improved in order to most effectively respond to the needs of managing still-innovative software development projects. Hence, there are so many methods that are still looking for new, more perfect, solutions; although, they undoubtedly already seem to be better adapted than the classic methods to dynamically changing project environments. Implementation based on iteration allows for effectively adapting signals that come from both inside the project and from the external environment.

Classifying specific types of tasks in software development projects will allow for determination of the requirements, the time necessary, and the expenditure that will be needed to implement them. This will allow for more effective work on further adaptive project planning methods. Linking task types with the roles of the project team will allow you to simulate project work, which will support project management by identifying bottlenecks in the software development process. In addition, it will allow for avoiding over employment and will allow support for quick "what if" simulations. At the same time, the introduction of the task classification in terms of business and technological components, in conjunction with the employee competency model, will allow for automatically selecting the composition of the project team and optimally managing the team; this has the highest priority in agile methodologies. This is of great importance for the optimization of software development process management and leads to the development of perfect methods in response to the dynamism of real-world changes. It will have replicative and predictive capabilities to plan the project, to simulate it during changes, and to detect bottlenecks during the entire process. In this article, experiments were carried out on many real IT projects to classify task types and to attempt to find the relationship between the nature of the planned project and the specificity and number of these tasks.

3. Materials and Methods

The model of task types was built on the basis of the data analysis of real software projects from the financial sector, managed by the Jira system. Table 1 contains summary data of these projects. In four projects, the team worked in accordance with traditional methodologies, to which more and more elements of the agile approach were introduced. In two projects, the teams worked in accordance with the scrum approach. It is quite a large data set that allows for the analysis of phenomena related to modern manufacturing processes. The total number of project issues exceeds 30,000, which makes it possible to use artificial intelligence methods. The total duration of the projects is approximately 22 years, which does not mean that historical records are dealt with. Project teams worked on most projects in parallel.

Table 1. Projects, the data of which were used for research on task types.

Project	Methodology	Number of Issues	Effort Person/Hour	Duration of the Project		Team Size		
				Years	Months	MIN	Avg	Max
P1	Hybrid	10,773	67,444	8.0	96.3	2	14	26
P2	Hybrid	2492	17,063	3.6	43.6	1	12	25
P3	Hybrid	7754	41,530	3.2	37.9	1	19	31
P4	Hybrid	1212	7453	2.6	30.6	1	8	21
P5	Scrum	5466	55,354	3.4	40.9	6	22	39
P6	Scrum	3949	27,397	1.9	22.8	1	23	41
	Total	31,646	216,241	22.7			98	

In Jira, teams use issues to describe and track specific tasks to be done. Issues are the basic building blocks of projects. The issue can be of a specific type. The most common type of issue is task, which in practice is used for many purposes. The second most frequent type is bug, which describes the defects found in the developed software and the work needed for fixing them. The full list of types used in the projects with their frequency is shown in Figure 1.

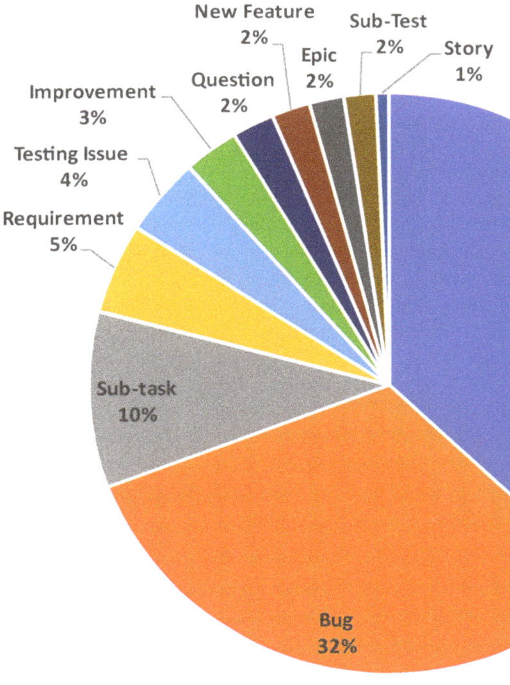

Figure 1. The occurrence frequencies of issue types in the researched projects.

The most commonly used issue types are task, bug, and subtask. The summary determines the idea of the actions to perform. The description field contains a detailed description of the work to do. A significant feature of the issue is its state. Tracking issue state changes allows for the recognition of the performed work. The workflow describes the values of the state field and the allowed transitions between them. Figure 2 shows a typical simplified workflow used in the analyzed projects.

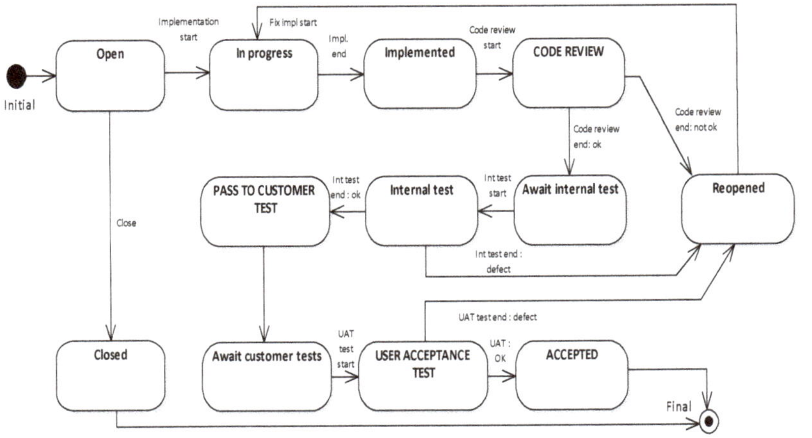

Figure 2. A typical simplified issue workflow used in projects.

As shown in Figure 2, transitions between issue states correspond to performing basic actions in implementing new system functions or fixing bugs. For example, the DevOps (development and operations) methodology helps to establish cooperation between developers and operators to automate the continuous delivery of new software, which is expected to contribute to shortening the development cycle and to creating high-quality software [44,45]. Another development of DevOps is the concept of development, security, and operations (DevSecOps), which at the same time is designed to integrate security methods with the software development process, where security measures are built in to ensure the integrity and availability of the application [46].

A valuable feature of the Jira system is storing the history of changes in the value of issue elements, including the state. Storing history supports the tracking of issue execution. Not all issues use the state to track work. The Jira system allows employees to register working hours devoted to work on an issue. For some types of work, there is no need to keep track of their state, as they are repetitive works performed as needed. In this case, the possibility of registering working hours is sufficient.

The presented approach is based on a precise division of the development process into activities and roles in line with the RUP methodology, adapted to hybrid and agile processes [47,48]. The previous series of articles described the agent–object model of the manufacturing process (AGOMO), which was first used to assess the maturity of RUP processes, then to plan hybrid water–scrum–fall processes [49,50]; it became an inspiration for the research presented here.

As we showed in the previous section, the types of Jira issues are insufficient to clearly define the work's purpose and type. The model was created in order to fill the gap that prevents linking the work carried out with the composition and competencies of the project team. The combination of these two areas will allow for a more precise quantitative analysis of the projects' works and the detection of the causes of the observed phenomena. It is a way to optimize the efficiency of development processes and to increase the level of maturity of project teams and organizations [51].

Software projects consist of tasks that a project team performs to build an IT system [52,53]. The task represents the Jira issue. Each of the tasks performed has a clearly defined goal. This goal can be, for example, implementing a requirement, testing a component or the entire module, administering environments, or managing a project; it determines its type. Members of the project team perform the tasks. Roles define the competencies and responsibilities of those carrying out the tasks. One person can have many roles; one role can have many people.

The task of implementing system functions requires performing some essential subtasks. They include implementation, i.e., the creation of component source code, code review, creation of unit tests, testing and verification of functions, and acceptance tests. Such tasks correspond to the issue, the state of which changes according to the workflow shown in Figure 2. The implementation tasks that consist of subtasks are named stateful tasks in the model.

Each task and subtask contains the number of project team members' working hours. Task and subtask types enable association efforts with the roles of project team members. They help to recognize bottlenecks or over-employment in completed IT projects and to avoid them during project planning.

Issues whose states do not change during the project are represented in the model by recurring tasks. They have been called recurring because a single task of this type can be performed as needed for the project's duration. For example, this might be a project development management task performed by an administrator when defects arise or when new software components need to be configured. For recurring tasks, you can calculate the average labor intensity in a period and, on this basis, assume what the fixed costs of the project are [54]. Genuinely recurring tasks are, for example, meetings such as daily stand-up meetings, workshops with clients, or steering group meetings.

The authors analyzed the tasks of the studied projects in terms of their type. The results, in the form of graphs showing the percentage number and the sizes of stateful and recurring tasks, are shown in Figure 3. The chart on the left shows the ratio of the number of stateful tasks to the recurring tasks in the projects. The chart on the right shows the ratio of the effort of stateful tasks to recurring tasks.

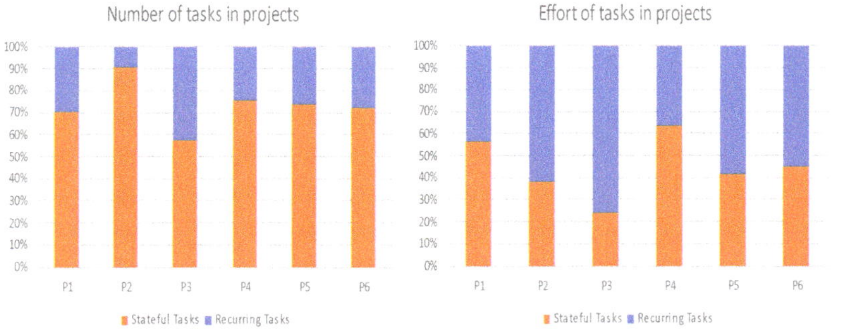

Figure 3. The number and effort of stateful and recurring tasks in the researched projects.

The case of the P2 project is significant, where the number of stateful tasks is greater than the recurring tasks (amounting to over 90%), while the effort of these tasks is slightly over 40%. It is very interesting, because stateful tasks are responsible for implementing the functions of the built system and for fixing the detected defects in it. From the developer's point of view, this workload should probably be the greatest. From the point of view of a researcher of software development processes, this phenomenon arouses great curiosity. In order to satisfy it, it is first necessary to precisely define the types of tasks in the projects, which this article implements on the basis of actual and implemented projects from practical activity.

4. Results

The classification of task types is based on the division of tasks into stateful and recurring tasks. By definition, stateful tasks represent work related to the implementation of system components and the fixing of defects. Stateful tasks are processed by a subtasking algorithm. Recurring tasks are responsible for the remaining works. Recurring tasks can be classified on the basis of summary and extended-text descriptions included in the issue. Initially, these tasks were classified manually. Subsequently, simple classification algorithms were created based on keyword searches. Recurring tasks are also broken down into subtasks that correspond to the registered work.

Figure 4 shows a diagram of the classification algorithm activity. The algorithm first checks how many times the issue has changed its state; if at least three times (more than open and close), the workflow is the basis for dividing the task into subtasks. The split algorithm tries to create subtasks (e.g., development, testing, code review). For this, it uses the value of the state before and after, the time of the change, the role of the person, and the registered works and then assigns completed work to the created subtask. The result is a stateful task.

If the task did not change state or if the algorithm did not detect any subtasks, the task type is determined by searching for keywords in the text description. When it fails, the tasks go to a spreadsheet, where they are manually classified. Then, the algorithm checks who was working on the task. If many people worked on a task on one day, it is a recurring group task (e.g., stand-up and other meetings). If one person worked on a task on one day, it is a recurring individual. The way of dividing the recurring task into subtasks depends on the individual/group classification. If the programmer and the tester alternately execute a task, it is stateful; the algorithm divides them into subtasks according to the roles of the team members.

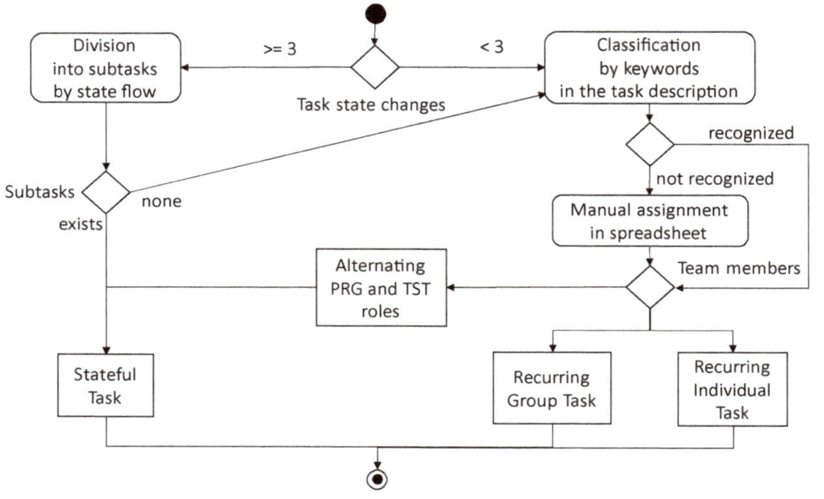

Figure 4. Algorithm of task classification.

The algorithm for breaking stateful tasks into subtasks is very complex due to the many ways that users use Jira to work on a project issue. Searching for keywords is not very accurate and supporting it by manual classification makes it impossible to use the classification in practice. Therefore, it is planned to replace these solutions with the NLP (natural language processing) classification. However, there are some good points to manual ranking. During the work on the algorithm, the process of the manual analysis of recurring tasks detected more than 50 types of tasks that were aggregated into three groups containing 14 main types of tasks.

Stateful tasks consist of tasks for implementing new features and for fixing bugs reported by customers and testers. Recurring tasks are divided into three main groups: implementation, meetings, and organizational tasks. A complete list of the main task types is provided in Table 2. The task types listed in the table form a hierarchical structure divided into types and groups.

The task classification algorithm—classifying tasks performed alternately by programmers and testers as stateful tasks—increased the number and effort of state tasks. An updated version of the graphs in Figure 3 is included in Figure 5. The changes are significant. For example, for project P3, the percentage of stateful tasks increased from 58% to 66% and the percentage of stateful task efforts increased from 24% to 43%.

The development of task types and an algorithm enabling automatic classification of the researched projects allowed for a detailed analysis of the tasks of the researched projects. Below, we present effort charts (Figure 6) for six groups of task types in the researched projects.

The P1 project was implemented in a hybrid methodology. It is the longest project among the researched, with a duration of more than 8 years. The project went through many phases of implementation, delivery, and maintenance. Therefore, the values and the ratio of effort in the project were averaged. They can serve as a benchmark when compared with other hybrid and traditional designs.

The P2 project was carried out in the hybrid methodology. It is characterized by a large number of hours used for various types of meetings. On the other hand, the very small scope of work in the management field suggests that the project may have been managed collectively. The very little work involved in fixing defects found by clients indicates that time spent in meetings was well spent.

Table 2. Types of stateful and recurring tasks used by the model.

Kind	Group	Type
Stateful	DEV requirement, function, feature	DEV-PRG: analysis, design, programming
		DEV-TST: testing, verification
	BUG-DEV defect from internal tests	BUG-DEV-PRG: programming
		BUG-DEV-TST: testing, verification
	BUG-CLI defect found by customer	BUG-CLI-PRG: programming
		BUG-CLI-TST: testing, verification
Recurring	R-DEV development	A&D: analysis and design
		BLD: build versions, find the causes of errors, create scripts
		CFG: software configuration
		CUST: training, technical support, commuting
		MIGR: data migration from previous systems
		RQM: requiremets engineering
		RVW: code review
		TST: subsystem, module tests, manual tests
	R-MEET meetings	M-CUST: other customer meetings
		M-DEV: development team meetings with external teams
		M-INT: development team internal meetings
		M-STDNP: daily stand-up meetings
		M-WRK: requirements workshop with the customer
	R-ORG organizational	ADM: administration of environments and servers
		PM: project management

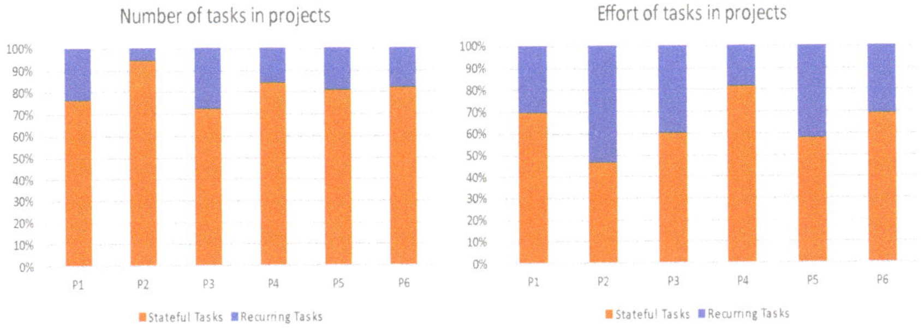

Figure 5. Updated numbers and effort of stateful and recurring tasks.

The P3 project was carried out in a hybrid methodology. It has high administrative and management costs. The ratio of repairing defects found by customers to repairing defects found by testers is interesting. There is no such tendency in the other projects, except perhaps for the P4 project. This may indicate inaccurately defined requirements or difficult contact with the customer.

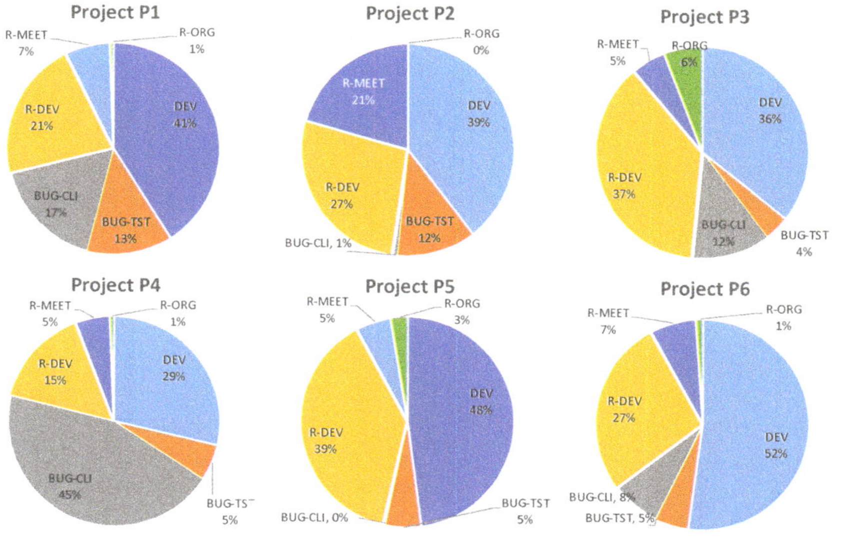

Figure 6. Effort of the six main groups of task types in projects.

The P4 project was implemented in the hybrid methodology. The distribution of the workload of tasks in the P4 project differs from others in terms of a very large amount of work to fix the defects detected by customers. The cost of this work is one-and-a-half times greater than that for implementation tasks and five times greater than the cost of repairing defects detected by the development team. The situation is similar to the P3 project, only the management and development costs are much lower. It is possible that the P4 project is in the maintenance phase of a project, with a large number of defects.

The P5 project was implemented using the scrum methodology. At the high level of abstraction given by the task type group analysis, no difference can be seen between this project and the hybrid projects. What is important is the lack of resources to rectify defects reported by customers. It is very possible that the project did not enter the customer implementation phase and was not put into production.

The P6 project was also implemented using the scrum methodology. More than half of the work was devoted to product implementation. The product has likely been delivered to the customer, as indicated by 8% of the efforts to fix defects reported by customers. The meetings have a significant share in the work on the project, which is consistent with the scrum methodology. The outlays for the maintenance and management of the project are small.

Graphs of total efforts by groups of task types give a very synthetic view of the projects. We can get a deeper look at the differences between projects by focusing attention on the detailed effort of cyclic task types. Figure 7 shows radar charts of recurring task effort, shown as a percentage of total recurring effort. The details of the R-DEV group from Figure 6 are shown here. The left side of the graph shows the labor intensity of meetings, cooperation with the client, administrative work, and management. The right side of the chart shows the expenditure on recurring development tasks. The charts differ from each other to reflect the characteristics of the projects. The charts show the "fingerprints" of the projects, making it possible to identify their detailed characteristics and to compare them with each other.

In most of the charts, the left organizational and management side dominates over the right developer side. The P1, P2, P3, and P6 projects follow a similar pattern in the recurring works chart, which indicates greater expenditure on meetings and management than on development work. Analysis and design play a large role in the P3, P5, and P6 projects.

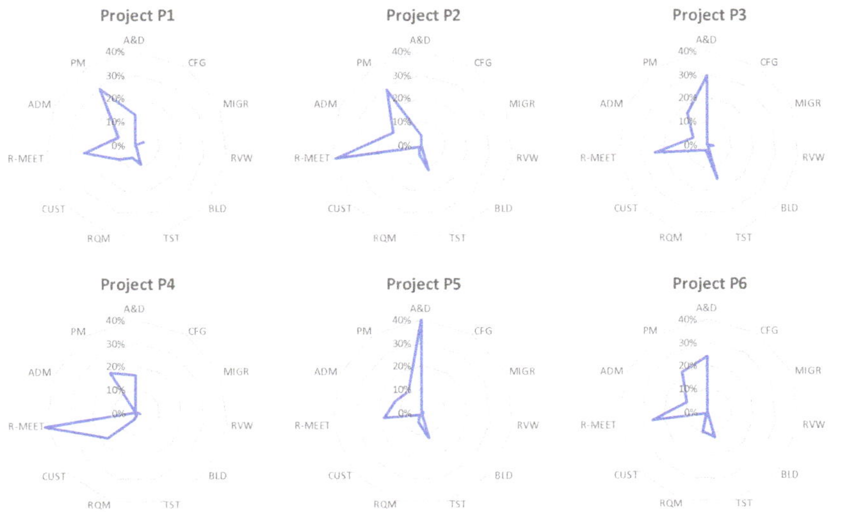

Figure 7. Structure of expenditure on recurring tasks in the researched projects.

Comparing the details of recurring work in projects allows you to determine the minimum, average, and maximum values of recurring work, which will enable the use of linguistic variables in the work for project planning. On the basis of static summary graphs, the nature of the project can be determined and projects can be compared with each other. It is also important to determine the area of expenditure of recurring tasks in projects and to determine their minimum, average, and maximum values. This is important for project planning. The entered types of tasks can be useful to answer the question of how tasks are created and performed during the development process.

The previous chapters introduced the division of tasks performed in the software process into stateful tasks related to the implementation of new tasks and recurring ones, with works performed periodically. This division may lead to the assumption that stateful tasks related to new features are mainly created at the beginning of the project. As for bugs, it would be prudent to assume that they arise after implementing certain requirements or features. The very name of recurring tasks (e.g., daily stand-ups) suggests that they are performed in equal intensity throughout the manufacturing process. Creating new tasks is very important when planning the development process, because the implementation of tasks cannot be started when they have not yet been created. This fact limits the size of the planned project team.

The model of task types and the research carried out on actual projects show what this case looks like in reality. Figure 8 shows when the state tasks in the researched projects were created. The charts do not show the number of created tasks, rather their effort, which better reflects the total size of tasks created in a month. DEV—new functions; BUG-DEV—defects detected by testers; BUG-CLI—defects detected by the customer (see Table 2 for details).

Most of the charts show that new features are developed throughout the life of the project. This phenomenon may come as a big surprise, especially since the P1-P4 designs were produced according to a hybrid approach in which traditional practices had a large share. The situation in the long-term P1 project is understandable, because it consists of many phases and, in each of them, new functions were created to be implemented. The P2 project is an exception among the examined projects, because new functions are created during the first 8 months at the beginning of the development process and then, for 24 months, they are implemented, and repairs of defects detected by the development team are created.

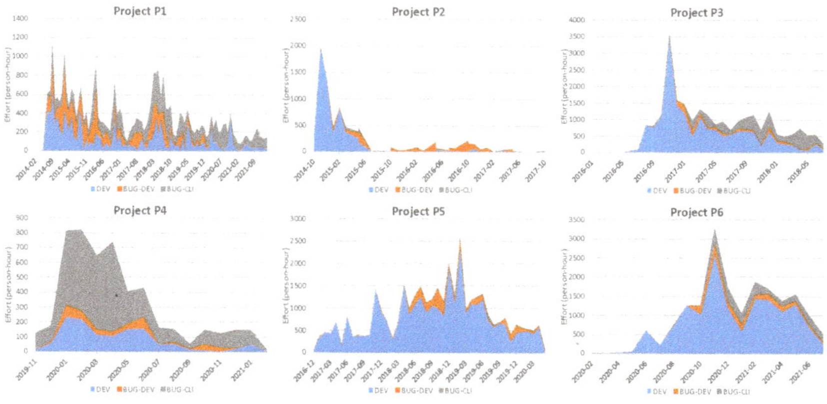

Figure 8. Stateful tasks created in the development process of the researched projects.

The graphs of the P3–P6 projects show, however, that new functions are created until the very end of the manufacturing process, although to a lesser extent. The reasons for this are interesting. Does it result from a long-term process of acquiring new requirements parallel to the development process? Or, maybe the reason is getting to know the details of the requirements obtained earlier? Unfortunately, the data placed in the Jira system do not answer these questions directly, because they do not take into account the requirements engineering processes. An interesting phenomenon is also the periodic increase and decrease in both the work on new functions and the repair of detected defects.

Work on the implementation of stateful tasks proceeds in a different rhythm than the creation of new stateful tasks. The number of man-hours used per month for stateful assignments depends on the number of people on the project team and their assigned roles. It should be taken into account that the team also performs recurring tasks. Figure 9 shows the monthly expenditures on the execution of stateful works in the researched projects.

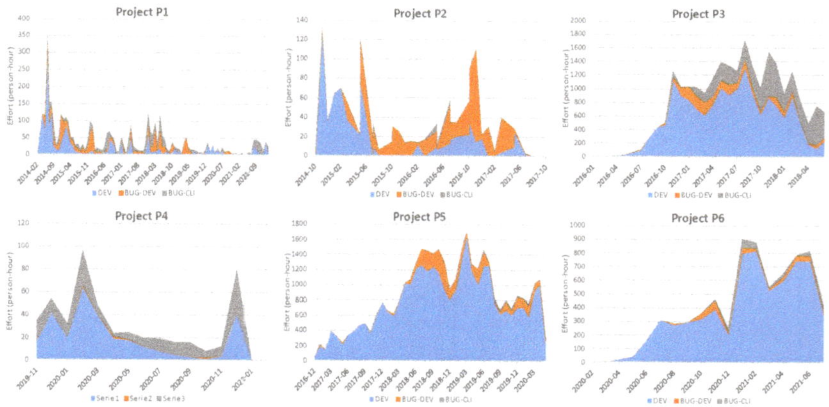

Figure 9. Work on stateful tasks in the researched projects.

The charts show that in most projects, bug fixes detected by the development team are delayed in relation to the implementation of the functions of the developed software. Even more delayed is the repair of errors detected by the client, because they are the last detected. This phenomenon is best seen in the graphs of the P3 and P6 projects. Comparing the work charts with the charts of the created tasks (see Figure 8) gives better insight into the project. For example, in the P2 project, after creating many thousands of new features, there is a break for several months, and then defects detected by the project team are created. The P2 project work graph shows that there was no break in the project, the work was less intensive, but at that time defect fixes and then the implementation of new features were ongoing.

The answer to the question of what the effort of recurring work in the software development process looks like can be found in Figure 10. The graphs show a similar periodicity as the graphs of the effort of stateful tasks. The source of these periodic disturbances is very interesting. Probably, to some extent, the outlays for stateful and recurring work are complementary, i.e., increases in the first graph correspond to decreases in the second.

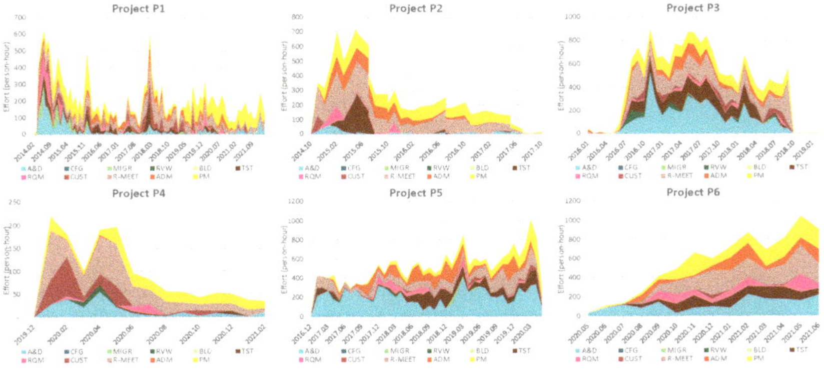

Figure 10. Work on recurring tasks in the researched projects.

The model of task types, in conjunction with the roles of the project team members, allows for the analysis of the work carried out in the project. On this basis, it is possible to trace the implementation of tasks and to recreate the composition of the project team. The next step in the development will be the possibility of simulating the work in the software development process or of planning the composition of the project team based on task plans.

The project team consists of the roles and the number of jobs of people employed in a given role. The analysis of actual project data is the source of the model, hence the lack of certain roles in the team, e.g., the role of an analyst. The current set of roles is defined as follows: ADM—administrator, PM—team leader, PRG—developer (who also deals with design and collection of requirements), TST—tester. The team consists of roles and the number of positions for a given role.

The adopted set of roles is not consistent with the agile approach represented by the scrum methodology. The scrum team chiefly consists of three roles: the scrum master, the product owner, and the development team. Developers are everyone belonging to the development team who are involved in software development. However, in practice [21], it is worth distinguishing the role of a tester (whose main activities are software testing and quality assurance), a programmer (who creates production code), and an administrator (who manages development and production environments and tools supporting the development and implementation of emerging software).

Figure 11 shows the concept of the relationship between task types and the roles of project team members. It consists of task types, roles, and two kinds of connections: simple and proportional. A simple link between a task type and a role determines that tasks of a specific type are performed by members of the project team with that role. For example, DEV-PRG tasks are performed by people with the PRG role and BUG-CLI-TST tasks are performed by people with the TST role.

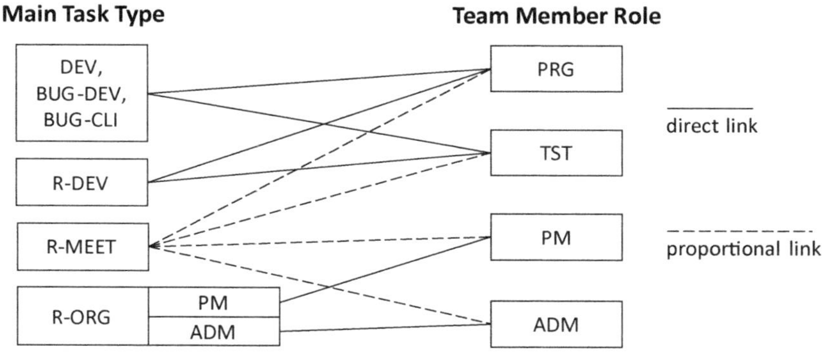

Figure 11. Relationship between the main types of tasks and the roles of project team members performing them.

A proportional link exists between meeting tasks and all roles of the project team. The idea of proportional connection is to divide the man-hours allocated to meetings among people from the project team in proportion to the number of people performing a given role. For example, there are 100 person-hours of meetings recorded in a month and programmers completed 72% of the work per month, so 72 meeting person-hours per month are added to the workload of the developers' tasks.

The workload of tasks performed monthly by roles is divided by the adopted average number of hours worked by a person per month. The score is the number of positions for that role. The number of positions is quantized to 1/4 and rounded up. The adopted average number of 165 h of work per month takes into account only non-working days. It does not take into account holidays and possible dismissals due to the employee's illness. This number can be changed freely.

Linking task types to the roles of project team members enables the approximation of the team composition needed to complete the project. Table 3 presents the monthly work of the P6 project, broken down by the main types of tasks and the composition of the project team reconstructed on their basis.

Table 3. Monthly work of the P6 project and reconstructed composition of the project team.

Date	Actual Monthly Effort				Team Number of Jobs			
	DEV	R-DEV	R-MEET	R-ORG	ADM	PRG	TST	PM
2020-05	155	-	-	11	0.25	1	0.25	0.25
2020-06	307	3	-	8	0.25	1.75	0.25	0.25
2020-07	284	41	21	106	0.5	2	0.5	0.25
2020-08	290	113	65	105	0.5	2.5	0.5	0.5
2020-09	361	191	133	128	0.25	3	1.25	1
2020-10	467	200	129	293	0.25	3.5	1.25	2
2020-11	255	130	256	158	0.5	2.75	0.5	1
2020-12	901	212	202	292	1	6.25	1.5	1.25
2021-01	881	193	208	328	1.25	6.25	1.5	1.25
2021-02	555	138	152	293	0.75	3.25	1.75	1.5
2021-03	650	151	198	333	0.75	4.25	1.5	1.75
2021-04	782	269	289	326	0.5	6.5	1.25	2
2021-05	821	110	196	404	1	5.75	0.75	1.75
2021-06	423	54	114	184	0.25	2.75	0.75	1.25

The number of people on the project team goes up and down in line with monthly stateful and recurring tasks. The team has been growing since the beginning of the project. In July and October 2020, it was at its highest; the number of positions was 10.25. After that, the team shrunk and the project ended in December 2020. There was a maximum number of 6.5 programmers and 1.75 tester positions per project. It is interesting that, in April and October 2020, there were two positions for the project manager in the team. Often, projects employ people to perform organizational and support work, for example, managing issues in the Jira system, maintaining Kanban boards, or creating reports.

5. Model Verification

The classification algorithm, according to Figure 4, consists of a method of dividing into subtasks and assigning task types based on a set of keywords built from the manual classification of P3 project tasks. Since the total number of issues in all projects was large, it was difficult to classify them manually. The research is intended to serve as a proof of concept, so the same set of keywords was used to classify the tasks of the other projects.

The method of dividing tasks into subtasks is closely related to the classification of tasks, because the types of subtasks affect the distinction, for example, of whether implementation or testing has been performed. If a subtask is not correctly identified, labor intensity will not be assigned to it and it will not be included in the accuracy indicator (EA_p). Subtask division is complex, because people who record work in the JIRA system do it in many different ways. The method of subtask division developed most first for the P3 project, then was adapted to other projects. The primary indicator of the effectiveness of project task classification (EA_p) is the ratio of the labor intensity of the recognized subtasks (E_p^R) to the total labor intensity of the project (E_p):

$$EA_p = \frac{E_p^R}{E_p} \cdot 100\%$$

where p is the project; EA_p is the index of effectiveness of classification of labor intensity of project tasks p; E_p^R is labor intensity of correctly identified project subtasks p; E_p is total labor intensity of the project p.

To further verify the accuracy of task classification, a manual check of a sample of 100 randomly selected tasks for each project was conducted. As a result, the accuracy rate was obtained (NA_p^M) determining the percentage of correctly classified tasks in the sample (N_p^{MC}) to the number of tasks in the sample (N_p^M):

$$NA_p^M = \frac{N_p^{MC}}{N_p^M} \cdot 100\%$$

where p is the project; NA_p^M is the indicator of the effectiveness of the classification of the number of tasks of the project p in the sample; N_p^{MC} is the number of correctly identified project tasks p in the sample; N_p^M is the number of project tasks p in the sample, $N_p^M = 100$.

The labor-intensity accuracy rate (EA_p^M) is obtained, determining the percentage of the labor intensity of correctly classified tasks in the sample (E_p^{MC}) to the total labor intensity of the sample (EA_p^M):

$$EA_p^M = \frac{E_p^{MC}}{E_p^M} \cdot 100\%$$

where p is the project; EA_p^M is the index of effectiveness of classification of labor intensity of project tasks p in the sample; E_p^{MC} is the labor intensity of correctly identified project tasks p in the sample; E_p^M is the total labor intensity of the project p in the sample.

Table 4 shows the results of the verification of the effectiveness of project task and subtask classification and the manual verification of the samples of project tasks.

Table 4. Results of verification of the effectiveness of the classification of tasks and subtasks of projects.

Project	EA	NA^M	EA^M
P1	91.50%	93.00%	93.72%
P2	61.50%	67.00%	31.14%
P3	99.70%	98.00%	99.85%
P4	25.00%	24.00%	52.88%
P5	99.60%	87.00%	100.00%
P6	54.70%	68.00%	99.50%

The accuracy rates of the P3 project can be considered exemplary, since a classification algorithm was developed for this project. The differences between the values of numerical and labor-intensive accuracy indicators determined during the manual verification of small samples (from 1% to 4% of the number of tasks in the project) are due to the fact that not all correctly classified tasks have labor hours recorded. The low indicator values for projects P2, P4, and P6 are due to the frequent use of a language other than English in task descriptions. The set of keywords developed for project P3 consists of English words, hence the poor transferability of the classification algorithm to some projects. In addition, the manual verification of the P4 project detected the following: a lack of keywords in the description and a large number of tasks with no change in status, resulting in a lack of subtasks and tasks with no registered work. The results in Table 4 indicate the need to translate job descriptions from the JIRA system into English before starting classification based on keywords or using NLP models.

6. Conclusions

The data of actual projects are the basis of the presented research and model. On the one hand, they increase the possibility of practical applications, on the other hand, they limit the model to the types of tasks and roles present in the researched projects. With this in mind, we tried to make the model flexible and open to modifications.

Connecting the model with the Jira system enables easy data acquisition for analysis and increases its commercial potential. A separate abstract layer of the model, in combination with a dedicated database, supports the possibility of creating interfaces for other IT project management systems.

The classification algorithms presented in the article are based on the manual recognition of task types. With manual recognition, rules based on keywords are created, which allows automatic recognition of task types at subsequent occurrences. As is known from the literature and practice, such algorithms are not very elastic and not very accurate (45% accuracy) [55]. However, with a growing base of manually recognized tasks, it will be easy to change to NLP models such as BERT [56]. This will allow fully automated operation of the task classification and subtask classification algorithm on a real-time basis. It will allow the analysis of the data collected in JIRA, the production of reports and charts to provide insight into the manufacturing process, and support for the project manager in decision making.

The division into state and cyclic tasks shows that state tasks are created during software development and their number and labor intensity depend on the size of the project. The rate of growth and completion of state tasks depends on the composition of the project team. The project's execution time depends on this rate. The number and labor intensity of cyclic tasks, on the other hand, depends on the duration of the project. Thus, the classification of tasks becomes the basis for constructing a generator of state and cyclic tasks to create software development plans. In turn, the creation of a development plan and the composition of the project team will allow the construction of a simplified simulation of the work in the project.

The ability to create a project plan and to select the appropriate composition of the project team, and, then, thanks to the simulation, to check how the work will proceed, will allow for comprehensive support of the management of the development process. Thanks to simulation, it will be possible to estimate whether the composition of the team is suitable for the specifics of the project. Simulation can show that, for example, programmers have implemented the requirements and that the team is waiting for testers to perform tests. In this way, the project manager can recognize the risk of a bottleneck in the project and prevent it in advance.

Project planning is useful not only before starting; real-time automatic task classification will allow analysis and will use the calculated task statistics to plan the next sprints or stages of the development process with increasing accuracy.

The introduction of an additional classification of tasks in terms of business and technological components, in conjunction with the employee competency model, would automatically collect information about employees' experiences in business and technology areas. This would allow the assessment of the level of employees' competences, selecting the composition of the project team and perhaps managing the team so that the competences are duplicated and dispersed among team members.

Author Contributions: Conceptualization, W.W. and I.M.; methodology, W.W.; software, W.W.; I.M. and M.M.; validation, W.W., I.M. and M.M.; formal analysis, W.W., I.M. and M.M.; investigation W.W. and I.M.; resources, W.W. and I.M.; data curation, W.W. and I.M.; writing—original draft preparation, W.W. and I.M.; writing—review and editing, W.W., I.M. and M.M.; visualization, W.W. and I.M.; supervision, W.W., I.M. and M.M.; project administration, W.W., I.M. and M.M.; funding acquisition, W.W. and I.M. All authors have read and agreed to the published version of the manuscript.

Funding: The project was financed within the framework of the program of the Minister of Science and Higher Education in Poland under the name "Regional Excellence Initiative" in the years 2019–2022, project number 001/RID/2018/19, the amount of financing PLN 10,684,000.00.

Data Availability Statement: Data are contained within the article.

Acknowledgments: Many thanks to Marcin Korzeń, Jakub Swacha, and Leon Dorozik for scientific support.

Conflicts of Interest: The authors declare no conflict of interest.

References

1. Wysocki, J. The use of information technologies in the enterprise. In *Science about the Enterprise: Selected Issues*; Lichniak I.: Warsaw, Poland, 2009; pp. 347–377. Available online: http://hdl.handle.net/20.500.12182/905 (accessed on 21 June 2022).
2. Zakrzewska, M.; Miciuła, I. Using e-government services and ensuring the protection of sensitive data in EU member countries. *Procedia Comput. Sci.* **2021**, *192*, 3457–3466. Available online: https://doi.org/10.1016/j.procs.2021.09.119 (accessed on 11 July 2022).
3. Açikgöz, A.; Günsel, A. Individual Creativity and Team Climate in Software Development Projects: The Mediating Role of Team Decision Processes. *Creat. Innov. Manag.* **2016**, *25*, 445–463. [CrossRef]
4. Mieśmieńska, L. Shaping the involvement of IT employees in the personnel strategy of company X. In *Man and Work in a Changing Organization. Towards Respecting the Interests of Employees*; Gableta, M., Pietroń-Pyszczek, A., Eds.; Scientific Papers No. 223; Wrocław University of Economics: Warsaw, Poland, 2011.
5. Olsen, T.L.; Tomlin, B. Opportunities and Challenges for Operations Management. *Manuf. Serv. Oper. Manag.* **2019**, *22*, 113–122. [CrossRef]
6. Azizyan, G.; Magarian, M.; Kajko-Mattsson, M. The Dilemma of Tool Selection for Agile Project Management. In Proceedings of the The Seventh International Conference on Software Engineering Advances ICSEA 2012, Lisbon, Portugal, 16–20 October 2022.
7. Powell, T.C.; Dent-Micallef, A. Information Technology as Competitive Advantage: The Role of Human, Business and Technology Resources. *Strateg. Manag. J.* **1997**, *18*, 375–405. [CrossRef]
8. Kędziora, A.F. SCRUM Methodology in Small and Medium IT Projects. 2011. Available online: http://min.wmi.amu.edu.pl/wpcotent/uploads/2011/04/PMKEDZIORASCRUM.pdf (accessed on 14 August 2022).
9. Elssamadisy, A. *Agile Patterns of Implementing Agile Practices*; Helion: Gliwice, Poland, 2010.
10. Shore, J.; Warden, S. *The Art of Agile Development*, 2nd ed.; O'Reilly Media, Inc.: Sebastopol, CA, USA, 2008.
11. Bielec, J. The same technology—Different results. Key elements of the success of an IT project implementation. In Proceedings of the 13th PLOUG Kościelisko Conference, Post-Conference Materials, Kościelisko, Poland, 17–20 October 2007.
12. Spałek, S. *Critical Success Factors in Project Management*; Publishing House of the Silesian University of Technology: Gliwice, Poland, 2004.
13. Spolsky, J. *IT Project Management. Subjective View of a Programmer*; Helion Publishing House: Warsaw, Poland, 2005.
14. Wróbleski, P. *IT Project Management for Practitioners*; Helion Publishing House: Warsaw, Poland, 2005.
15. Abrahamsson, P.; Salo, O.; Ronkainen, J.; Warsta, J. *Agile Software Development Methods: Review and Analysis*; VTT Publications: Espoo, Finland, 2002; pp. 1–112.

16. Dima, A.M.; Maassen, M.A. From Waterfall to Agile software: Development models in the IT sector, 2006 to 2018. Impacts on company management. *J. Int. Stud.* **2018**, *11*, 315–326. [CrossRef] [PubMed]
17. Łabuda, W. Agile and traditional approach to software development projects. *Zesz. Nauk. WWSI* **2015**, *9*, 57–87.
18. Highsmith, J. *APM: Agile Project Management. How to Create Innovative Products*; Mikom: Warsaw, Poland, 2005.
19. Bhatt, P. An influence model for factors in outsourced software maintenance. *J. Softw. Maint. Evol. Res. Pract.* **2006**, *18*, 385–423. [CrossRef]
20. Kruchten, P. *The Rational Unified Process: An Introduction*; Addison-Wesley Professional: Boston, MA, USA, 2004.
21. Manifest Agile. 2001. Available online: https://agilemanifesto.org/iso/pl/principles.html (accessed on 17 June 2022).
22. *Agile Project Management Handbook*; Version 1.1; Dynamic Systems Development Method Limited: Katowice, Poland, 2013.
23. Andersen, E. Warning: Activity planning is hazardous to your project's health! *Int. J. Proj. Manag.* **1996**, *14*, 89–94. [CrossRef]
24. Bamel, U.K. Organizational climate and managerial effectiveness: An Indian perspective. *Int. J. Organ. Anal.* **2011**, *21*, 198–218. [CrossRef]
25. Beck, K. *Efficient Programming eXtreme Programming*; MIKOM Publishing House: Warsaw, Poland, 2001.
26. Bieliński, J. *The Development of Sectors in the Modern Economy*; Publishing House of the University of Gdańsk: Gdańsk, Poland, 2006.
27. Jędrzejowicz, P. *IT Management Systems*; WSM in Gdynia: Gdynia, Poland, 2001.
28. Łabuda, W. How to implement a successful IT project. In Proceedings of the Conference Materials Summarizing the Project Program for the Development of the Teaching Offer and Improving the Competences of Lecturers at the Warsaw University of Information Technology, Warsaw, Poland, 20–21 September 2011.
29. Kisielnicki, J. *Management Infrastructure—Poland in Europe*; Master of Business Administration: Poznań, Poland, 2002.
30. Raunak, M.S.; Binkley, D. Agile and Other Trends in Software Engineering. In Proceedings of the IEEE 28th Annual Software Technology Conference (STC), Gaithersburg, MD, USA, 25–28 September 2017.
31. Turk, D.; France, R.; Rumpe, B. Limitations of Agile Software Processes. *arXiv* **2014**, arXiv:1409.6600.
32. Crispin, L.; Gregory, J. *Agile Testing: A Practical Guide for Testers and Agile Teams*; Addison-Wesley Professional: Boston, MA, USA, 2009.
33. Czerska, M.; Rutka, R. Assessment of the Designers' Motivation System in the IT Industry. In *Man and Work in a Changing Organization*; Gableta, M., Pietroń-Pyszczek, A., Eds.; Naukowe No. 43; University of Economics in Wrocław: Wrocław, Poland, 2009.
34. PMI. Foundation for Business Agility | Disciplined Agile. 2020. Available online: https://www.pmi.org/disciplined-agile,dostęp (accessed on 21 August 2022).
35. Wysocki, W. A hybrid software processes management support model. *Procedia Comput. Sci.* **2020**, *176*, 2312–2321. [CrossRef]
36. Miłosz, M.; Borys, M.; Plechawska-Wójcik, M. *Contemporary Information Technologies*; Agile methodologies of software development; Lublin University of Technology: Lublin, Poland, 2011.
37. Lievens, A.; Blažević, V. A service design perspective on the stakeholder engagement journey during B2B innovation: Challenges and future research agenda. *Ind. Mark. Manag.* **2021**, *95*, 128–141. [CrossRef]
38. Wojtaszek, H.; Miciuła, I. Analysis of Factors Giving the Opportunity for Implementation of Innovations on the Example of Manufacturing Enterprises in the Silesian Province. *Sustainability* **2019**, *11*, 5850. [CrossRef]
39. Krzos, G. Measures of the success of the project manager and projects co-financed by the EU. In *Selected Aspects of Managerial Work*; Cyfert, S., Ed.; Scientific Papers No. 187; Publishing House of the University of Economics in Poznań: Poznań, Poland, 2011.
40. Miciuła, I.; Wojtaszek, H. Automatic hazard identification information system (AHIIS) for decision support in inland waterway navigation. *Procedia Comput. Sci.* **2019**, *159*, 2313–2323. [CrossRef]
41. Orzechowski, R. Business-IT Alignment in Poland, *E-Mentor (E-Biznes)* Number 1 (23). February 2008. Available online: https://www.e-mentor.edu.pl/artykul/index/numer/23/id/520 (accessed on 27 October 2022).
42. Pawlak, M. *Project Management*; Polish Scientific Publishers PWN: Warsaw, Poland, 2006.
43. Wysocki, W.; Orlowski, C. A multi-agent model for planning hybrid software processes. *Procedia Comput. Sci.* **2019**, *159*, 1688–1697. [CrossRef]
44. Zaydi, M.; Nassereddine, B. DevSecOps practices for an agile and secure it service management. *J. Manag. Inf. Decis. Sci.* **2020**, *23*, 1–16.
45. Akbar, M.A.; Smolander, K.; Mahmood, S.; Alsanad, A. Toward successful DevSecOps in software development organizations: A decision-making framework. *Inf. Softw. Technol.* **2022**, *147*, 106894. [CrossRef]
46. Myrbakken, H.; Colomo-Palacios, R. DevSecOps: A multivocal literature review. In *Proceedings of the International Conference on Software Process Improvement and Capability Determination*; Springer: Berlin/Heidelberg, Germany, 2017; pp. 17–29.
47. Wysocki, W. Agents of RUP Processes Model for IT Organizations Readiness to Agile Transformation Assessment. In *Intelligent Information and Database Systems*; Nguyen, N.T., Ed.; Springer International Publishing: Berlin/Heidelberg, Germany, 2017; pp. 777–786.
48. Piwowar-Sulej, K. Project management as a desirable managerial competence. In *Selected Aspects of Managerial Work*; Cyfert, S., Ed.; Scientific Papers No. 187, Wyd; Poznań University of Economics: Poznań, Poland, 2011.
49. West, D. *Water-Scrum-Fall Is the Reality of Agile for Most Organizations Today*; Forrester Research: Cambridge, MA, USA, 2011; Volume 26.
50. Attri, R.; Grover, S.; Dev, N.; Kumar, D. Analysis of barriers of total productive maintenance (TPM). *Int. J. Syst. Assur. Eng. Manag.* **2013**, *4*, 365–377. [CrossRef]
51. Akbar, M.A.; Sang, J.; Nasrullah; Khan, A.A.; Mahmood, S.; Qadri, S.F.; Hu, H.; Xiang, H. Success factors influencing requirements change management process in global software development. *J. Comput. Lang.* **2019**, *51*, 112–130. [CrossRef]
52. Poppendieck, M.; Poppendieck, T. *Lean Software Development: An Agile Toolkit*; Addison Wesley: Boston, MA, USA, 2013.

53. Zolnowski, A.; Anke, J.; Gudat, J. Towards a cost-benefit-analysis of data-driven business models. In Proceedings of the 13th International Conference on Wirtschaftsinformatik, St. Gallen, Switzerland, 12–15 February 2017.
54. Iriarte, C.; Bayona, S. IT projects success factors: A literature review. *Int. J. Inf. Syst. Proj. Manag.* **2020**, *8*, 49–78. [CrossRef]
55. Mccallum, A.; Kamal, N. *Text Classification by Bootstrapping with Keywords, EM and Shrinkage*; ACL: Stroudsburg, PA, USA, 2001.
56. Devlin, J.; Ming-Wei, C.; Kenton, L.; Toutanova, C. BERT: Pre-Training of Deep Bidirectional Transformers for Language Understanding. *arXiv* **2019**, arXiv:1810.04805. [CrossRef]

electronics

MDPI

Article

Knowledge Mining of Interactions between Drugs from the Extensive Literature with a Novel Graph-Convolutional-Network-Based Method

Xingjian Xu *, Fanjun Meng and Lijun Sun

College of Computer Science and Technology, Inner Mongolia Normal University, Hohhot 010022, China
* Correspondence: xingjian@imnu.edu.cn

Abstract: Interactions between drugs can occur when two or more drugs are used for the same patient. This may result in changes in the drug's pharmacological activity, some of which are beneficial and some of which are harmful. Thus, identifying possible drug–drug interactions (DDIs) has always been a crucial research topic in the field of clinical pharmacology. As clinical trials are time-consuming and expensive, current approaches for predicting DDIs are mainly based on knowledge mining from the literature using computational methods. However, since the literature contain a large amount of unrelated information, the task of identifying drug interactions with high confidence has become challenging. Thus, here, we present a novel graph-convolutional-network-based method called DDINN to detect potential DDIs. Combining cBiLSTM, graph convolutional networks and weight-rebalanced dependency matrix, DDINN is able to extract both contexture and syntactic information efficiently from the extensive biomedical literature. At last, we compare our DDINN with some other state-of-the-art models, and it is proved that our work is more effective. In addition, the ablation experiments demonstrate the advantages of DDINN's optimization techniques as well.

Keywords: knowledge mining; drug–drug interaction; graph convolutional network; self-attention; deep learning

Citation: Xu, X.; Meng, F.; Sun, L. Knowledge Mining of Interactions between Drugs from the Extensive Literature with a Novel Graph-Convolutional-Network-Based Method. *Electronics* **2023**, *12*, 311. https://doi.org/10.3390/electronics12020311

Academic Editors: Agnieszka Konys and Agnieszka Nowak-Brzezińska

Received: 29 November 2022
Revised: 4 January 2023
Accepted: 6 January 2023
Published: 7 January 2023

1. Introduction

When treating patients with drugs, doctors often use multiple drugs at the same time because the effectiveness of one drug is limited. Particularly in the case of severe and chronic diseases, many different drugs have to be used at the same time to treat lesions, relieve pain, prevent complications or are used for other medical reasons. As drugs are taken together, complex biochemical reactions may take place in vivo, resulting in unpredictable results, which are called drug–drug interactions (DDIs) [1]. In terms of their side effects, DDIs can be basically divided into two types: beneficial and adverse [2]. A beneficial drug interaction can improve patient outcomes, whereas adverse drug interactions can pose serious threats to patients' health, reducing the effectiveness of drugs, prolonging the course of disease, and even putting patients' lives at risk. Therefore, the identification of possible DDIs has always been a crucial research topic in clinical pharmacology [3]. A number of databases were constructed by researchers in order to document the DDIs found, such as DrugBank [4], DDInter [5], TwoSides [6] and SFINX [7].

The traditional method of obtaining DDIs involves the use of clinical trials, and these are time-consuming, expensive, and often have serious ethical implications [8]. In spite of the fact that in vivo trials remain the most accurate method for identifying DDIs, the disadvantages described above severely limit the pace at which DDIs can be identified. In recent years, many biomedical research papers have been published at high frequencies, which led researchers to study how meaningful information can be extracted from these papers. Clearly, manually curation is not feasible, so machine learning or other knowledge-mining-based methods must be employed [9]. The two examples in Figure 1 illustrates

DDI extraction from drug-related text sentences, for example, the published literature or drug descriptions. For sentence S1, the DDI type of Fluoxetine and Phenelzine is "Advice" (see Section 3.1 for a description of the specific DDI types). For sentence S2, the DDI type of PGF2alpha and Oxytocin is "Effect". Although these automated prediction methods may output false-positive and true-negative DDI predictions, they nevertheless became a mainstream approach for the DDI prediction task due to their efficacy. If it is necessary, researchers may then validate these high-confidence DDIs produced by automated DDI prediction methods clinically [10].

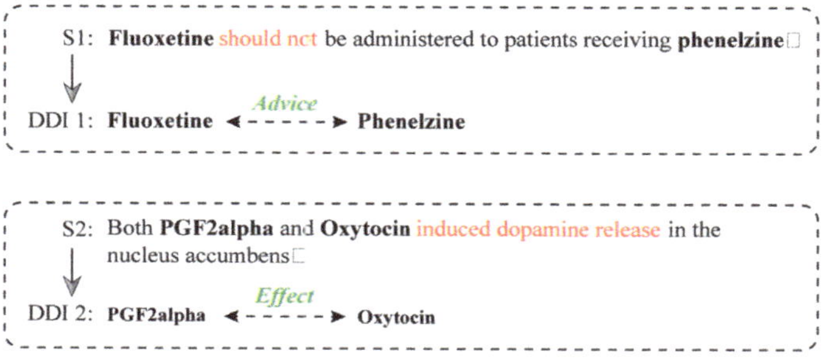

Figure 1. Two examples illustrating DDI extraction from drug-related text sentences.

Initially, there are mainly two kinds of traditional machine learning methods for automatically extracting DDI: pattern-based and feature-based ones. In pattern-based methods, experts with extensive domain knowledge are required to propose some recognizable patterns based on their own experiences [10]. Later, a number of feature-based methods are proposed, among which the best-performing ones are based on support vector machine (SVM), for example, FBK-irst [11] and NIL_UCM [12]. In general, machine learning methods that are based on features have experienced great success and are more portable than those that rely on patterns [13]. There is, however, an inherent disadvantage to these methods, which is that they heavily rely on tedious feature engineering and redundant feature selection, and defining the feature set in a supervised manner will also limit the identification of other valuable patterns. Moreover, as these methods are based on traditional machine learning models and are not capable of extracting deep features from input data, they will become much less effective when dealing with large data sets [14].

Deep learning can solve the above problems well, and it has been applied widely and successfully in a variety of other fields as well, such as in the field of computer vision, natural language processing (NLP) and speech recognition [15,16]. Deep learning methods based on graph structure have been proposed and successfully applied to the DDI prediction task [17,18]. The first wave of popular deep-learning-based DDI detection methods rely primarily on sequence-based networks, for example, the convolutional network (CNN) and recurrent neural networks (RNNs) [19]. In most cases, these methods can achieve better results than methods based on traditional machine learning models. However, the main drawback of this approach is that they cannot handle long or complex sentences in the literature's text or other information sources, mainly because of the inherent characteristics of CNN or RNN. The researchers then proposed dependency-based methods, which can be used to extract corpora that contain multiple long and complex sentences, incorporating structural information into a neural architecture for DDI prediction. As many DDI extraction corpora contain a large number of long sentences ($\geq$150 words) [20], dependency-based methods obviously have advantages over sequence-based ones. In regard to all these methods, there are still some challenges to overcome: (1) These methods only use the literature's text as input data and lack relevance to other information extraction sources; (2) due to the difficulty of parallelizing existing dependencies-based methods, such as tree-LSTM, they are often inefficient and have a disappointing runtime performance; (3) as their network is essentially linear, most of these methods are only capable of predicting the interaction of one pair of drugs at a time, which severely limits their practical usage.

In order to resolve the issues outlined above, we propose DDINN (DDI Neural Network) for the DDI prediction task, which is a novel graph-convolutional-network-based method featured by the self-attention mechanism for pruning. Our method utilizes contextual features of sentences as vertices and syntactic features as edges to construct a graph, which will be fed to GCN layers sequentially. DDINN can capture more neighborhood information of the graph more effectively by stacking the convolution layer. In particular, we rebalance the weights of each edge via a self-attention mechanism. Thus, DDINN is able to exploit both the context and structure of the input sentence to the maximum extent possible. Our final step was to train and evaluate the DINN model on the dominant DDI extraction dataset from SemEval-2013 Task 9 of the DDIExtraction 2013 dataset [21]. Validation experiments and ablation study show the effectiveness of DDINN and its superiority compared to other similar methods. Performance assessments are also conducted on the DDINN model's components to show the improvement compared with other traditional methods.

To summarize, we can state the following as our main contribution:

- DDINN: Combining graph convolutional networks with recurrent networks, we propose a novel deep learning method, DDINN, that can effectively utilize the contextual and syntactic information of input literature text at the same time.
- Weight-rebalanced dependency matrix: On the basis of dependency-aware embedding representation and self-attention-based pruning strategy, we propose a method for rebalancing the weights of all edges in the dependency matrix for GCN.
- Extensive experiments: The experimental results show that our model can predict DDI with the best F-score and has a better performance in comparison with state-of-the-art models.

Following is the outline of the remainder of this paper. In Section 2, we review the characteristics of existing DDI extraction approaches and briefly summarize the improvements made in the DDINN method proposed here to overcome their shortcomings. Section 3 describes the implementation specifics of DDINN in detail. Then, the experiments and analysis of their results are presented in Sections 4 and 5. As a final point, in Section 6, our conclusions regarding the entire work of DDINN is presented.

2. Related Works

Currently, there are three main types of DDI extraction methods: feature-based, kernel-based, and deep learning neural-network-based methods. The representative methods below will serve as the baseline for further experimental validation.

2.1. Feature-Based Methods

Feature-based methods aim to find a way to distinctively represent data characteristics using some feature representation techniques, which are called feature engineering. This process involves transforming the original data into feature vectors that can better express the essence of the problem. Then, classifiers are trained based on various linguistic features extracted from the data. For example, UTurku [22] uses dependency graph features to mine entity associations and it achieved an F-value of 59.4% in the DDIExtraction 2013 competition. WBI-DDI [23] proposes a two-stage method that first classifies the results using multiple methods including APG (all path graph), Moara, SL (shallow linguistic), and TEES (urku event extraction system) separately, and then it votes on these classification results to obtain the best classification result, which achieved an F-value of 60.9%. FBK-irst [11] constructs a combined kernel classifier by combining the feature kernel, shallow linguistic kernel and closure tree kernel for binary classification, deleting negative examples and then constructing a combined kernel classifier to achieve multi-classification, which scored 65.1% in the DDIExtraction 2013 competition F-value.

2.2. Kernel-Based Methods

The purpose of kernel-based methods is to find and learn the mutual relationships in a set of data. Widely used kernel methods include support vector machines, Gaussian processes, etc. Kernel-based methods are an effective way to solve nonlinear pattern analysis problems. The core idea is as follows: First, the original data are embedded into a suitable high-dimensional feature space by some nonlinear mapping; then, the patterns are analyzed and processed in this new space using a generic linear learner. Feature- and kernel-based DDI extraction can achieve better results than the rule-based extraction, and these methods have been the mainstream method for DDI extraction for a long period of time. The disadvantage is that they are time-consuming and laborious for performing multiple complex feature extractions, so the extraction's performance is bottlenecked and cannot be improved significantly. In 2015, Kim et al. [13] constructed kernel functions by employing a set of lexical and syntactic features based on a series of lexical and syntactic features with an F-value of 67% in DDIExtraction 2013. In 2016, Zheng et al. [24] constructed kernel functions for a graph kernel with an F-value of 68.4%. This method became the best model among the current methods using feature-based and kernel functions. It is similar to our approach in that semantic and syntactic information is integrated. However, the performance of previous studies has not been satisfactory since they have only looked at the shortest dependency path (SDP).

2.3. Neural-Network-Based Methods

Neural networks have an extremely strong feature representation capability. Thus, deep learning methods have a significant advantage over other machine learning methods in terms of accuracy and do not require a complex pre-processing process. In classification tasks, neural networks can be treated as classifiers capable of automatically extracting features. With the rapid development of deep learning, many neural-network-based DDI extraction methods emerged in recent years and have excellent performances in DDI extraction task over traditional feature- or kernel-based methods. The relationship between drug entities can be extracted using neural networks in two basic ways: sequence-based and dependency-based methods.

Different neural architectures, including CNNs and RNNs, are used in sequence-based models. Quan et al. [25] proposed a multichannel convolutional neural network (MCCNN) for automated biomedical relation extraction. As a result of MCCNN's performance on the DDIExtraction 2013 challenge dataset, MCCNN was reported to achieve an overall F-score of 70.2% compared to the linear SVM-based standard system (e.g., 67.0%). Sun et al. [26] proposed a recurrent hybrid convolutional neural network (RHCNN) for DDI extraction from the biomedical literature in which semantic embeddings and position embeddings are both used to represent the texts mentioning two drug entities. RHCNN is reported to achieve DDI automatic extraction with a micro F-score of 75.48%. In addition to CNN-based models, RNN-based ones have also been adopted for extracting DDI effectively. For example, in GGNN [27], textual drug pairs are encoded with convolutional neural networks, while molecule pairs are encoded with graph convolutional networks. DDI relations are then extracted by concatenating the outputs of these two networks. Sahu et al. [28] present three long short-term memory (LSTM) network models for mining DDI relation from biomedical text, namely B-LSTM, AB-LSTM and Joint AB-LSTM. The experimental results on the DDIExtraction2013 dataset show that the Joint AB-LSTM model produces reasonable performances with an F-score of 69.39%.

Dependency-based neural network architectures are constructed using structural information of a given sentence. It is common for the DDI extraction corpus (literature text or drug description, etc.) to contain multiple long and complex sentences, and the longest sentence may contain over 150 words, so using only sequence-based networks for extraction is extremely challenging. It is therefore very helpful to introduce structural knowledge (such as dependency trees) into the DDI extraction task. For example, Zhao et al. [29] present a syntax convolutional neural network (SCNN) for DDI extraction. In SCNN, a new syntax word embedding method is proposed that incorporates syntactic sentence information.

2.4. Improvements Made by DDINN

In order to address the shortcomings of the approaches discussed above, we made considerable improvements with respect to DDINN for the DDI extraction task:

1. To avoid the lack of representation depth caused by using traditional sequence-based or dependency-based networks alone, DDINN combines the contextual features of sentences and syntactic features together to construct a graph, which will be fed to GCN layers sequentially. By stacking the convolution layer in GCN, DDINN is able to capture more neighborhood information about the graph.
2. Traditional GCN model only allows edges between nodes with a weight of 0 or 1. There are many complex interactions between drugs in the DDI extraction task that cannot be adequately described in this manner. Thus, we propose a new method to rebalance the weights of all edges in the dependency matrix of GCN based on the dependency-aware embedding representation, so that the weights can take values ranging from 0 to 1.
3. Full dependency trees are used to avoid losing key information during the extraction of syntactic features. Specifically, we propose an attention-based pruning mechanism to minimize the loss of important cues in the full dependency tree. Unlike the rule-based or SDP-based pruning algorithms used in previous studies, this pruning strategy can be used to achieve selective pruning with different weight ratios and to reflect the different strengths of the relatedness between nodes.

3. Materials and Methods

3.1. Problem Definition

Words in the literature's text can be denoted as $\mathbf{X} = [\mathbf{x_1}, \mathbf{x_2}, \cdots, \mathbf{x_i}, \cdots, \mathbf{x_n}] \in \mathbb{R}^{d \times n}$, where n denotes the total number of words and $\mathbf{x_i} \in \mathbb{R}^d$ denotes the d-dimensional i-th embedded token. Drugs described in this text can be denoted as $\mathscr{D} = \{D_k \mid k \in [1, n]\}$. The mapping relationship between words and drugs is already known, and it can be represented as $R_{xd}(\mathbf{x_i}, D_k), R_{xd} \subset \{0, 1\}$. If $R_{xd}(\mathbf{x_i}, D_k) = 0$, it means that there is no relationship between $\mathbf{x_i}$ and D_k; otherwise, it shows a positive relationship.

All drug entities can be annotated with the following five drug–drug interaction relationship types [21]:

1. Advice: Describes recommendations when two drugs are used together;
2. Mechanism: Describes the pharmacokinetic mechanisms of two drug entities;
3. Effect: The result of the interaction of two drugs is clearly stated;
4. Int: Indicates some relationship between the two drugs, but it does not define the specific type of relationship.
5. Negative: Indicates that there is no interaction between the two drugs.

Thus, in the problem of DDI relation extraction, $\mathscr{C}$ represents the overall prediction classes as follows.

$$\mathscr{C} = \{Advice, Mechanism, Effect, Int, Negative\} \tag{1}$$

Now, the problem of DDI predication can be defined as follows. Given $\mathbf{X}$ and the R_{xd}, our DDINN method will predict drug relation set $\mathscr{R}_D$.

$$\mathscr{R}_D = \{R_{dd}(D_a, D_b) \mid a \in [1, n], b \in [1, n], a \neq b\}, \tag{2}$$

$$R_{dd}(D_a, D_b) \in \mathscr{C} \tag{3}$$

3.2. Overview of Architecture

The outline of the overall architecture of our novelly proposed DDINN model is illustrated in Figure 2. Firstly, each word in the input literature text is transformed into a token vector that consists of the embeddings of the word itself, its dependency, part of speech, and distance in sentences. These embedding vectors are concurrently sent to cBiLSTM and the weight-rebalanced dependency parser to extract the contextual and syntactic features, respectively. Then, DDINN constructs a graph, which is fed to the GCN layers, by converting contextual features into graph vertices and syntactic features into graph edges. Consequently, the representations of drug pairs and sentences consisting of other remaining words are obtained by masking the output of GCN layers. At the last step, the PPI prediction classifier, which is the final output of DDINN, is generated by concatenating the representations above sequentially and passing them to the softmax and linear layers. Below, we will provide a detailed description of the process for building the DDINN model.

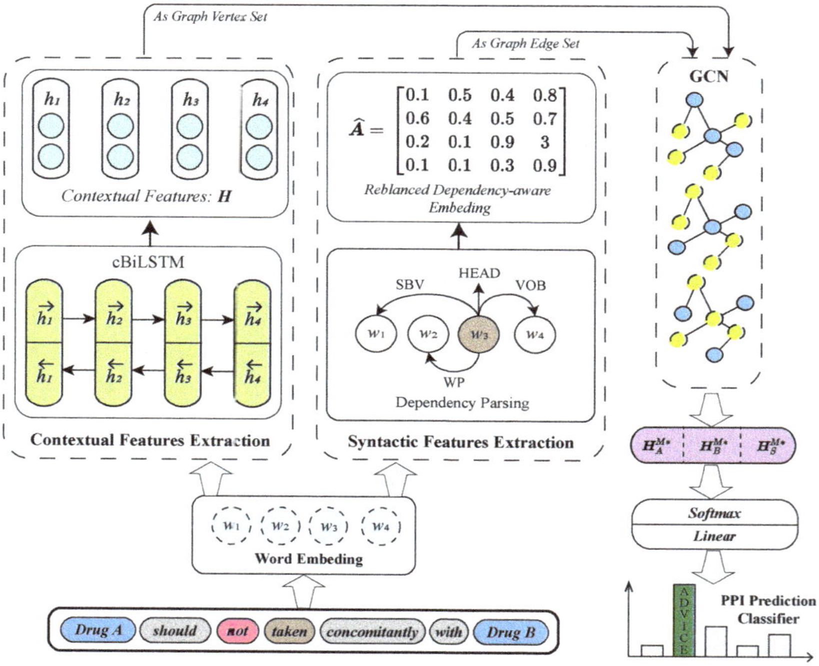

Figure 2. Architecture overview of our proposed DDINN method.

3.3. Contextual Feature Representations

In our work, the contextual and syntactic representation of sentences is used to analyze the literature's text. The concept of a bag-of-words model is often used in traditional sentiment analysis, where a document is viewed as a collection of terms or combinations of short compound words regardless of grammatical and word order. As a result, when processing sentences, word vectors are often used. It is very common for obtaining word embeddings by pre-training, and the word representation obtained in this way is often independent of the sentence's context. However, due to polysemy, the word itself can have different meanings in different contexts. Therefore, it is impossible to accurately describe the contextual meaning of the word itself in a certain context only by using the word vector. The use of context-sensitive vectors can enhance the representations of semantic relations between sentences [30].

Our solution to these issues involves the use of contextual bidirectional long short-term memory recurrent neural networks (cBiLSTM). In cBiLSTM, the contextual information extraction problem is viewed as a sequence classification problem, and a type of pooling will be performed to obtain sentence-level polarity after using RNNs as discriminative binary classifiers. There are two separate layers of LSTM in cBiLSTM. As for word token $\mathbf{x_i}$, these two LSTM layers are responsible for capturing both forward and reverse contextual information, respectively. By estimating the probability of a word based on its complete left and right contexts, the networks process the bi-directional period adjacent to the position of a word in the sentence. Therefore, the cBiLSTM is able to understand the contextual meaning of words more effectively than traditional network models.

3.3.1. Word Embedding

The first step is the vectorization of words to obtain $\mathbf{X}$. Considering that the word T_i in it does not necessarily have a mapping relationship in $\mathbf{X}$, in this case, this paper will use a uniform distribution on interval $[-0.5, 0.5]$ for its random initialization. Let $\mathbf{x}(T_i)$ denote the vector of word T_i; this representation rule is described as follows:

$$\mathbf{x}(T_i) = \begin{cases} \mathbf{x}_i, & T_i \in \mathbf{X}, \\ Uniform([-0.5, 0.5])^d, & T_i \notin \mathbf{X}. \end{cases} \tag{4}$$

3.3.2. Construct cBiLSTM

Later, word vector $\mathbf{x}$ will be processed by cBiLSTM, which will produce the forward $\overrightarrow{h_i}$ and backward $\overleftarrow{h_i}$ for word vector $\mathbf{x_i}$.

$$\overrightarrow{h_i} = LSTM(x_i, \overrightarrow{h_{i-1}}) \tag{5}$$

$$\overleftarrow{h_i} = LSTM(x_i, \overleftarrow{h_{i-1}}) \tag{6}$$

Then, we can calculate the contextual feature, h_i, of word vector $\mathbf{x_i}$ by concatenating $\overrightarrow{h_i}$ and $\overleftarrow{h_i}$ as follows.

$$h_i = [\overrightarrow{h_n}; \overleftarrow{h_i}] \in \mathbb{R}^d \tag{7}$$

At the final step of this section, all contextual information (denoted as $\boldsymbol{H}$) of the sentences will be fed to the later networks for parsing.

$$\boldsymbol{H} = (h_1, h_2, \cdots, h_n) \in \mathbb{R}^{n*d} \tag{8}$$

3.4. *Syntactic Feature Representations*

Dependent syntactic analyses aim to parse the text into a dependent syntactic tree. This is performed by obtaining the dependencies and association paths between words. Thus, the method gives the model a better understanding of natural language by extracting text features based on sentence structure. In addition to contextual information, syntactic information is also important. In fact, contextual and syntactic features complement each other. Here, we adopt the graph convolutional network (GCN) [31,32] to extract syntactic information. The syntactic structure of texts is more similar to that of graph data. For such non-Euclidean spatial data, traditional deep learning models do not effectively exploit or may even corrupt its intrinsic information. By extending convolution to graph-structured data, GCN is proposed, which has the ability to model common graph data in reality, and then it explores the complex relationships in it. In this paper, we use the full dependency tree as the input of the graph convolutional network and introduce the attention mechanism during the training process so as to selectively focus on the dependency substructure.

3.4.1. Construct Dependency Matrix

Based on the dependency structure, we first generate the corresponding adjacency matrix $A \in \mathbb{R}^{n \times n}$. Most traditional dependency-tree-based networks do not employ full dependency trees to convey syntactic information from sentences. These methods often use 1 or 0 to encode syntactic dependencies between words, which indicate that the elements in the adjacency matrix A take values of only 1 or 0. However, this approach ignores the impact of different dependencies on the target task and introduces other redundant features. As a result of such strategies, which are normally determined by rule-based preprocessing, crucial information may also be lost [33,34].

To address the problems above, we introduce two more steps: a dependency-aware embedding representation method based on dependency relations in the layers and self-attention-based pruning. The dependency-aware embedding representation not only focuses on the dependency correlations between words but also considers the dependency tag types and the semantics of the words associated with the tags. The following paragraphs provide the implementation details of the dependency-aware embedding representation method.

For $\mathbf{X}$, if there is a dependency relationship between word i and j and the dependency type is φ, the corresponding dependency-type embedded vector is $\aleph_\varphi \in \mathbb{R}^{d_\varphi \times 1}$, and the dependency relationship between these two words can be embedded represented as follows:

$$a_{ij} = Sigmod(Avg[\mathbf{x_i}, \mathbf{x_j}] \times \omega_\varphi \times N_\varphi + b_\varphi) \tag{9}$$

where ω_φ and b_φ are trainable parameters, Avg denotes the average value function, $Sigmod$ denotes the activation function and $\aleph_\varphi$ is initialized before the model's training and will be updated during the training process. Thus, if words i and j have syntactic dependency, the elements in matrix A can be represented as $A_{ij} = a_{ij}$; otherwise, $A_{ij} = 0$.

3.4.2. Self-Attention-Based Pruning

Then, in order to exploit syntactic dependencies more fully, self-attention-based pruning is employed to assign weights to all edges in the dependency graph. By incorporating the self-attention mechanism, we transform A into a soft adjacent matrix $\widehat{A}$. Self-attention has the advantage of noticing the relationship between different positions in a single sequence. Thus, the edge weights of all node pairs in the graph are reassigned regardless of whether they are directly or indirectly connected. This is why we call output $\widehat{A}$ as a *soft* adjacent matrix.

In the specific calculation process, we use query and key pairs of $\mathbf{x}_i$ as self-attention function parameters. By employing multi-head attention [35,36], we were able to capture a different context from multiple perspectives. In particular, the soft adjacent matrix, $\widehat{A}$, can be calculated as follows:

$$\widehat{A} = Softmax\left(\frac{Q{W_h}^Q \times (K{W_h}^K)^T}{\sqrt{d}}\right) \tag{10}$$

where $Softmax$ is the activation function and Q and K are the features of the previous convolutional layer $h^{(l-1)}$. ${W_h}^Q \in \mathbb{R}^{d*d}$ and ${W_h}^K \in \mathbb{R}^{d*d}$ are used for projection parameters, where h denotes the h-th head in H, which is defined in Equation (8).

3.4.3. Construct GCN

Then, contextual information H, which is the output of Equation (8), and adjacency matrix $\widehat{A}$ will be fed into the l-level GCN:

$$H^{(l)} = Relu(\widehat{D}^{-\frac{1}{2}}\widehat{A}_D\widehat{D}^{-\frac{1}{2}}H^{(l-1)}W^{(l-1)} + b_l) \tag{11}$$

where *Relu* is the activation function, $\widehat{A}_D$ is the edge matrix of $\widehat{A}$, $\widehat{D}$ denotes the degree matrix of $\widehat{A}_D$, $H^{(l-1)}$ denotes the node features of the $(l-1)$-th level GCN (when $l = 1$, $H^{(l-1)} = H$) and $W^{(l-1)}$ denotes the weight matrix of the $(l-1)$-th level GCN.

Finally, in order to further enhance the generalization capability of the model, the output of the GCN layers above will be processed by a pooling layer, dropout layer, and Relu layer:

$$H^* = \omega \times Relu(Dropout(Pooling(H^{(l)}))) + b \tag{12}$$

where H^* is the final output of GCN, which holds the contextual and syntactic feature information of text $\mathbf{X}$ at the same time.

3.5. Extract DDI

3.5.1. Extract Masked Representations

After completing the above steps, we have hidden representations of each word in the input literature text, which can be simply denoted as $\mathbf{w}_i$ for word i. The problem in this step can be defined as follows: Within the input word representations $[\mathbf{w}_1, \cdots, \mathbf{w}_n]$, drug A is mapped to $\mathbf{w}_a$, and drug B is mapped to $\mathbf{w}_b$; we want to extract the relationship between drug A and B. In order to achieve this, we first calculate the masked representations of drug A, drug B, and the sentence including other words (i.e., words except for $\mathbf{w}_a$ and $\mathbf{w}_b$), which are denoted as H_A^{M*}, H_B^{M*}, and H_S^{M*}, respectively. The calculation process is as follows:

$$H_S^{M*} = MaxPooling(Mask_S(H^*)) \tag{13}$$

$$H_A^{M*} = MaxPooling(Mask_A(H^*)) \tag{14}$$

$$H_B^{M*} = MaxPooling(Mask_B(H^*)) \tag{15}$$

where H^* is the output of Equation (12), $MaxPooling$ denotes an activation function that can transform n output vectors to only one vector, i.e., $MaxPooling \in \mathbb{R}^{n \times d} \to \mathbb{R}^d$. $Mask_A$, $Mask_B$ and $Mask_S$ denote functions that can select only representations for drug A, drug B and sentences formed by the remaining words, respectively.

3.5.2. Construct DDI Classifier

Finally, we can predict the DDI by using a classifier. Firstly, we concatenate the masked representations above and then feed them to a fully connected layer [37]. The final result of this classifier is denoted as H_{Final}, which is calculated as follows:

$$H_{Final} = FC(Concat(H_A^{M*}, H_B^{M*}, H_S^{M*})) \tag{16}$$

where *FC* is the fully connected layer, and *Concat* is the function that concatenates all its parameters. H_{Final} will then be fed into a linear layer and a softmax layer to output the probability distribution for the DDI relationship between these drugs [38,39]:

$$P = Softmax(Linear(H_{Final})) \tag{17}$$

4. Experiments

4.1. Dataset

In this paper, we evaluate DDINN on the DDIExtraction2013 dataset [20], which is most widely used when comparing the performances of different DDI extraction algorithms. Prior to 2011, there were relatively few studies related to the DDIExtraction task due to the lack of standard datasets, and almost all of those studies were rule-based. These rules have to be formulated by professionals, and the DDI extraction is achieved by matching the DDI expressions in the sentences with the formulated rules. This approach is more effective for composing simple sentences. However, for long and complex sentences, especially those with many subordinate clauses, the performance of this method is much less effective. In 2011, the SemEval 2011 competition established the DDIExtraction subtask and provided the standard DDIExtraction dataset for the first time. Subsequently, in 2013, the SemEval 2013 competition supplemented and improved the dataset, which can be referred to as DDIExtraction2013.

The text corpus of this dataset has two sources: (1) literature abstracts in the discipline of drug interactions downloaded from the MedLine (https://medline.com/, accessed on 20 February 2022) medical literature retrieval system and (2) articles studying drug interactions downloaded from the DrugBank (https://drugbank.com/, accessed on 23 February 2022) online database. A total of 18,491 pharmacological substances and 4999 drug–drug interactions were manually annotated in this DDI corpus, which consists of 1017 documents (784 paragraphs from DrugBank and 233 abstracts from MedLine). All documents contain 5806 sentences and 127,653 tokens. The details of the DDIExtraction2013 dataset are listed in Table 1.

Table 1. The statistics information of DDIExtraction 2013 dataset.

Type	Training	Test	Total
Advice	826	221	1047
Mechanism	1319	302	1621
Effect	1687	360	2047
Int	188	96	284
Negative	23,772	4737	28,509

4.2. Training

In the training process, cross entropy cost function and L^2 regularization are used as the optimization objective. The cross entropy is defined as follows:

$$l_i = -\ln {Y_i}^T P_i \tag{18}$$

where Y_i denotes the one-hot representation of the i-th instance label, and P_i is the model output, which is defined in Equation (17). For a mini batch $\mathcal{M} = [\mathbf{X}_1, \mathbf{X}_2, \cdots, \mathbf{X}_M]$, we defined the optimization objective as follows:

$$\mathcal{J}(\theta) = \frac{1}{|\mathcal{M}|} \sum_{i=1}^{|\mathcal{M}|} l_i + \lambda \|\theta\|_2^2 \tag{19}$$

where θ includes all the parameters in our model. At the final step, parameter θ in the objective function, $\mathcal{J}(\theta)$, is optimized with Nadam [40], which is an algorithm that performs first-order gradient optimization on an efficient stochastic objective function.

The models are randomly initialized at the beginning, so if a higher learning rate is selected at this point, the model may become unstable or oscillate, while a lower learning rate will result in a slower convergence speed. The learning rate scheduler with exponential decay [41] is used to control the dynamic change of the learning rate during the training process (see Figure 3). It can slow down overfitting in the initial stages and maintain the stability of the deep layer. Upon the completion of training, the model that can predict interactions between two drugs is obtained.

4.3. Experiment Setup

The DDINN is implemented with PyTorch (https://pytorch.org/, accessed on 10 January 2022) and open-sourced at Github (https://github.com/xingjianxu/DDINN, accessed on 10 January 2022). We use pre-trained word embeddings from GloVe [42] combined with PMCVec [43,44], which is based on unlabeled biomedical texts from PubMed (https://pubmed.ncbi.nlm.nih.gov/, accessed on 10 January 2022) and PubMed Central (https://www.ncbi.nlm.nih.gov/pmc/, accessed on 10 January 2022). In order to obtain the dependency tree, dependency label, and POS tag of each word, we use the Stanford Parser (https://nlp.stanford.edu/software/lex-parser.shtml, accessed on 10 January 2022). All experiments are conducted with two RTX 3090 GPUs. The detailed parameters are listed in Table 2.

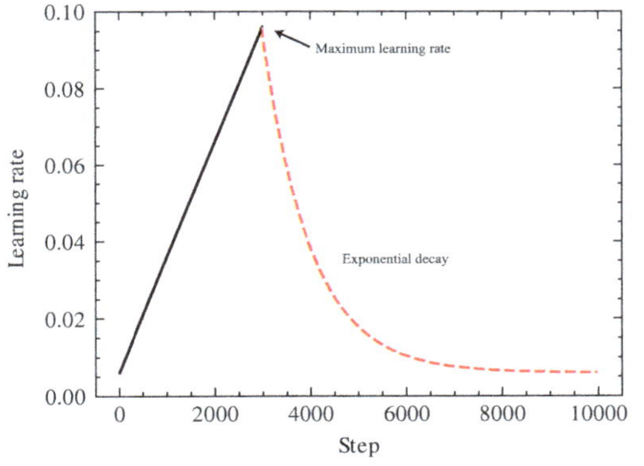

Figure 3. Learning rate exponential decay.

Table 2. The main hyperparameter settings used in DDINN implementations and evaluation experiments.

Module	Parameter	Value
Word embedding	Size (per word)	200
Stanford Parser	Word dimension	200
	Dependency dimension	20
	Distance embedding	20
cBiLSTM	Dimension	400
	Dropout	0.4
GCN	Dimension	300
	Layer	3
Attention	Dimension	300
	Dropout	0.2
Training	Epoch	30
	L^2 regularization	5×10^{-5}
	Batch	30
	Maximum learning rate	0.005

4.4. Assessment Metrics

In order to evaluate the quality of prediction results, micro-precision, micro-recall, and micro-F score are employed as assessment metrics, which are denoted as P_{micro}, R_{micro}, and F_{micro}, respectively. As described in Table 1, we can define the prediction classes. We set $\mathscr{D}$ as

$$\mathscr{D} = \{Advice, Mechanism, Effect, Int, Negative\} \tag{20}$$

and these metrics above can be calculated as follows:

$$P_{micro} = \frac{\overline{TP}}{\overline{TP} + \overline{FP}} = \frac{\sum_{i=1}^{n} TP_i}{\sum_{i=1}^{n} TP_i + \sum_{i=1}^{n} FP_i} \tag{21}$$

$$R_{micro} = \frac{\overline{TP}}{\overline{TP} + \overline{FN}} = \frac{\sum_{i=1}^{n} TP_i}{\sum_{i=1}^{n} TP_i + \sum_{i=1}^{n} FN_i} \tag{22}$$

$$F_{micro} = \frac{2 \times P_{micro} \times R_{micro}}{P_{micro} + R_{micro}} \tag{23}$$

where TP_i denotes the true positives in the prediction class $i \in \mathscr{D}$, FP_i denotes the false positives and FN_i denotes the false negatives.

4.5. Baselines

The following two kinds of methods are selected as the baseline for evaluating the performance of DDINN in this paper:

- Traditional statistical-model-based methods, including UTurku [22], FBK-irst [11] and WBI-DDI [23]: Such kinds of methods mainly use features and kernel functions to predict the DDI relationship.
- Deep learning neural-network-model-based methods, including MCCNN [25], Joint AB-LSTM [28], GGNN [27], RHCNN [26] and GCNN [45]: The application of neural networks significantly improved prediction performances compared to methods based on traditional statistical models.

5. Results and Discussion

5.1. Performance Comparison

As shown in in Table 3, we compare the performance of our DDINN method to those of the other eight baseline methods. For each method, the F_{micro} score for four kinds of DDI types and the overall precision, recall and F_{micro} score are listed. The performance statistics are obtained by conducting test experiments on the DDIExtraction2013 dataset, except for UTurku, GGNN and GCNN, which are directly cited from their original papers. This is because we cannot find available codes or runnable binaries for these methods, and they all conducted the performance test on the DDIExtraction2013 dataset. The highest values in each test are marked in bold, and the second best ones are marked underlined.

In comparison with all baseline methods, except for the PPI type of Int, DDINN exhibited the highest performance scores. The main reason for this is that DDINN requires a relatively large amount of training data, and training data with the Int PPI type only rarely (1.68% in total training data) appears in the DDIExtraction2013 dataset (see Table 1). The experimental results proved that the series of optimization used in DDINN finally worked and successfully improved the quality of the results of the DDI prediction task.

The training process of this model on the DDIExtraction2013 dataset is shown in Figure 4, which shows the changes in the precision, recall, and the F_{micro} score values over the epoch. From the figure, it can be seen that all these values improve faster in the early stage of the training, and then they fluctuate continuously to find the local optimal value; finally, they gradually converge to smooth values.

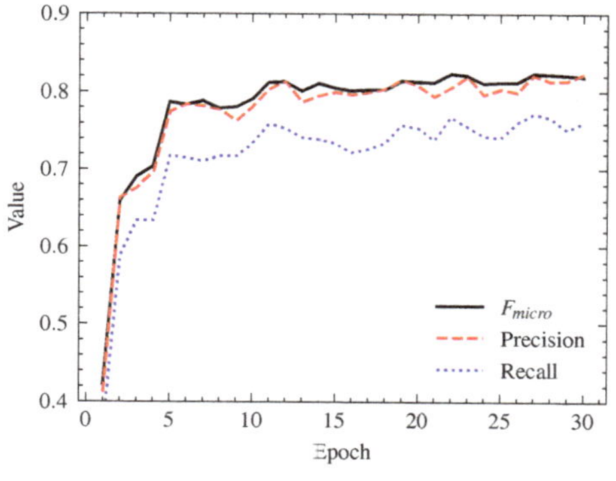

Figure 4. Precision, recall, and F_{micro} value on entire test dataset in the training process.

5.2. Error Analysis

Figure 5 shows the confusion matrix of the model in this paper. Each column of the matrix represents an instance prediction of a class, while each row represents an actual instance of the class. The darker color in the figure indicates a larger proportion of error. To clearly highlight the misclassification of the DDI predicted by our model, the values in the confusion matrix are normalized.

Table 3. Performance comparisons with other DDI prediction methods.

Method		F-Score for Each DDI Type				Overall		
		Advice	Mechanism	Effect	Int	Precision	Recall	F-Score
Traditional models	UTurku	0.621	0.586	0.601	0.504	0.728	0.489	0.599
	FBK-irst	0.695	0.671	0.626	0.554	0.651	0.651	0.658
	WBI-DDI	0.627	0.611	0.608	0.510	0.650	0.563	0.606
Deep learning networks	MCCNN	0.785	0.719	0.683	0.510	0.759	0.652	0.702
	Joint AB-LSTM	0.796	0.761	0.674	0.461	0.733	0.698	0.715
	GGNN	0.817	0.735	0.710	0.460	0.734	0.719	0.726
	RHCNN	0.806	0.780	0.735	**0.578**	0.773	0.735	0.753
	GCNN	0.834 [1]	0.798	0.759	0.514	0.800	0.738	0.769
	DDINN (ours)	**0.863** [2]	**0.820**	**0.772**	0.566	**0.822**	**0.761**	**0.816**

[1] The second best value of the column is marked by underline style. [2] The best value of the column is marked by bold style.

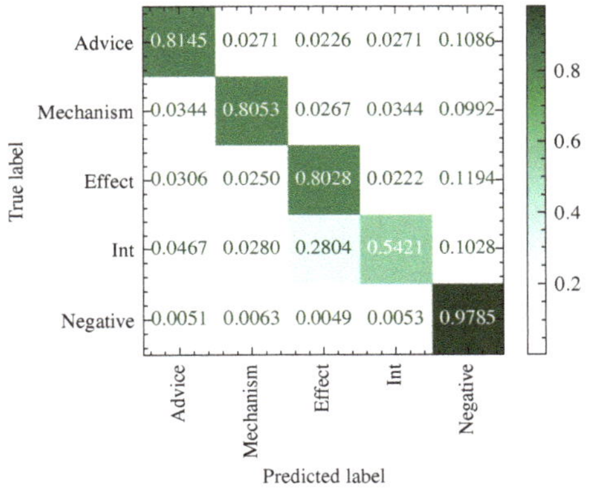

Figure 5. Confusion matrix with L1 normalization.

From Figure 5, we can see that there are two main types of classification errors for the model: (1) the class of relations with the Int type is often incorrectly classified as the Advice type; (2) the four positive classes of relations (Advice, Mechanism, Effect and Int) are often incorrectly classified in the negative class.

For the first type of error, which is already briefly discussed in Section 5.1, the reason is that the number of Int DDI type is too small, with only 96 instances in the training set, and we observed in this paper that the instances of DDI type Int and Effect in the dataset have similar semantics, resulting in the model's inability in classifying these two categories well. The second type of error is also mainly caused by the dataset, where the number of negative categories in the dataset is 28,509, while the number of remaining positive examples is only 4999, which inevitably allows a small number of DDI types to be misclassified into the negative DDI type.

5.3. Ablation Study

Additional ablation experiments are conducted in order to evaluate the influence of different modules or optimizations on DDI prediction. Firstly, the impact of contextual representation methods has been investigated. The corresponding results are shown in Table 4, in which method "GCN only" refers to the model without any contextual representation engagement, and the others are models using GRU, LSTM and cBiLSTM to extract contextual representations, respectively. From Table 4, we can see that cBiLSTM improves the F-score of the GCN-only model by 6.1%, and the cBiLSTM model is indeed more suitable for DDI prediction tasks than some other RNN models.

Table 4. Ablation study on different contextual representation methods.

Method	Precision	Recall	F-Score
GCN only	0.761	0.722	0.777
+GRU	0.784	0.735	0.783
+LSTM	0.796	0.741	0.803
+cBiLSTM	0.822	0.761	0.816

We also investigate the influence of the self-attention pooling strategy used in the construction of the weight-rebalanced dependency matrix, and the results are listed in Table 5. "Full tree" means the method without any pruning strategy. "LAC ($k = n$)" means using the LCA strategy [46] to conduct the tree pruning, and the subtree only includes tokens with the range of n words. From Table 5, we can see that the self-attention-based pruning strategy improved the F-score by 5.4% compared with the full tree strategy. Self-attention adds some complexity to the model, but it is worth it.

Table 5. Ablation study on different syntactic dependency extraction methods.

Method	Precision	Recall	F-Score
Full tree	0.762	0.749	0.762
LCA ($k = 0$)	0.727	0.694	0.725
LCA ($k = 1$)	0.749	0.703	0.738
LCA ($k = 2$)	0.747	0.711	0.744
LCA ($k = 3$)	0.761	0.729	0.759
Self-attention	0.822	0.761	0.816

6. Conclusions

In this paper, we proposed a novel graph-convolutional-network-based method for the knowledge mining of interactions between drugs from the extensive literature, which is called DDINN. Our method makes full use of cBiLSTM to capture the contextual information of input sentences and target drug entities. Additionally, the self-attention mechanism is used to maximize the acquisition of syntactic information related to the DDI extraction task and discard irrelevant information. At last, the output of cBiLSTM and weight-rebalanced dependency matrix will be fed into GCN layers to obtain the DDI type classifier.

The evaluation experiments prove that the DDINN model in this paper achieved higher performance results compared to other state-of-the-art DDI prediction methods in the DDIExtraction2013 dataset. In future work, we will consider data augmentation and other schemes to improve the performance of the DDINN relative to the imbalanced dataset. Additionally, we hope to improve the interpretability [47,48] of deep learning networks in DDINN, which will enhance its utility in the medical field.

Author Contributions: Conceptualization, X.X.; funding acquisition, X.X. and F.M.; project administration, X.X.; software, X.X. and L.S.; validation, L.S. and F.M.; writing—original draft, X.X. All authors have read and agreed to the published version of the manuscript.

Funding: This work was supported by grants from the Fundamental Research Funds for Inner Mongolia Normal University (2022JBQN105), Inner Mongolia JMRH Project (JMRKX202201) and Fundamental Research Funds for Inner Mongolia Normal University (2022JBQN109).

Institutional Review Board Statement: Not applicable.

Informed Consent Statement: Not applicable.

Data Availability Statement: The datasets and codes used in this paper to produce the experimental results are publicly available at GitHub (https://github.com/xingjainxu/DDINN, accessed on 29 December 2022). The project code of biolitNER is also open sourced and accessible at GitHub under the GPLv3 license.

Acknowledgments: We thank Sun for their help in setting up the experiment's server node.

Conflicts of Interest: The authors declare no conflict of interest.

Abbreviations

The following abbreviations are used in this manuscript:

DDI	Drug–drug interaction;
GCN	Graph convolutional network;
cBiLSTM	Contextual bidirectional long short-term memory recurrent neural networks;
RNN	Recurrent neural network;
GRU	Gate recurrent Unit;
POS	Part of speech.

References

1. Becker, M.L.; Kallewaard, M.; Caspers, P.W.J.; Visser, L.E.; Leufkens, H.G.M.; Stricker, B.H.C. Hospitalisations and emergency department visits due to drug-drug interactions: A literature review. *Pharmacoepidemiol. Drug Saf.* **2007**, *16*, 641–651. [CrossRef] [PubMed]
2. Chee, B.W.; Berlin, R.; Schatz, B. Predicting Adverse Drug Events from Personal Health Messages. *AMIA Annu. Symp. Proc.* **2011**, *2011*, 217–226.
3. van der Heijden, P.G.M.; van Puijenbroek, E.P.; van Buuren, S.; van der Hofstede, J.W. On the assessment of adverse drug reactions from spontaneous reporting systems: The influence of under-reporting on odds ratios. *Stat. Med.* **2002**, *21*, 2027–2044. [CrossRef] [PubMed]
4. Wishart, D.S.; Feunang, Y.D.; Guo, A.C.; Lo, E.J.; Marcu, A.; Grant, J.R.; Sajed, T.; Johnson, D.; Li, C.; Sayeeda, Z.; et al. DrugBank 5.0: A major update to the DrugBank database for 2018. *Nucleic Acids Res.* **2018**, *46*, D1074–D1082. [CrossRef] [PubMed]
5. Xiong, G.; Yang, Z.; Yi, J.; Wang, N.; Wang, L.; Zhu, H.; Wu, C.; Lu, A.; Chen, X.; Liu, S.; et al. DDInter: An online drug–drug interaction database towards improving clinical decision-making and patient safety. *Nucleic Acids Res.* **2022**, *50*, D1200–D1207. [CrossRef] [PubMed]
6. Tatonetti, N.P.; Ye, P.P.; Daneshjou, R.; Altman, R.B. Data-Driven Prediction of Drug Effects and Interactions. *Sci. Transl. Med.* **2012**, *4*, 125ra31. [CrossRef]
7. Böttiger, Y.; Laine, K.; Andersson, M.L.; Korhonen, T.; Molin, B.; Ovesjö, M.L.; Tirkkonen, T.; Rane, A.; Gustafsson, L.L.; Eiermann, B. SFINX-a drug-drug interaction database designed for clinical decision support systems. *Eur. J. Clin. Pharmacol.* **2009**, *65*, 627–633. [CrossRef]
8. Zhang, L.; Zhang, Y.D.; Zhao, P.; Huang, S.M. Predicting Drug–Drug Interactions: An FDA Perspective. *AAPS J.* **2009**, *11*, 300–306. [CrossRef]
9. Zhao, L.; Au, J.L.S.; Wientjes, M.G. Comparison of methods for evaluating drug-drug interaction. *Front. Biosci. Elite Ed.* **2010**, *2*, 241–249.
10. Roblek, T.; Vaupotic, T.; Mrhar, A.; Lainscak, M. Drug-drug interaction software in clinical practice: A systematic review. *Eur. J. Clin. Pharmacol.* **2015**, *71*, 131–142. [CrossRef]
11. Chowdhury, M.F.M.; Lavelli, A. *FBK-irst: A Multi-Phase Kernel Based Approach for Drug-Drug Interaction Detection and Classification that Exploits Linguistic Information*; Association for Computational Linguistics: Atlanta, GA, USA, 2013; pp. 351–355.
12. Bokharaeian, B.; Díaz, A. NIL_UCM: Extracting Drug-Drug interactions from text through combination of sequence and tree kernels. In *Proceedings of the Second Joint Conference on Lexical and Computational Semantics (*SEM), Volume 2: Proceedings of the Seventh International Workshop on Semantic Evaluation (SemEval 2013)*; Association for Computational Linguistics: Atlanta, GA, USA, 2013; pp. 644–650.
13. Kim, S.; Liu, H.; Yeganova, L.; Wilbur, W.J. Extracting drug-drug interactions from literature using a rich feature-based linear kernel approach. *J. Biomed. Inform.* **2015**, *55*, 23–30. [CrossRef]
14. Vilar, S.; Friedman, C.; Hripcsak, G. Detection of drug-drug interactions through data mining studies using clinical sources, scientific literature and social media. *Brief. Bioinform.* **2018**, *19*, 863–877. [CrossRef] [PubMed]
15. LeCun, Y.; Bengio, Y.; Hinton, G. Deep learning. *Nature* **2015**, *521*, 436–444. [CrossRef] [PubMed]
16. Cho, K.; Courville, A.; Bengio, Y. Describing Multimedia Content Using Attention-Based Encoder-Decoder Networks. *IEEE Trans. Multimed.* **2015**, *17*, 1875–1886. [CrossRef]

17. Karim, M.R.; Cochez, M.; Jares, J.B.; Uddin, M.; Beyan, O.; Decker, S. Drug-Drug Interaction Prediction Based on Knowledge Graph Embeddings and Convolutional-LSTM Network. In Proceedings of the 10th ACM International Conference on Bioinformatics, Computational Biology and Health Informatics, Niagara Falls, NY, USA, 7–10 September 2019; Association for Computing Machinery: New York, NY, USA, 2019; pp. 113–123. [CrossRef]
18. Nicholson, D.N.; Greene, C.S. Constructing knowledge graphs and their biomedical applications. *Comput. Struct. Biotechnol. J.* **2020**, *18*, 1414–1428. [CrossRef]
19. Kumar Shukla, P.; Kumar Shukla, P.; Sharma, P.; Rawat, P.; Samar, J.; Moriwal, R.; Kaur, M. Efficient prediction of drug-drug interaction using deep learning models. *IET Syst. Biol.* **2020**, *14*, 211–216. [CrossRef]
20. Herrero-Zazo, M.; Segura-Bedmar, I.; Martínez, P.; Declerck, T. The DDI corpus: An annotated corpus with pharmacological substances and drug–drug interactions. *J. Biomed. Inform* **2013**, *46*, 914–920. [CrossRef]
21. Segura-Bedmar, I.; Martínez Fernández, P.; Herrero Zazo, M. *SemEval-2013 Task 9: Extraction of Drug-Drug Interactions from Biomedical Texts (DDIExtraction 2013)*; Accepted: 2015-04-23T13:42:53Z; Association for Computational Linguistics: Atlanta, GA, USA, June 2013.
22. Björne, J.; Kaewphan, S.; Salakoski, T. UTurku: Drug Named Entity Recognition and Drug-Drug Interaction Extraction Using SVM Classification and Domain Knowledge. In *Proceedings of the Second Joint Conference on Lexical and Computational Semantics (*SEM), Volume 2: Proceedings of the Seventh International Workshop on Semantic Evaluation (SemEval 2013)*; Association for Computational Linguistics: Atlanta, GA, USA, 2013; pp. 651–659.
23. Thomas, P.; Neves, M.; Rocktäschel, T.; Leser, U. WBI-DDI: Drug-Drug Interaction Extraction using Majority Voting. In *Proceedings of the Second Joint Conference on Lexical and Computational Semantics (*SEM), Volume 2: Proceedings of the Seventh International Workshop on Semantic Evaluation (SemEval 2013)*; Association for Computational Linguistics: Atlanta, GA, USA, 2013; pp. 628–635.
24. Zheng, W.; Lin, H.; Zhao, Z.; Xu, B.; Zhang, Y.; Yang, Z.; Wang, J. A graph kernel based on context vectors for extracting drug–drug interactions. *J. Biomed. Inform.* **2016**, *61*, 34–43. [CrossRef]
25. Quan, C.; Hua, L.; Sun, X.; Bai, W. Multichannel Convolutional Neural Network for Biological Relation Extraction. *BioMed Res. Int.* **2016**, *2016*, e1850404. [CrossRef]
26. Sun, X.; Dong, K.; Ma, L.; Sutcliffe, R.; He, F.; Chen, S.; Feng, J. Drug-Drug Interaction Extraction via Recurrent Hybrid Convolutional Neural Networks with an Improved Focal Loss. *Entropy* **2019**, *21*, 37. [CrossRef]
27. Asada, M.; Miwa, M.; Sasaki, Y. Enhancing Drug-Drug Interaction Extraction from Texts by Molecular Structure Information. In Proceedings of the 56th Annual Meeting of the Association for Computational Linguistics (Volume 2: Short Papers), Melbourne, Australia, 15–20 July 2018; Association for Computational Linguistics: Melbourne, Australia, 2018; pp. 680–685. [CrossRef]
28. Sahu, S.K.; Anand, A. Drug-drug interaction extraction from biomedical texts using long short-term memory network. *J. Biomed. Informatics* **2018**, *86*, 15–24. [CrossRef] [PubMed]
29. Zhao, Z.; Yang, Z.; Luo, L.; Lin, H.; Wang, J. Drug drug interaction extraction from biomedical literature using syntax convolutional neural network. *Bioinformatics* **2016**, *32*, 3444–3453. [CrossRef]
30. Camacho-Collados, J.; Pilehvar, M.T. From word to sense embeddings: A survey on vector representations of meaning. *J. Artif. Intell. Res.* **2018**, *63*, 743–788. [CrossRef]
31. Zhang, S.; Tong, H.; Xu, J.; Maciejewski, R. Graph convolutional networks: A comprehensive review. *Comput. Soc. Netw.* **2019**, *6*, 11. [CrossRef]
32. Santoro, A.; Raposo, D.; Barrett, D.G.; Malinowski, M.; Pascanu, R.; Battaglia, P.; Lillicrap, T. A simple neural network module for relational reasoning. In Proceedings of the 31st International Conference on Neural Information Processing Systems, Long Beach, CA, USA, 4–9 December 2017; Curran Associates Inc.: Red Hook, NY, USA, 2017; NIPS'17, pp. 4974–4983.
33. Lin, J.; Sun, X.; Ma, S.; Su, Q. Global Encoding for Abstractive Summarization. In Proceedings of the 56th Annual Meeting of the Association for Computational Linguistics (Volume 2: Short Papers), Melbourne, Australia, 15–20 July 2018; Association for Computational Linguistics: Melbourne, Australia, 2018; pp. 163–169. [CrossRef]
34. Yu, A.W.; Dohan, D.; Luong, T.; Zhao, R.; Chen, K.; Le, Q. QANet: Combining Local Convolution with Global Self-Attention for Reading Comprehension. *arXiv* **2018**, arXiv:1804.09541.
35. Vaswani, A.; Shazeer, N.; Parmar, N.; Uszkoreit, J.; Jones, L.; Gomez, A.N.; Kaiser.; Polosukhin, I. Attention is all you need. In Proceedings of the 31st International Conference on Neural Information Processing Systems, Long Beach, CA, USA, 4–9 December 2017; Curran Associates Inc.: Red Hook, NY, USA, 2017; pp. 6000–6010.
36. Hacene, G.B.; Lassance, C.; Gripon, V.; Courbariaux, M.; Bengio, Y. Attention Based Pruning for Shift Networks. IEEE Computer Society. In Proceedings of the 2020 25th International Conference on Pattern Recognition (ICPR), Milan, Italy, 10–15 January 2021; pp. 4054–4061. [CrossRef]
37. Lee, K.; He, L.; Lewis, M.; Zettlemoyer, L. End-to-end Neural Coreference Resolution. In Proceedings of the 2017 Conference on Empirical Methods in Natural Language Processing, Copenhagen, Denmark, 7–11 September 2017; Association for Computational Linguistics: Copenhagen, Denmark, 2017; pp. 188–197. [CrossRef]
38. Mohammed, A.A.; Umaashankar, V. Effectiveness of Hierarchical Softmax in Large Scale Classification Tasks. In Proceedings of the 2018 International Conference on Advances in Computing, Communications and Informatics (ICACCI), Bangalore, India, 19–22 September 2018; pp. 1090–1094. [CrossRef]

39. Qi, X.; Wang, T.; Liu, J. Comparison of Support Vector Machine and Softmax Classifiers in Computer Vision. In Proceedings of the 2017 Second International Conference on Mechanical, Control and Computer Engineering (ICMCCE), Harbin, China, 8–10 December 2017; pp. 151–155. [CrossRef]
40. Kingma, D.P.; Ba, J. Adam: A Method for Stochastic Optimization. *arXiv* **2017**, arXiv:1412.6980.
41. You, K.; Long, M.; Wang, J.; Jordan, M.I. How Does Learning Rate Decay Help Modern Neural Networks? *arXiv* **2019**, arXiv:1908.01878.
42. Pennington, J.; Socher, R.; Manning, C.D. GloVe: Global Vectors for Word Representation. In Proceedings of the Empirical Methods in Natural Language Processing (EMNLP), Doha, Qatar, 26–28 October 2014; pp. 1532–1543.
43. Moen, S.; Ananiadou, T.S.S. Distributional semantics resources for biomedical text processing. *Proc. LBM* **2013**, 39–44.
44. Gero, Z.; Ho, J. PMCVec: Distributed phrase representation for biomedical text processing. *J. Biomed. Inform.* **2019**, *100*, 100047. [CrossRef]
45. Yi, Z.; Li, S.; Yu, J.; Tan, Y.; Wu, Q.; Yuan, H.; Wang, T. Drug-Drug Interaction Extraction via Recurrent Neural Network with Multiple Attention Layers. In Proceedings of the Advanced Data Mining and Applications; Lecture Notes in Computer Science, Singapore, 5–6 November 2017; Cong, G., Peng, W.C., Zhang, W.E., Li, C., Sun, A., Eds.; Springer International Publishing: Cham, Switzerland, 2017; pp. 554–566. [CrossRef]
46. Zhang, Y.; Qi, P.; Manning, C.D. Graph Convolution over Pruned Dependency Trees Improves Relation Extraction. In Proceedings of the 2018 Conference on Empirical Methods in Natural Language Processing, Brussels, Belgium, 31 October–4 November 2018; Association for Computational Linguistics: Brussels, Belgium, 2018; pp. 2205–2215. [CrossRef]
47. Murdoch, W.J.; Singh, C.; Kumbier, K.; Abbasi-Asl, R.; Yu, B. Definitions, methods, and applications in interpretable machine learning. *Proc. Natl. Acad. Sci. USA* **2019**, *116*, 22071–22080. [CrossRef]
48. Meng, C.; Trinh, L.; Xu, N.; Enouen, J.; Liu, Y. Interpretability and fairness evaluation of deep learning models on MIMIC-IV dataset. *Sci. Rep.* **2022**, *12*, 7166. [CrossRef] [PubMed]

electronics

MDPI

Article

An RG-FLAT-CRF Model for Named Entity Recognition of Chinese Electronic Clinical Records

Jiakang Li [1,2], Ruixia Liu [1], Changfang Chen [1], Shuwang Zhou [1,3], Xiaoyi Shang [1] and Yinglong Wang [1,*]

1 Shandong Artificial Intelligence Institute, Qilu University of Technology (Shandong Academy of Sciences), Jinan 250353, China; lijiakang178@163.com (J.L.); liurx@sdas.org (R.L.); chenchangfang012@163.com (C.C.); zhoushw@sdas.org (S.Z.); shangxy@sdas.org (X.S.)
2 Shandong Computer Science Center (National Supercomputer Center in Jinan), Qilu University of Technology (Shandong Academy of Sciences), Jinan 250353, China
3 College of Computer Science and Engineering, Shandong University of Science and Technology, Qingdao 266590, China
* Correspondence: wangylscsc@126.com

Abstract: The goal of Clinical Named Entity Recognition (CNER) is to identify clinical terms from medical records, which is of great importance for subsequent clinical research. Most of the current Chinese CNER models use a single set of features that do not consider the linguistic characteristics of the Chinese language, e.g., they do not use both word and character features, and they lack morphological information and specialized lexical information on Chinese characters in the medical field. We propose a RoBerta Glyce-Flat Lattice Transformer-CRF (RG-FLAT-CRF) model to address this problem. The model uses a convolutional neural network to discern the morphological information hidden in Chinese characters, and a pre-trained model to obtain vectors with medical features. The different vectors are stitched together to form a multi-feature vector. To use lexical information and avoid the problem of word separation errors, the model uses a lattice structure to add lexical information associated with each word, which can be used to avoid the problem of word separation errors. The RG-FLAT-CRF model scored 95.61%, 85.17%, and 91.2% for F1 on the CCKS 2017, 2019, and 2020 datasets, respectively. We used statistical tests to compare with other models. The results show that most p-values less than 0.05 are statistically significant.

Keywords: clinical named entity recognition; Chinese medical text; pre-trained model

Citation: Li, J.; Liu, R.; Chen, C.; Zhou, S.; Shang, X.; Wang, Y. An RG-FLAT-CRF Model for Named Entity Recognition of Chinese Electronic Clinical Records. *Electronics* **2022**, *11*, 1282. https://doi.org/10.3390/electronics11081282

Academic Editors: Agnieszka Konys and Agnieszka Nowak-Brzezińska

Received: 9 March 2022
Accepted: 15 April 2022
Published: 18 April 2022

1. Introduction

Informatization has penetrated all aspects of social life. In the medical field, more and more hospitals are building information systems to improve their service level and core competitiveness, effectively use limited medical resources, and provide patients with high-quality treatment. These information systems can not only improve doctors' efficiency but also enhance internal management, making information communication among departments more efficient and simplifying and standardizing the medical treatment process. Medical staff can be released from tedious and repetitive work, with extra time and energy being used to provide better patient services.

Existing medical systems have generated countless medical data, and if the data cannot be used effectively, it will be a waste of professional knowledge. As a medical record, Electronic Medical Record (EMR) has received great attention in scientific research [1] because it contains complete and detailed clinical information generated by patients during each visit. EMR refers to the digital information such as words, symbols, charts, graphics, data, images, and so on, generated by medical personnel using the information system of medical institutions in medical activities. EMR contains various information such as text and medical images. Medical images are mainly the results of laboratory tests of patients, such as CT and B-ultrasound. These medical images can currently be analyzed using pattern recognition and machine learning methods, but EMR also contains much textual data. To make use of the text data, Natural Language Processing (NLP) technology is essential. Electronic medical records cover all patient information from admission to discharge, including admission time, symptoms, body parts, examination methods, medication, and other physical information [2]. Medical services may consider providing patients with the facility to submit inquiries in the form of comments [3].

EMR information extraction is to identify various medical entities from texts and establish relationships among them. The information extraction of EMR was first carried out on English medical records, and many achievements have been achieved, while domestic research on Chinese EMR is still in its infancy. Therefore, it is our top priority.

Named Entity Recognition (NER) is the foundation of text data mining and information processing. For entity recognition in the medical field, it refers to identifying entities such as symptoms, body parts, examinations, etc. Identifying this information and analyzing the relationship among different entity information plays an indispensable role in establishing a knowledge map in the medical field, building an auxiliary diagnosis model, and providing data support for clinical decision-making.

Early NER systems are mainly rule-based approaches. This method extracts the target entity through the preset rule template and has achieved certain results. Although for some uncommon fields, experts need to write rules, which is demanding, time-consuming, and limited, rule-based approaches are not outdated but are still an important complement to other approaches.

Feature-based Supervised Learning Approaches transform NER tasks into classification tasks or sequence labeling tasks. Conditional Random Fields (CRF) and Hidden Markov Models (HMM) [4] are two common algorithms.

With the rapid expansion of deep learning, Gated Recurrent Unit (GRU) and Long Short-Term Memory (LSTM) are applied to the CNER tasks [4]. Alam et al. [5] proposed a new framework based on association rule mining for prognostic factor identification in malignant mesothelioma. At present, the integration of LSTM and CRF is a common method. However, there are limitations. Transformer [6] proposes self-attention, enabling the LSTM networks to solve long-distance dependencies. Transformers gradually replaced LSTM as the mainstream feature extractor in NLP.

Unsupervised pre-trained models are suitable for general domains but not appropriate for the medical domain.

In addition, Chinese NER is related to word segmentation. Since Chinese entities are generally composed of words, word segmentation errors will lead to errors in Chinese NER. The character-based Chinese NER model cannot fully utilize the information of words. The Lattice-LSTM proposed by Zhang et al. [7] improves the accuracy of this task by adding dictionary information to the model. However, due to the complexity of the lattice structure, it does not support parallel computing. Li et al. [8] proposed a Flat Lattice Transformer (FLAT), which uses a flatten lattice structure and transformer to realize parallel processing. At the same time, FLAT uses the calculation method of relative position in the Transformer-XL model [9], and by adding additional position information in the Transformer structure, it solves the modeling of long text and captures ultra-long distance dependencies.

Also, unlike other languages, Chinese is a pictograph. Chinese characters contain rich semantic information. Many words with similar meanings are similar in composition and structure of Chinese characters, which is especially obvious in the medical field. The glyph information of Chinese characters is also of significant reference value. Glyce, proposed by Meng [10], can extract the glyph vectors of Chinese characters. It attempts to extract the semantics of Chinese characters from various ancient and modern Chinese characters and various writing styles, and the performance is improved.

To solve problems, we propose a RoBerta Glyce-Flat Lattice Transformer-CRF (RG-FLAT-CRF) model suitable for Chinese CNER tasks. First, the glyph vector is obtained by Glyce, the character vector and word vector are obtained by Word2vec [11], and the character vector obtained by RoBerta is spliced with the glyph vector and the word vector obtained by Word2vec. At the same time, the Flat-lattice structure is used, word information is added, the head position code and tail position code are constructed for each character and vocabulary, and the relative position code is calculated. The concatenation of vectors and the corresponding position encoding are sent to a transformer to extract the context information of every Chinese character. Finally, we jointly decode the labels of the entire sentence using CRF. Our main contributions are as follows:

Contribution

In our contributions, we have:

1. A RoBerta Glyce-Flat Lattice Transformer-CRF model is proposed, which can make full use of the glyph information and language features of Chinese medical texts, has strong coding and text representation capabilities, and can accurately identify various types of Chinese electronic clinical Entity records.
2. According to the particularity of medical entities and the language characteristics of Chinese, a multi-feature fusion vector is constructed. The pre-trained model is used to obtain vectors that conform to medical characteristics. At the same time, to strengthen the semantic representation of medical entities, convolutional neural networks are used to extract the glyph features of medical entities, and different character vectors are spliced together to form a composite character vector.
3. Use of a lattice structure to add potential lexical information to each word to avoid word segmentation errors. The relative position vector in the improved transformer directly captures the dependencies between words and vocabulary, makes full use of case information, and can be implemented in parallel.

The rest of this article is organized as follows. Section 2 provides a brief review of related work of NER. The proposed model is presented in Section 3. The relevant content of the experiment is described in detail in Section 4. Finally, Section 5 gives the conclusions.

2. Related Work

We include the following studies: (1) How to enhance the semantic representation of Chinese word vectors. (2) Feature extraction networks more applicable to the Chinese language. (3) The characteristics and difficulties of named entity recognition in Chinese electronic medical records. (4) Related Evaluation Metrics [12]. We used multiple strings such as "Chinese electronic medical record named entity recognition", "Chinese named entity recognition", and "medical named entity recognition" to retrieve peer-reviewed articles using Multiple databases, including Scopus, ACM Digital Library, IEEE Xplore, ScienceDirect, SpringerLink, and Google Scholar [13].

This section primarily provides a brief introduction to rule-based and dictionary-based methods, machine learning-based methods, and deep learning-based methods. Then, the representation method of the word vector is introduced.

2.1. Rule-and-Dictionary-Based Clinical Named Entity Recognition

Nowadays, Rule-and-Dictionary-Based CNER is commonly used, and these methods benefit from the development of professional medical dictionaries. Researchers complete the NER task by pattern matching according to the belonging list in the dictionary. Friedman et al. [14] developed a clinical document processor that recognized medical information in the medical record and mapped this information into a structured representation containing medical terms. Fukuda et al. [15] proposed a method to identify the names of substances such as proteins from biological papers, using the characteristics of proper noun descriptions in the professional field, which eliminates the need to prepare a professional term dictionary in advance. Names can be extracted with precision, whether they are known or newly defined or are single or compound words.

The completeness and accuracy of the dictionary and the accuracy of the matching algorithm can determine the accuracy of such methods. Therefore, dictionary-based methods are more suitable for fields where proper nouns are fixed and updated infrequently. In the biomedical field, there are problems such as the fast updating of proper nouns and different expressions of the same entity name. Experts need to spend much time and effort writing rules, and the cost is high. In addition, different rules are needed for different systems. They are of poor portability and are hard to reuse quickly.

2.2. Clinical Named Entity Recognition Based on Machine Learning

In the past, traditional machine learning based on CNER has been widely used, including HMM, CRF, Support Vector Machine (SVM) [16], Naive Bayesian Model (NBM) [17], etc. Settles [18] used combined feature sets with CRF in biomedical NER tasks. Tang [19] developed an SVM-based NER system for medical entities in the medical record. Roberts et al. [20] utilized SVM with a manually constructed dictionary to classify. Liu [21] evaluated the contribution of different features in the CRF-based CNER task.

Compared with the methods analyzed in Section 2.1, the method in Section 2.2 does not require the experimenter to master much language knowledge, thus saving time and effort. However, this type of method requires a lot of energy to design features. The effect of the model depends on the designed features. With deep learning modeling, the feature extraction problem in traditional machine learning can be addressed.

2.3. Deep-Learning-Based Clinical Named Entity Recognition

Recently, we have witnessed the great success of deep learning in the field of NLP, such as NER and event extraction tasks. Commonly used network models include Convolutional Neural Networks (CNN) [22], Recurrent Neural Networks (RNN) [23], and LSTM. Ma et al. [24] proposed the Bi-directional LSTM-CNNs-CRF model, character-level representations are extracted using CNN, Bi-directional LSTM (BiLSTM) is responsible for modeling the contextual information of each word. Xu et al. [25] combined bidirectional LSTM and CRF based, BiLSTM-CRF model can learn the information features of a given dataset and achieved a score of 0.8022 at NCBI, outperforming many widely used baseline methods. Yin et al. [26] used convolutional neural nets for Chinese character radical feature extraction and captured the correlation between characters using self-attentiveness. Kong et al. [27] proposed a Chinese medical named entity recognition based on a multi-layer CNN and attention mechanism, constructing a multi-layer CNN to extract short-term and long-term memories and using an attention mechanism to capture global information. However, the above deep neural network-based CNER methods cannot model the ambiguity of Chinese.

The BERT-BiLSTM-CRF model was proposed by Jiang et al. [28] to be applied to CNER. The semantic representation of words was enhanced with a BERT pre-trained language model, and the BiLSTM was to learn contextual information. Qin et al. [29] proposed a BERT-BiGRU-CRF model in the field of Chinese electronic medical records, which uses BERT to convert the electronic medical record text into low-dimensional vectors and BiGRU to obtain contextual features. Wu et al. [30] used a bi-directional LSTM model to learn a medical entity's partial head information using Roberta to learn medical features. Wang et al. [31] used information from medical encyclopedias as additional information to enhance the recognition of Chinese electronic medical record entities. However, these models do not fully consider the characteristics of medical domain data, and it is not very effective in medical entity extraction.

2.4. Research Status of Word Vector Representation Methods

If you want to reflect a word in a text and perform mathematical calculations, it must be done through word embedding. The bag-of-words model simply represents words without any semantic features. As the number of words increases, so does the dimension. Researchers propose a way to solve this problem using a pre-trained language model for word representation. Pre-training refers to obtaining a training model independent of subsequent tasks from a large-scale corpus using self-supervised learning. The model can be transferred to other tasks, thereby reducing the training burden of subsequent tasks. The Word2Vec model was proposed by Mikolov et al. to obtain vectors. The GloVe algorithm was proposed by Pennington et al. [32]. In recent years, pre-trained models have received increasing attention. Since this type of model is a context-independent word vector trained by static pre-training technology, it cannot accurately model the polysemy of a word. Therefore, Peters et al. [33] proposed the ElMo algorithm. The bidirectional LSTM network structure was used for context encoding, which could effectively capture context information.

2.4.1. Models for BERT and Its Variants

Devlin et al. [34] proposed Bidirectional Encoder (BERT). The emergence of Bert opened a new era of research in the field of NLP. Then some improved pre-training models based on BERT, mainly including ERNIE [35], BERT-WWM [36], RoBerta [37], and XLNet [38]. The ERNIE model is pre-trained using massive corpora in multiple fields, including encyclopedias, news, forums, etc. BERT-WWM's improvement over BERT is to replace a complete word with a Mask label instead of a subword. The RoBerta model uses a dynamic mask mechanism for pre-training, cancels the NSP task, and expands the batch size. As an auto-regressive model, the XLNet model can expand the language model and increase the prediction of bidirectional words, the above predicting the next word and the following predicting the previous words.

2.4.2. Research on Chinese Characters

The structure of Chinese characters is different from that of English. Chinese characters are pictographs, and their glyphs also contain rich meanings. Therefore, many scholars have carried out characterization studies on the glyph features of Chinese characters. Sun [39] proposed to learn the radical features of Chinese. Wang et al. [40] proposed a Chinese character root and stroke-enhanced embedding method for learning Chinese character roots from the internal information of semantics and form. Wei [41] proposed a visual embedding method for semantic association among visual words, segmented the glyph, spliced the average embedding vectors corresponding to each sub-region, and converted it into a fixed-length vector for keyword detection. Su [42] used convolutional autoencoders to learn glyph features from images of traditional Chinese characters and introduced glyph features during training using the corpus. Meng [6] proposed the Glyce model. It tried to extract the semantics of Chinese characters from various ancient and modern Chinese characters and various writing styles, and the performance was improved.

These are the characteristics of Chinese, which improve CNER tasks. However, the current mainstream CNER methods cannot integrate the pre-trained model with the Chinese glyph information.

3. Proposed Method

In the NER task, the character sequence of the input text is represented by $X = (x_1, x_2, \ldots, x_n)$. The labels of the input text are represented by $Y = (y_1, y_2, \ldots, y_n)$. The goal of a NER system is to predict the correct sequence Y of labels for the text given the known sequence of characters X of the text. The RG-FLAT-CRF model proposed in this chapter consists of three parts; the embedding layer, the encoding layer, and the decoding layer. The overall structure is shown in Figure 1.

The model first matches the latent words related to the character in the input text and splices the character information and words information into the embedding layer. The embedding layer consists of three parts, and the character vector is spliced after processing by RoBerta, Glyce, and Word2vec. The word vector is obtained using Word2vec, head and tail position encoding are constructed for each character and word, and the relative position encoding is calculated. The concatenation of word vectors and the corresponding position encoding are input into the encoding layer, consisting of a Transformer neural network that captures deep features and encodes the input sequence. Finally, the output of the encoding layer is input to the decoding layer, which predicts the final label sequence.

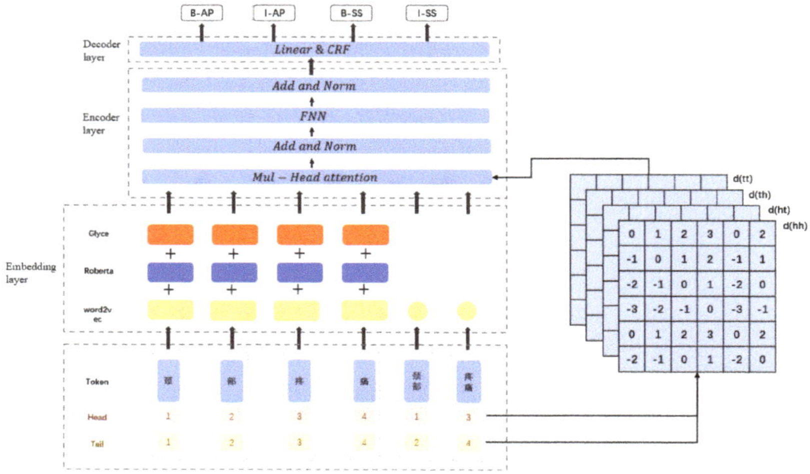

Figure 1. Model structure diagram of RG-FLAT-CRF.

This study uses NER to perform entity recognition on Chinese EMR. Specific steps are as follows:

(1) Electronic medical record data preprocessing, that is, the original electronic medical record text data set is processed, and the electronic medical record text set is represented as $J = (j_1, j_2, \ldots, j_n)$, where the i-th electronic medical record text is represented as j_i. The predefined entity category $C = (c_1, c_2, \ldots, c_m)$, is divided and annotated according to the character level, and the characters and predefined categories are separated by spaces when annotating.
(2) Establish a Chinese EMR text training dataset.
(3) Model training, that is, training the RGT-CRF model. Take the Chinese EMR test text set $J_{test} = (j_1, j_2, \ldots, j_N)$ as input and take the entity and its corresponding category pair as output: $\{\langle m_1, c_1\rangle, \langle m_2, c_2\rangle, \ldots, \langle m_p, c_p\rangle$. The entity m_i represents the entity that appears in the document, and b_i and e_i represent the start and end positions of m_i, respectively. There is no need to overlap between entities; that is, $e_i < b_i + 1$. C_{m_i} represents the predefined category of entity m_i, calculates the F1 score according to the precision and recall rate, and uses the F1 score as the comprehensive evaluation index of the model.

3.1. Embedding Layer

The embedding layer consists of three parts: RoBerta layer, Glyce layer, and Word2vec layer:

(1) RoBerta layer: the model adopts the better pre-training model RoBerta to capture the characteristics of medical text and converts each word of medical text into a low-dimensional vector form through RoBerta.
(2) Glyce layer: scan each word in the sentence to obtain the glyph vector corresponding to each word, and enhance the representation of the word.
(3) Word2vec layer: Using Word2vec, the vector representation of each word in the medical text and the vector representation of the latent words can be obtained to enrich the semantic representation.

The character vectors processed by RoBerta, Glyce, and Word2vec are spliced to obtain multi-feature word vectors, and then the character vectors and word vectors processed by Word2vec are spliced together.

3.1.1. RoBerta

Pretrained language models are often used in NER tasks to generate richer semantic representations. BERT and its variant RoBerta are widely used in research. We use RoBerta for text encoding instead of BERT. Compared with BERT, the model structure of RoBerta has not changed. They are all composed of 12 stacked transformers. Each layer has a hidden state of 768 dimensions. Each Transformer uses a 12-head self-attention mechanism. The only thing that has changed is the pre-training method. Dynamic masks and text encoding are adopted to remove the NSP task and use more data to train the model.

The vector is obtained through the RoBerta. The RoBerta structure is shown in Figure 2. The input text is $Z = \{Z_1, Z_2, \ldots, Z_x\}$. First, the sequence is vectorized. This part consists of token embedding, clause embedding, and position embedding. These three embedding layers are essentially equivalent to the static embedding layers, and the table lookup is performed by the embedding matrix. For the x-th token in the processed token sequence, the vector calculation is as follows:

$$e_x = W_t(E_{t_x}) + W_s(E_{s_x}) + W_p(E_x) \tag{1}$$

where W_t, W_s, W_p are the token embedding matrix, the clause embedding, and the position matrix.

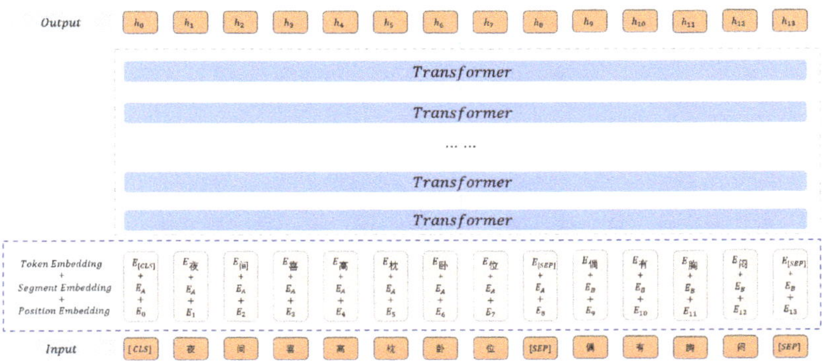

Figure 2. Structure diagram of RoBerta.

Token Embeddings represent the Embedding vector of each word. Segment Embeddings are used to distinguish different sentences before and after punctuation marks. Position Embeddings represent the embeddings of a word's position. The input feature of RoBerta is the sum of the above 3 embeddings. "[CLS]" is used as the starting symbol of the input, indicating that the feature can be used in the classification model. "[SEP]" indicates the clause symbol, which is used to cut off the clauses in the sentence.

The obtained vector is input into the stacked Transformer to extract features. The final output is the result of encoding the input sentence text. Finally, we obtained the sentence representation vector with the dependency information among words and words in the sentence text. The calculation is as follows:

$$H = Mul_{trans}(E) \tag{2}$$

where $Mul_{trans}(.)$ represents the stacked Transformer, outputting the text encoding of the entire sentence through the last layer H, which can be expressed as $H = h_0, h_1, \ldots, h_x$. Here h_x is the text representation vector to the xth token.

3.1.2. Glyce

Chinese characters are pictographs, and most Chinese characters are evolved from graphics. Chinese characters contain rich semantic information, especially in the medical field. Most of the words for diseases have the same parts. Therefore, we believe that adding glyph information to word vectors can enhance the representation of characters.

Glyce used different versions of the writing method, as well as different writing to enhance the representation of the characters.

Glyce is different from traditional CNN. There are about 100,000 Chinese characters, but only a few thousand are commonly used. Compared with classification on the ImageNet dataset. There're few training examples for Chinese characters. Compared with the size of Imagenet images, Chinese images are usually smaller, with a size of 12 × 12. Thus according to the Chinese writing habits, a 2 × 2 Tianzi lattice structure is used. As shown in Figure 3, this structure can reflect the glyph information of Chinese, including components such as radicals, which is suitable for the extraction of glyph information.

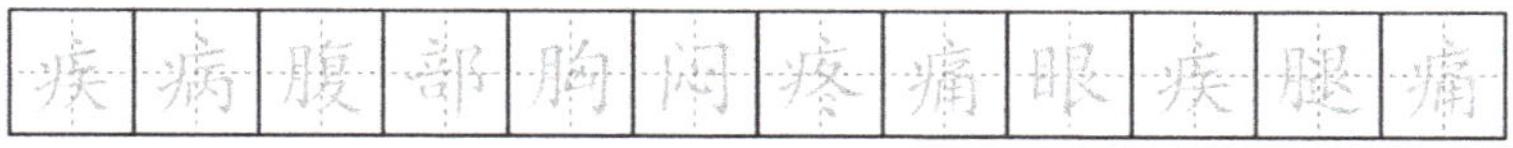

Figure 3. Schematic diagram of the Tianzi lattice.

The structure of Glyce Tianzi lattice-CNN is shown in Figure 4. The processing process is shown in Figure 5. To capture lower-level graph features, the input image approximation firstly passes through a convolutional layer with kernel size 5. In addition, the convolutional layer has to increase the number of feature channels to 1024. Then we apply a max-pooling layer with a pooling kernel of 4 × 4 to perform feature downsampling. After this, the resolution is reduced from 8 × 8 to 2 × 2. This 2 × 2 Tianzi lattice structure shows the glyph features of Chinese characters, and finally, we apply the group convolution operation to map the Tianzi lattice to the final output.

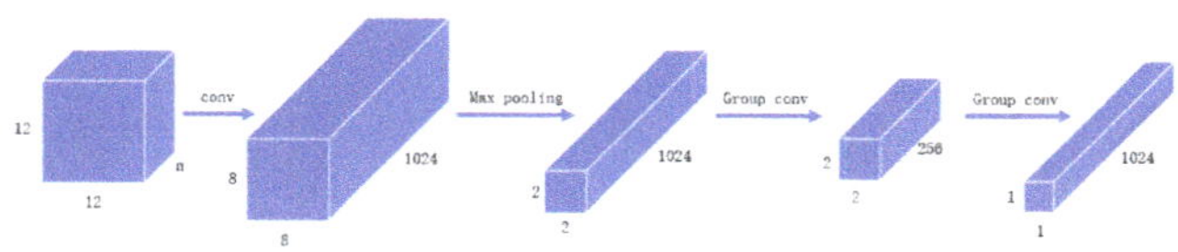

Figure 4. CNN structure diagram in Glyce.

layer	kernel	Output
input		n×12×12
Conv2d	5	1024×8×8
Relu		1024×8×8
Max pool	4	1024×2×2
8 group conv	1	256×2×2
16 group conv	2	1024×1×1

Figure 5. The Tianzi lattice—CNN structure.

For the input text $Z = \{Z_1, Z_2, \ldots, Z_x\}$, the glyph vector obtained by Glyce is $E_G = (e_{G0}, e_{G1}, \ldots, e_{Gx})$ as shown in Figure 6.

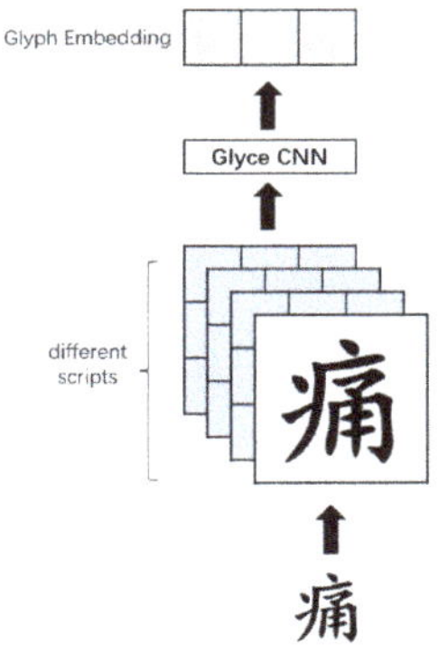

Figure 6. Glyce character embedding.

3.1.3. Word2vec

We use Word2vec to get word vectors, a typical representative of distributed representation. Compared with one-hot, Word2vec takes into account the relationships among words. In addition, Word2vec also optimizes the training efficiency of the model, so it is used more frequently.

3.2. Position Encoder

Chinese NER tasks are often considered sequence labeling tasks. By calculating the probability of each character corresponding to each entity type label, The label with the highest probability is used as the final identification result. There are usually two vectorization methods to vectorize Chinese characters into the model calculation: methods based on word vectors and methods based on character vectors.

The first task of the word vector-based model is to segment the text into the form of words. The improvement effect of word vectors on entities is significant. The word contains more semantic information, but if there is a false classification, it will affect the results of NER.

For instance, in Figure 7, this sentence can be divided into '济南人 (Jinan People)', '和 (and)', '山庄 (Mountain Villa)', and can also be divided into '济南 (Jinan People)', '人和山庄 (Renhe Mountain Villa)'. These two-word segmentation methods have a great impact on recognition.

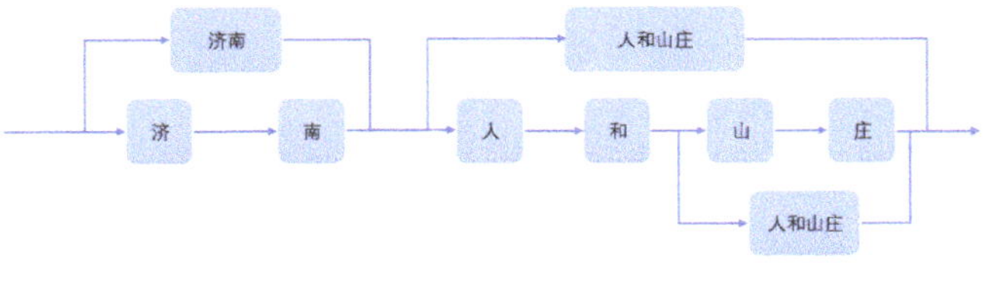

Figure 7. Structure diagram of Lattice.

Using character vector-based models avoids word segmentation error information but lacks lexical information. For example, '感冒 (cold)', separate the word '感 (feel)' and '冒 (emit)' represent different semantic information. '感 (feel)' means feeling, and '冒 (emit)' means to penetrate outward or rise upward. It is difficult to express the information of the word '感冒 (cold)' in medicine after '感 (feel)' and '冒 (emit)' are separated, which is especially obvious in the medical field.

To address the above problems, we adopted the FLAT-lattice structure, shown in Figure 8. This structure uses both character vectors and word vectors. Based on character vectors, the latent vocabulary of each character is matched, and the word vectors are added to the model. This method utilizes the semantic relationship of words and avoids the phenomenon of word segmentation errors.

Figure 8. Structure diagram of Flat-lattice.

After using the dictionary to obtain lattice information from the string, it is flattened, and the structure is shown in Figure 8.

These flat lattices can also be defined as spans. A span comprises a token, a head, and a tail. A token is a word or character, and the head represents the starting position of the token in the original sequence, and the tail represents the ending position of the token in the original sequence. For characters, the head and tail are the same. For the matched words, head indicates the start position of the word in the sequence, and tail indicates the end position of the word in the sequence. The flat lattice can preserve the original structure of the lattice and, at the same time, preserve the word order information of the original sentence.

According to the Flat-lattice structure, there are three interrelationships, intersection, involvement, and separation. We use relative position encoding to encode the positional relationship among each span. Relative position encoding does not directly model the interaction relationship but obtains a dense vector by computing a set of head and tail changes. Not only the interrelationships among spans can be represented, but more detailed sequence relationships can be shown, such as the distance among words and characters. Let $tail_x$ and $tail_x$, $head_y$ and $tail_y$ denote the head and tail positions of s_x and s_y, respectively. Four kinds of relative distances can be used to represent the relative relationship between s_x and s_y. Their calculation formulas are as follows:

$$r_{xy}^{hh} = head_x - head_y \tag{3}$$

$$r_{xy}^{ht} = head_x - tail_y \tag{4}$$

$$r_{xy}^{th} = tail_x - head_y \tag{5}$$

$$r_{xy}^{tt} = tail_x - tail_y \tag{6}$$

where r_{xy}^{hh} stands for the distance from the head of s_x to the head of s_y, r_{xy}^{ht} is the distance from the head of s_x to the tail of s_y, r_{xy}^{th} represents the distance from the tail of s_x to the head of s_y, r_{xy}^{tt} is the distance from the tail of s_x to the tail of s_y. The final relative position encoding is a nonlinear transformation of the four distances, which can be calculated like:

$$L_{xy} = ReLU\left(W_l\left(P_{r_{xy}^{hh}} \bigoplus P_{r_{xy}^{ht}} \bigoplus P_{r_{xy}^{th}} \bigoplus P_{r_{xy}^{tt}}\right)\right) \tag{7}$$

among them, W_l is a learnable parameter, $\oplus$ represents the connection operator, and the calculation method of P_r refers to the calculation method of the transformer. The calculation is as shown in the equation:

$$P_r^{2k} = \sin\frac{r}{1000^{\frac{2k}{d_{model}}}} \tag{8}$$

$$P_r^{2k+1} = \cos\frac{r}{1000^{\frac{2k}{d_{model}}}} \tag{9}$$

3.3. Encoder

The encoding layer consists of Transformers, which aim to extract semantic and temporal features from the context automatically.

Before the transformer appeared, most NER used BiLSTM as the model's encoder. However, BiLSTM has some problems: (1) The sequential nature of the recurrent neural network represented by LSTM hinders the parallelization of training samples; (2) The problem of long-term dependence cannot be completely solved.

Transformer avoids recurrent model structure and uses attention mechanism for modeling. The structure is shown in Figure 9. We used its encoding part, which consists of two parts, a feedforward network and a multi-head self-attention layer, both of which have a residual network. Multi-head self-attention consists of stacked self-attentions, all accompanied by a "layer normalization" step.

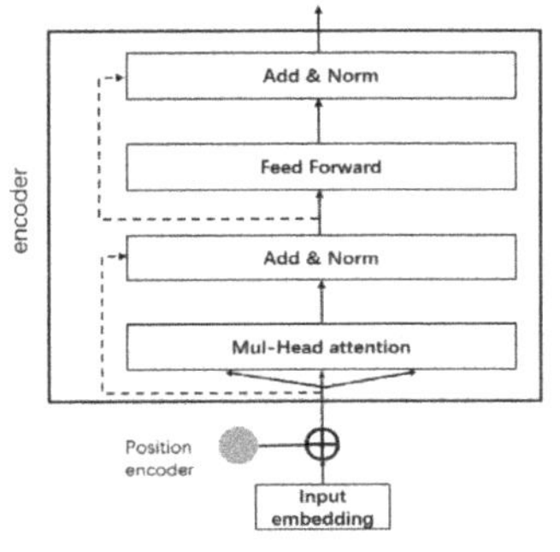

Figure 9. Structure diagram of Transformer.

When the encoder encodes this word, the self-attention mechanism can take other words in this sentence into consideration.

First, we send the vector output of the embedding layer and the corresponding relative position encoding to the encoding layer, using the encoding layer of the transformer. A Query vector, a Key vector, and a Value vector are created for each word by this self-attention mechanism. They are obtained through the vector multiplication by the three matrices we trained. Their calculation formula is as follows:

$$Q = Linear(X) = XW_q \tag{10}$$

$$K = Linear(X) = XW_k \tag{11}$$

$$V = Linear(X) = XW_k \tag{12}$$

The second step is to calculate the score, which will make the gradient more stable, and then it is divided by $\sqrt{d_{head}}$. The traditional Transformer model can capture contextual semantics by adding position information to the input, but there is a problem of sentence errors in the face of text segmentation input. Therefore, extra position information is added to the Transformer structure of the Transformer-XL model, and the absolute vector is converted into a relative vector. Solve the modeling of long text, capture ultra-long distance dependencies, and calculate the attention score vector among input vectors by the formula:

$$A^*_{x,y} = \frac{W_q^\top E_{s_x}^\top E_{s_y} W_{k,E} + W_q^\top E_{s_x}^\top L_{xy} W_{k,R} + u^\top E_{s_x} W_{k,E} + v^\top L_{xy} W_{k,R}}{\sqrt{d_{head}}} \tag{13}$$

where $W_q, W_{k,E}, W_{k,R}, u, v$ are learnable parameters,E_{s_x}, E_{s_y} are the embedded representations of s_x and s_y.

Then pass the result through softmax, which normalizes the scores for all words. For the weighted value vector, the output of the self-attention layer at that position is obtained, and the following is its formula:

$$Attention(A, V) = softmax(A)V \tag{14}$$

The multi-head attention mechanism consists of multiple self-attentions. Define multiple groups of different Q, K, and V, and let them focus on different contexts, respectively. The process of calculating Q, K, V is still the same, except that the matrix of linear transformation has changed from one set of (W_Q, W_K, W_K) to multiple sets of (W_Q, W_K, W_K).

For the input matrix X, each group of Q, K, V can get an output matrix Z. Concatenate the different matrices together and multiply with an additional matrix W_o.

The multi-head attention mechanism enhances the attention layer's performance in two aspects:

(1) It empowers the model with a closer focus on different locations.
(2) Multiple "representation subspaces" are given to the attention layer, and multi-head attention allows us to possess multiple sets of Q, K, and V matrices. After training, each group projects the output into a different representation subspace. The calculation formula is as (15):

$$MH_{att}(A, V) = Concat(head_1, \ldots, head_h)W_o \tag{15}$$

The resulting output is subjected to layer normalization and residual connections. The specific formula is as follows:

$$X_{MH_{att}} = X_{MH_{att}} + X \tag{16}$$

$$X_{MH_{att}} = LayerNorm(X_{MH_{att}}) \tag{17}$$

After the operation of Feedforward, the formulas are shown in equations:

$$X_{hidden} = Linear(ReLu(Linear(X_{attention}))) \tag{18}$$

$$X_{hidden} = X_{attention} + X_{hidden} \tag{19}$$

$$X_{hidden} = LayerNorm(X_{hidden}) \tag{20}$$

3.4. Decoder

The decoding layer consists of CRFs, whose purpose is to resolve the correlation between the output labels to obtain the globally optimal annotation sequence for the text.

For the input sequence $X = (x_1, x_2, \ldots, x_n)$, its predicted label is $Y = (y_1, y_2, \ldots, y_n)$. The score matrix P output by the encoding layer is n×k in size, n is the length of the input sequence, and q is the different types of labels defined. P_{i,y_i} represents the score of the ith character in the sentence on the y_i label. A state transition score matrix A represents the probability score of transition among different labels. $A_{y_i,y_{i-1}}$ represents the transition score from label y_i to label y_{i+1}. y_0, y_{n+1} represent the start tag and the end tag, respectively. Under the condition of the given sequence, the score $S(X, y)$ of the corresponding sequence tag is obtained. The functions can be described as follows:

$$SX, y = \sum_{i=0}^{n} A_{y_i,y_{i+1}} + \sum_{i=1}^{n} P_{i,y_i} \tag{21}$$

The predicted probability is $P(y|X)$. The calculation formula is shown in (22):

$$P(y|X) = \frac{e^{S(X,y)}}{\sum_{y' \in Y_X} e^{S(X,y')}} \tag{22}$$

The loss function, as shown in the formula:

$$-\log(P(y|X)) = \log \sum_{y' \in Y_X} e^{S(X,y')} - S(X,y) \tag{23}$$

In the last, we adopted the Viterbi algorithm to get the optimal path, that is, a more reasonable predicted label of the input sequence. The calculation formula is as follows (24):

$$y^* = arg_{y' \in Y_X} max\ S(X,y') \tag{24}$$

3.5. Time Complexity Analysis

We discuss the time complexity of the model.

$$O\left(n^2 \cdot d + n \cdot d^2 + \sum_{l=1}^{N} M_l^2 \cdot K_l^2 \cdot C_{l-1} \cdot C_l + n \cdot k^2\right)$$

where n is the sequence length and d is the dimension of embedding. n is the number of convolutional kernels the neural network has; l is the lth convolutional layer of the neural network; C is the number of output channels of the lth convolutional layer of the neural network; and for the lth convolutional layer, the number of input channels C_n is the number of output channels of the l-1st convolutional layer. k is the number of labels as

4. Experiment Design

This section presents the following aspects: the dataset used for the experiments, the labeling rules, the evaluation metrics, and an introduction to the comparative experimental model.

4.1. Dataset

Our proposed RG-FLAT-CRF model is validated with real datasets of three clinical NER tasks.

These three datasets are all from the CCKS competition dataset. The following is the introduction to these datasets.

CCKS-2017 data is adopted for the experiment. Since we did not participate in the competition, we only found some open-source data. The CCKS-CNER2017 dataset. Provides 300 electronic clinical record texts with 29,865 annotated instances (7816 sentences). It is annotated with five entity types: symptoms and signs, diseases and diagnosis, body parts, examinations and tests, and treatment. Table 1 lists its detailed statistics. The proportion of each part of the data is shown in Figure 11.

Table 1. Entity statistics of the three datasets.

Entity Type	CCKS2017	CCKS2019	CCKS2020	Total
symptoms and signs	7831	-	-	7831
diseases and diagnosis	721	5488	5628	11,837
body parts/anatomical parts	10,719	11,468	11,420	33,607
examinations and tests	9546	-	-	9546
treatment	1048	-	-	1048
examinations	-	1302	1626	2928
surgery	-	1182	1136	2318
tests	-	1678	2081	3759
drugs	-	2266	2814	5080
Total	29,865	23,384	24,705	77,954

CCKS-2019 contains 23,384 annotated instances (10,179 sentences). They are annotated with six entity types, namely diseases and diagnosis, examinations, tests, surgery, drugs, and anatomical parts. The elaborated statistics are shown in Table 1. The proportion of each part of the data is shown in Figure 10.

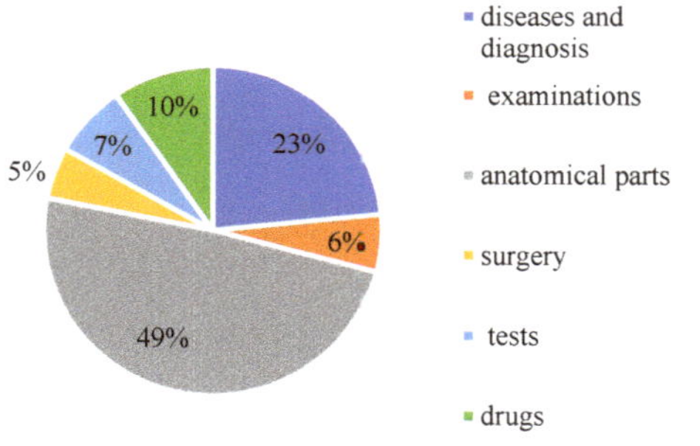

Figure 10. The proportion of medical entities on CCKS2019.

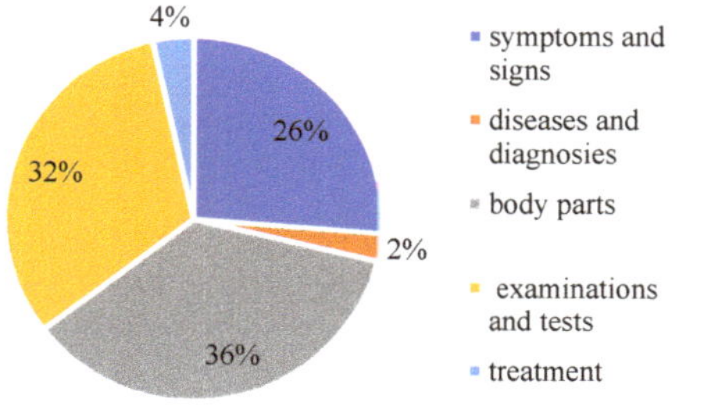

Figure 11. The proportion of medical entities on CCKS2017.

CCKS-2020 contains 24,341 annotated instances (13,308 sentences) with six entity types: diseases and diagnosis, examinations, tests, surgery, drugs, and anatomical parts. Table 1 shows the specific statistics. The proportion of each part of the data is shown in Figure 12.

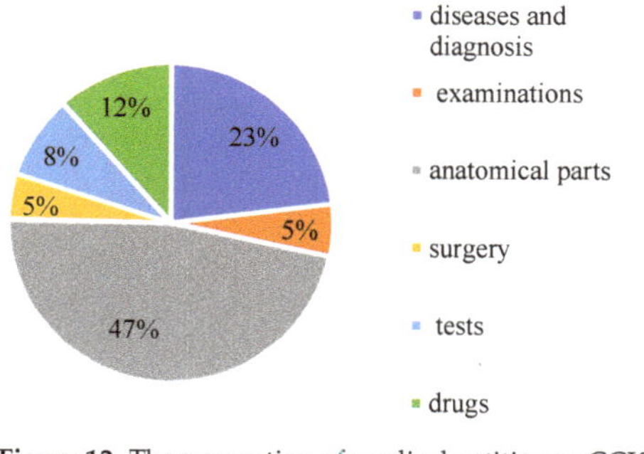

Figure 12. The proportion of medical entities on CCKS2020.

4.2. Labeling Rules

We adopt the BOI rule, where the entity's beginning is represented by B, I is the interior, and O stands for the other categories.

Annotation methods of five entity categories in CCKS2017: SS for symptoms and signs, DD for disease and diagnosis, AP for body parts, EE for inspection and examination, TM for treatment.

Annotation methods of six entity types in CCKS2019 and 2020: DD for disease and diagnosis, GEXA for examination, AP for the anatomical site, SU for surgery, EEXA for the test, and DR for the drug.

4.3. Evaluation Indicators

This paper uses the most common evaluation metrics in the NER field Precision, Recall, and F1 scores are used as the evaluation indicators of the model to evaluate the performance of the evaluation model comprehensively. TP is the number of positive samples predicted as positive samples, FN is the number of positive samples predicted as negative samples, and FP is the number of negative samples predicted as positive samples. They are widely used to evaluate classification and sequence annotation tasks [43].

Precision: The ratio of the number of recognized entities to the number of recognized entities is recorded as Precision, abbreviated as P. The calculation formula is Equation (25).

Recall: The percentage of correctly identified entities out of the number of entities in the sample. The calculation formula is Equation (26).

Both take values between 0 and 1, and the closer the value is to 1, the higher the precision or recall. Precision and recall are sometimes contradictory; a weighted harmonic mean that needs to be considered, and the F_1-score is a combination of the two. The higher the F1 score, the more robust the classification model is. The calculation formula is Equation (27).

$$Precision = \frac{TP}{TP + FP} \tag{25}$$

$$Recall = \frac{TP}{TP + FN} \tag{26}$$

$$F_1\text{-}score = \frac{2 \times Precision \times Recall}{Recall + Precision} \tag{27}$$

4.4. Experimental Parameters

The parameters of the RG-FLAT-CRF were tuned by Adam, and a hierarchical lr mechanism introduced. For the pre-trained RoBerta model, a learning rate of 3×10^{-5} is used, and for the other parts a learning rate of 2×10^{-4} is used. For the RG-FLAT-CRF model, the batch size used is 12. Details are shown in Table 2.

Table 2. Parameter settings.

Parameter	Value
lattice embedding	50
bigram embedding	50
glyce embedding	768
linear projection dim	768
dropout	0.1
Transformer head dim	20
Transformer head num	8

5. Results and Analysis

This part is divided into two parts: performance comparison with existing models, and ablation research.

5.1. Performance Comparison with Existing Models

To verify the effect of the RG-FLAT-CRF-model, the RGT-CRF model is compared to the existing state-of-the-art models. Evaluated on CCKS2017, CCKS2019, and CCKS2020 datasets, respectively. The comparison model is as follows:

(1) RoBerta: Liu et al. [37] improved the BERT model and proposed the RoBerta model. RoBerta performed better than BERT on NLP downstream tasks, and used RoBerta to enhance semantic representation and complete NER tasks.

(2) RoBerta-BiLSTM-CRF: Xu et al. [25] combined the bi-directional LSTM and CRF, which has become a classic model, and combined the RoBerta model with BiLSTM-CRF on this basis. Use RoBerta trained vectors and then use the BiLSTM-CRF model to extract entities.
(3) RoBerta-BiGRU-CRF: Qin et al. [29] proposed a BERT-BiGRU-CRF model in the field of Chinese electronic medical records, where the pre-trained model was replaced with an improved RoBerta.
(4) Ra-RC: Wu et al. [30] used RoBerta to obtain medical semantic features while using a bidirectional long short-term memory network to learn the radical features of Chinese characters.
(5) AR-CCNER: Yin et al. [26] used a convolutional neural network to extract radical features while using a self-attention mechanism to capture the dependencies between characters.
(6) ACNN: Kong et al. [27] used a multi-layer CNN structure to capture short-term and long-term contextual relations. CNN can also solve the problem that LSTM is difficult to exploit GPU parallelism, and the model uses an attention mechanism that can obtain global information.
(7) BE-Bi-CRF-JN: Wang et al. [31] cite additional medical knowledge information to correlate the original text in the named entity recognition task with its encyclopedic knowledge and enhance the ability of entity recognition by building a connection network.

Tables 3–5 show the precision, recall, and F1 results detailing various medical entities and all medical entities. From the comparison results of Table 6, the performance of the RGT-CRF model proposed in this chapter has achieved the best results on the three datasets, and the improvement on CCKS2017 is about 2~5%. The improvement is about 0.3~8% on CCKS2019 and about 3~9% on CCKS2020.

Table 3. Results of different models on CCKS2017.

Model	Evaluation Index	Entity Type					Comprehensive Value
		Symptoms and Signs	Diseases and Diagnosis	Body Parts	Examinations and Tests	Treatment	
RoBerta	P	96.95	74.21	88.09	95.59	76.39	91.99
	R	98.05	82.52	88.44	96.36	78.95	93.01
	F1	97.49	78.14	88.26	95.976	77.64	92.49
RoBerta-BiGRU-CRF	P	97.6	82.97	88.38	95.75	75.81	92.56
	R	97.99	81.82	88.69	96.58	77.99	93.09
	F1	97.79	82.39	88.53	96.16	76.88	92.82
RoBerta-BiLSTM-CRF	P	97.02	79.22	88.39	95.59	80.66	92.41
	R	98.39	85.31	90.65	96.53	81.81	94.11
	F1	97.7	82.15	89.5	96.05	81.23	93.25
Ra-RC	P	95	89.44	89.29	95.73	61.25	94.14
	R	98.11	89.17	90.79	97	69.01	92.39
	F1	96.53	89.31	90.03	96.36	64.9	93.26
ACNN	P	93.2	79.67	87.81	94.25	74.26	90.19
	R	97.92	79.39	84.41	95.9	75.7	90.78
	F1	95.5	79.52	86.08	95.07	74.97	90.49
AR-CCNER	P	96.53	74.07	89.38	94.78	82.91	92.27
	R	97.83	73.17	90.41	97.22	82.91	93.73
	F1	97.18	73.62	89.89	95.98	82.91	93
RGT-CRF	P	98.48	82.79	91.72	98.58	81.37	95.47
	R	98.66	85.31	90.16	97.2	82.99	95.76
	F1	98.56	84.03	90.93	97.88	82.17	95.61

Table 4. Results of different models on CCKS2019.

Model	Evaluation Index	Entity Type						Comprehensive Value
		Diseases and Diagnosis	Examinations	Anatomical Parts	Surgery	Tests	Drugs	
RoBerta	P	74.81	83.23	82.62	80.38	71.71	81.74	79.85
	R	75.27	83.48	84.14	78.4	63.79	82.08	79.85
	F1	75.03	83.35	83.37	79.37	67.51	81.9	79.85
RoBerta-BiGRU-CRF	P	67.15	65.47	82.94	76.1	65.58	64.62	72.95
	R	77.17	84.64	82.78	74.69	67.93	80.21	79.78
	F1	71.81	73.83	82.85	75.38	66.73	71.57	76.21
RoBerta-BiLSTM-CRF	P	75.76	84.97	83.06	78.26	69.73	77.49	79.7
	R	79.22	85.22	83.66	77.78	66.72	81.04	80.75
	F1	77.45	85.09	83.35	78.01	68.19	79.22	80.22
Ra-RC	P	78.74	81.23	82.67	79.61	74.28	93.27	83.31
	R	79	85.71	85.23	77.56	69.96	92.27	82.44
	F1	78.87	83.41	83.93	78.57	72.06	92.77	82.87
ACNN	P	75.84	85.37	88.2	76.27	68.23	92.34	83.07
	R	87.3	90.52	89.37	83.33	71.69	91.96	87.29
	F1	81.17	87.87	88.78	79.65	69.92	92.15	85.13
BE-Bi-CRF-JN	P	83.79	84.14	82.02	83.82	82.4	87.43	83.16
	R	83.66	89.71	87.55	90.48	89.8	85.11	86.67
	F1	83.73	86.83	84.7	87.02	85.94	86.25	84.88
RGT-CRF	P	78.3	88.84	85.58	84.23	78.5	88.44	85.36
	R	80.18	88.09	86.98	80.16	72.62	85.04	84.99
	F1	79.22	88.46	86.27	82.14	75.44	86.7	85.17

Table 5. Results of different models on CCKS2020.

Model	Evaluation Index	Entity Type						Comprehensive Value
		Diseases and Diagnosis	Examinations	Anatomical Parts	Surgery	Tests	Drugs	
RoBerta	P	85.28	78.05	90.03	88.03	69.28	87.77	86.68
	R	86.76	95.17	88.23	93.21	86.18	87.49	88.2
	F1	86.01	85.76	89.12	90.54	76.81	87.62	87.43
RoBerta-BiGRU-CRF	P	69.23	70.08	88.5	70.66	67.96	87.26	76.74
	R	84.65	91.5	89.4	91.4	86.59	87.92	88.09
	F1	76.16	79.37	88.94	79.7	76.15	87.58	82.02
RoBerta-BiLSTM-CRF	P	84.8	78.3	90.27	90.04	72.54	87.93	87.05
	R	84.8	92.56	88.57	94.12	83.74	87.27	87.67
	F1	84.8	84.83	89.41	92.03	77.73	87.59	87.35
BE-Bi-CRF-JN	P	81.88	81.25	94.64	83.7	76.82	94.64	82.52
	R	81.32	85.71	92.44	86.52	87.88	92.44	85.05
	F1	81.6	83.42	93.53	85.08	81.98	93.53	83.76
RGT-CRF	P	88.04	84.06	92.16	91.38	81.08	90.73	90.85
	R	87.33	94.05	91.81	92.4	88.96	90.35	91.57
	F1	87.68	88.77	91.98	91.88	84.83	90.53	91.2

The effect of ACNN is unstable in CCKS2017 and CCKS2019. Compared with other models, ACNN does not use BERT or an improved model based on BERT to enhance semantic representation, but multi-layer CNN and attention mechanisms play a certain positive role. From the three datasets, most of the models use BERT or an improved pre-training model based on BERT to enhance semantic representation and have achieved good experimental results. RoBerta-BiLSTM-CRF performs better than RoBerta-BiGRU-CRF on the three datasets. Although BiGRU has a simpler structure than BiLSTM, it is clear that BiLSTM is more suitable for Chinese electronic medical record NER. At the same time, these two models perform moderately well on the three datasets, as the feature extraction networks of the two models are variations of recurrent neural networks and cannot solve the long-range dependency problem. AR-CCNER and Ra-RC performed better on the

CCKS2017 and CCKS2019 datasets overall. Although AR-CCNER did not use a BERT-based pre-training model to enhance semantic representation, both AR-CCNER and Ra-RC were based on the characteristics of Chinese. BiLSTM and CNN are used to extract and use radical features, respectively, which utilize the glyph information of Chinese characters to a certain extent, but do not consider the information of learning the overall glyph structure of Chinese characters, and the model also lacks medical vocabulary information. BE-Bi-CRF-JN also achieved good results, proving that the use of external corpus in Chinese electronic medical records NER is effective. The above analysis shows that the RGT-CRF model is more suitable for Chinese electronic medical record named entity recognition electronic medical record recognition. This is mainly because the model adds glyph information while introducing lexical information based on words.

Table 6. Comparison of the results of different F1 of each model on different datasets.

Model	Dataset (F1)		
	CCKS2017	**CCKS2019**	**CCKS2020**
RoBerta	92.49	79.85	87.43
RoBerta-BiGRU-CRF	92.82	76.21	82.02
RoBerta-BiLSTM-CRF	93.25	80.22	87.35
Ra-RC	93.26	82.87	-
ACNN	90.49	85.13	-
AR-CCNER	93	-	-
BE-Bi-CRF-JN	-	84.88	83.76
RGT-CRF	95.61	85.17	91.2

From the perspective of entity type, the overall recognition effect of different medical entities is compared longitudinally. From Figures 13–15, it can be seen that the recognition results of different models on CCKS2017 show disease and diagnosis. Poor, because there are many long entities like '右股骨颈骨折髋关节股骨头表面置换术 (Right femoral neck fracture hip femoral head resurfacing)' in the two types of entities in the CCKS2017 dataset, and the boundaries of each entity cannot be clearly identified. The recognition results of different models on CCKS2019 and CCKS2020 show disease and diagnosis. The recognition results of these two types of entities are poor because the two types of entities in the CCKS2019 dataset and CCKS2020 dataset are similar to 'CA125', 'CEA'. Many entities coexist with English and numbers, such as 'CA199', which will also cause the model to fail to identify the boundaries of each entity.

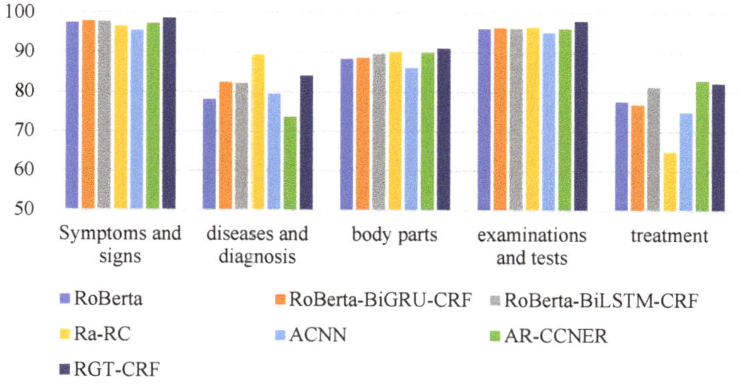

Figure 13. F1 values of different entities on CCKS2017 for different models.

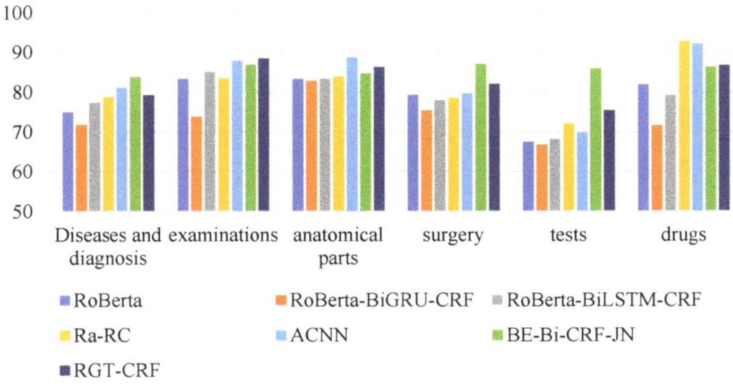

Figure 14. F1 values of different entities on CCKS2018 for different models.

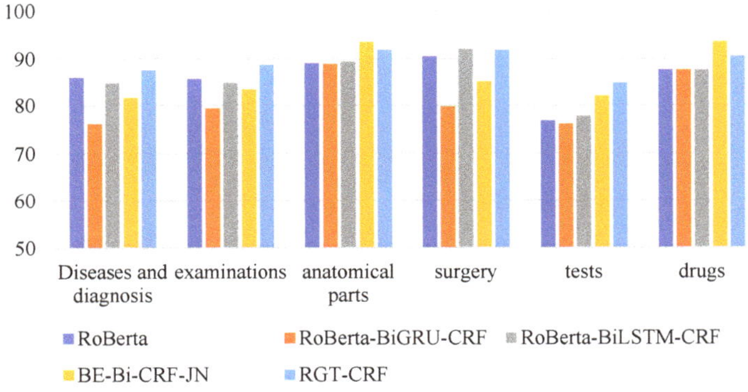

Figure 15. F1 values of different entities on CCKS2020 for different models.

To make the comparative results more convincing, a further hypothesis test was performed by calculating p-values using the t-test method, and *p*-values smaller than the significance level (usually 0.05) were considered statistically significant. Table 7 shows the statistical comparison of the proposed method with other methods. Most of the results are significant.

Table 7. Comparison results with different models on different datasets.

Model	CCKS2017	CCKS2019	CCKS2020
RoBera	0.0217	0.0019	0.0424
RoBerta-BiGRU-CRF	0.0382	0.0043	0.0049
RoBerta-BiLSTM-CRF	0.0029	0.0055	0.2025

5.2. Ablation Research

We design a set of ablation experiments to verify the contribution of each part to the model, where RGT-CRF-NG indicates that the model does not add glyph information. RGT-CRF-NF shows that the model does not add lexical information and its corresponding positional encoding. Finally, it is compared with RoBerta-BiLSTM-CRF and RGT-CRF on three datasets, and the results are shown in Table 8.

Table 8. Performance of different variants on three datasets.

Model	Dataset (F1)		
	CCKS2017	CCKS2019	CCKS2020
RoBerta-BiLSTM-CRF	93.25	80.22	87.35
RGT-CRF-NG	93.87	82.65	88.13
RGT-CRF-NF	94.92	84.03	89.83
RGT-CRF	95.61	85.17	91.2

The experimental results of RGT-CRF-NF and RGT-CRF-NG are better than the RoBerta-BiLSTM-CRF model regarding the three datasets, indicating that the glyph information and the use of lattice structure to add lexical information are effective for Chinese electronic medical record named entity recognition. The result of RGT-CRF-NG is slightly worse than that of RGT-CRF-NF, indicating that adding medical glyph information to the Chinese electronic medical record NER task is more effective than word information. This comparison can also be found in the above experiments using glyph information. Similarly, the final model with radical information is better than the model without radical information. This is because many Chinese characters in medical entities have the same glyph structure, so their meanings are also similar.

For example, '疼 (pain) ', '痛 (pain)', '病 (sick)', '腹 (belly)', '腰 (waist)', '肝 (liver)', '脾 (spleen)', '呕 (vomit)', '吐 (threw up)', '咳 (cough)', '嗽 (cough)', '胰 (pancreatic)', '肠 (intestinal)', '肿 (swell)', '胀 (swell)'. And this is very common in medical entities.

6. Conclusions

In this paper, an RG-FLAT-CRF model is proposed for Chinese CNER, which can learn the glyph features of medical fonts, and at the same time introduces word information to enhance word boundaries, and finally achieves good performance on three datasets. The RG-FLAT-CRF model obtains character vectors through RoBerta, Glyce, word2vec, and word vectors through word2vec. The word information is fused using the Flat-lattice structure and then encoded by the transformer network. In line with the output of the encoding layer, the label of each input character is predicted by the CRF layer. It addresses problems like word segmentation errors and lack of lexical information, given the characteristics of Chinese medical characters and the vector of multi-feature fusion. The final experimental results demonstrate that our proposed model outperformed the baseline models.

Several issues require further research. At this stage, deep learning requires a large amount of annotated data to train the model, as does our proposed model, but large-scale annotated data in the Chinese electronic medical record domain requires medical experts to annotate, which can be time-consuming. Therefore, our next research investigates how to perform named entity recognition on medical record texts with sparse data.

Author Contributions: J.L.: Conceptualization, Methodology, Software, Writing—original draft. Y.W.: Supervision, Project administration. R.L., C.C., and X.S.: Investigation, Writing—review & editing. S.Z.: Data curation, Resources. All authors have read and agreed to the published version of the manuscript.

Funding: This research was funded by the National Key R&D Program (Demonstration of R&D and Application of Integrated Science and Technology Service Platform for Central Plains Urban Agglomeration), grant number 2018YFB1404500.

Data Availability Statement: We used the CCKS open-source Chinese electronic medical record named entity recognition dataset and cite it in the article.

Conflicts of Interest: The authors declare no conflict of interest.

References

1. Chowdhury, S.; Dong, X.; Qian, L.; Li, X.; Guan, Y.; Yang, J.; Yu, Q. A multitask bi-directional RNN model for named entity recognition on Chinese electronic medical records. *BMC Bioinform.* **2018**, *19*, 75–84. [CrossRef] [PubMed]
2. Wang, Q.; Zhou, Y.; Ruan, T.; Gao, D.; Xia, Y.; He, P. Incorporating dictionaries into deep neural networks for the Chinese clinical named entity recognition. *J. Biomed. Inform.* **2019**, *92*, 103133. [CrossRef] [PubMed]
3. Shaukat, K.; Shaukat, U. Comment extraction using declarative crowdsourcing (CoEx Deco). In Proceedings of the 2016 International Conference on Computing, Electronic and Electrical Engineering (ICE Cube), Quetta, Pakistan, 11–12 April 2016; pp. 74–78.
4. Li, J.; Sun, A.; Han, J.; Li, C. A survey on deep learning for named entity recognition. *IEEE Trans. Knowl. Data Eng.* **2022**, *34*, 50–70. [CrossRef]
5. Alam, T.M.; Shaukat, K.; Hameed, I.A.; Khan, W.A.; Sarwar, M.U.; Iqbal, F.; Luo, S. A novel framework for prognostic factors identification of malignant mesothelioma through association rule mining. *Biomed. Signal Processing Control* **2021**, *68*, 102726. [CrossRef]
6. Vaswani, A.; Shazeer, N.; Parmar, N.; Uszkoreit, J.; Jones, L.; Gomez, A.N.; Kaiser, Ł.; Polosukhin, I. Attention is all you need. In *Advances in Neural Information Processing Systems 30*; Curran Associates Inc.: Red Hook, NY, USA, 2017; pp. 5998–6008.
7. Zhang, Y.; Yang, J. Chinese NER using lattice LSTM. *arXiv* **2018**, arXiv:1805.02023.
8. Li, X.; Yan, H.; Qiu, X.; Huang, X. FLAT: Chinese NER using flat-lattice transformer. *arXiv* **2020**, arXiv:2004.11795.
9. Dai, Z.; Yang, Z.; Yang, Y.; Carbonell, J.; Le, Q.V.; Salakhutdinov, R. Transformer-xl: Attentive language models beyond a fixed-length context. *arXiv* **2019**, arXiv:1901.02860.
10. Meng, Y.; Wu, W.; Wang, F.; Li, X.; Nie, P.; Yin, F.; Li, M.; Han, Q.; Sun, X.; Li, J. Glyce: Glyph-vectors for chinese character representations. *arXiv* **2019**, arXiv:1901.10125.
11. Mikolov, T.; Sutskever, I.; Chen, K.; Corrado, G.S.; Dean, J. Distributed representations of words and phrases and their compositionality. In *Advances in Neural Information Processing Systems 26*; Curran Associates Inc.: Red Hook, NY, USA, 2013; pp. 3111–3119.
12. Shaukat, K.; Luo, S.; Varadharajan, V.; Hameed, I.A.; Xu, M. A survey on machine learning techniques for cyber security in the last decade. *IEEE Access* **2020**, *8*, 222310–222354. [CrossRef]
13. Shaukat, K.; Luo, S.; Varadharajan, V.; Hameed, I.A.; Chen, S.; Liu, D.; Li, J. Performance comparison and current challenges of using machine learning techniques in cybersecurity. *Energies* **2020**, *13*, 2509. [CrossRef]
14. Friedman, C.; Alderson, P.O.; Austin, J.H.; Cimino, J.J.; Johnson, S.B. A general natural-language text processor for clinical radiology. *J. Am. Med. Inform. Assoc.* **1994**, *1*, 161–174. [CrossRef] [PubMed]
15. Fukuda, K.; Tamura, A.; Tsunoda, T.; Takagi, T. Toward information extraction: Identifying protein names from biological papers. *Pac. Symp. Biocomput.* **1998**, *707*, 707–718.
16. McCallum, A.; Freitag, D.; Pereira, F.C. Maximum entropy Markov models for information extraction and segmentation. *ICML* **2000**, *17*, 591–598.
17. Možina, M.; Demšar, J.; Kattan, M.; Zupan, B. Nomograms for visualization of naïve Bayesian classifier. In *European Conference on Principles of Data Mining and Knowledge Discovery*; Springer: Berlin/Heidelberg, Germany, 2004; pp. 337–348.
18. Settles, B. Biomedical named entity recognition using conditional random fields and rich feature sets. In Proceedings of the International Joint Workshop on Natural Language Processing in Biomedicine and Its Applications (NLPBA/BioNLP), Geneva, Switzerland, 28–29 August 2004; pp. 107–110.
19. Tang, B.; Cao, H.; Wu, Y.; Jiang, M.; Xu, H. Recognizing clinical entities in hospital discharge summaries using Structural Support Vector Machines with word representation features. *BMC Med. Inform. Decis. Mak.* **2013**, *13*, S1. [CrossRef]
20. Roberts, K.; Shooshan, S.E.; Rodriguez, L.; Abhyankar, S.; Kilicoglu, H.; Demner-Fushman, D. The role of fine-grained annotations in supervised recognition of risk factors for heart disease from EHRs. *J. Biomed. Inform.* **2015**, *58*, S111–S119. [CrossRef]
21. Liu, K.; Hu, Q.; Liu, J.; Xing, C. Named entity recognition in Chinese electronic medical records based on CRF. In Proceedings of the 2017 14th Web Information Systems and Applications Conference (WISA), Liuzhou, China, 11–12 November 2017; pp. 105–110.
22. LeCun, Y.; Boser, B.; Denker, J.S.; Henderson, D.; Howard, R.E.; Hubbard, W.; Jackel, L.D. Backpropagation applied to handwritten zip code recognition. *Neural Comput.* **1989**, *1*, 541–551. [CrossRef]
23. Mikolov, T.; Karafiát, M.; Burget, L.; Cernocky, J.; Khudanpur, S. Recurrent neural network based language model. *Interspeech. Makuhari* **2010**, *2*, 1045–1048.
24. Ma, X.; Hovy, E. End-to-end sequence labeling via bi-directional lstm-cnns-crf. *arXiv* **2016**, arXiv:1603.01354.
25. Xu, K.; Zhou, Z.; Hao, T.; Liu, W. A bidirectional LSTM and conditional random fields approach to medical named entity recognition. In Proceedings of the International Conference on Advanced Intelligent Systems and Informatics 2017; Springer: Cham, Swizerland, 2017; pp. 355–365.
26. Yin, M.; Mou, C.; Xiong, K.; Ren, J. Chinese clinical named entity recognition with radical-level feature and self-attention mechanism. *J. Biomed. Inform.* **2019**, *98*, 103289. [CrossRef]
27. Kong, J.; Zhang, L.; Jiang, M.; Liu, T. Incorporating multi-level CNN and attention mechanism for Chinese clinical named entity recognition. *J. Biomed. Inform.* **2021**, *116*, 103737. [CrossRef]
28. Zhang, W.; Jiang, S.; Zhao, S.; Hou, K.; Liu, Y.; Zhang, L. A BERT-BiLSTM-CRF model for Chinese electronic medical records named entity recognition. In Proceedings of the 2019 12th International Conference on Intelligent Computation Technology and Automation (ICICTA), Xiangtan, China, 26–27 October 2019; pp. 166–169.

29. Qin, Q.; Zhao, S.; Liu, C. A BERT-BiGRU-CRF Model for Entity Recognition of Chinese Electronic Medical Records. *Complexity* **2021**, *2021*, 6631837. [CrossRef]
30. Wu, Y.; Huang, J.; Xu, C.; Zheng, H.; Zhang, L.; Wan, J. Research on Named Entity Recognition of Electronic Medical Records Based on RoBERTa and Radical-Level Feature. *Wirel. Commun. Mob. Comput.* **2021**, *2021*, 2489754. [CrossRef]
31. Wang, Q.; Haihong, E. Bi-directional Joint Embedding of Encyclopedic Knowledge and Original Text for Chinese Medical Named Entity Recognition. In Proceedings of the 2021 2nd International Conference on Electronics, Communications and Information Technology (CECIT), Sanya, China, 27–29 December 2021; pp. 304–309.
32. Pennington, J.; Socher, R.; Manning, C.D. Glove: Global vectors for word representation. In Proceedings of the 2014 Conference on Empirical Methods in Natural Language Processing (EMNLP), Doha, Qatar, 25–29 October 2014; pp. 1532–1543.
33. Peters, M.E.; Neumann, M.; Iyyer, M.; Gardner, M.; Clark, C.; Lee, K.; Zettlemoyer, L. Deep contextualized word representations. *arXiv* **2018**, arXiv:1802.05365.
34. Devlin, J.; Chang, M.W.; Lee, K.; Toutanova, K. Bert: Pre-training of deep bidirectional transformers for language understanding. *arXiv* **2018**, arXiv:1810.04805.
35. Sun, Y.; Wang, S.; Li, Y.; Feng, S.; Chen, X.; Zhang, H.; Tian, X.; Zhu, D.; Tian, H.; Wu, H. Ernie: Enhanced representation through knowledge integration. *arXiv* **2019**, arXiv:1904.09223.
36. Cui, Y.; Che, W.; Liu, T.; Qin, B.; Yang, Z. Pre-training with whole word masking for chinese bert. *IEEE ACM Trans. Audio Speech Lang. Process.* **2021**, *29*, 3504–3514. [CrossRef]
37. Liu, Y.; Ott, M.; Goyal, N.; Du, J.; Joshi, M.; Chen, D.; Levy, O.; Lewis, M.; Zettlemoyer, L.; Stoyanov, V. Roberta: A robustly optimized bert pretraining approach. *arXiv* **2019**, arXiv:1907.11692.
38. Yang, Z.; Dai, Z.; Yang, Y.; Carbonell, J.; Salakhutdinov, R.R.; Le, Q.V. Xlnet: Generalized autoregressive pretraining for language understanding. In *Advances in Neural Information Processing Systems 32*; Curran Associates Inc.: Red Hook, NY, USA, 2019; p. 32.
39. Sun, Y.; Lin, L.; Yang, N.; Ji, Z.; Wang, X. Radical-enhanced chinese character embedding. In *International Conference on Neural Information Processing*; Springer: Cham, Switzerland, 2014; pp. 279–286.
40. Wang, S.; Zhou, W.; Zhou, Q. Radical and Stroke-Enhanced Chinese Word Embeddings Based on Neural Networks. *Neural Process. Lett.* **2020**, *52*, 1109–1121. [CrossRef]
41. Wei, H.; Zhang, H.; Gao, G. Word image representation based on visual embeddings and spatial constraints for keyword spotting on historical documents. In Proceedings of the 2018 24th International Conference on Pattern Recognition (ICPR), Beijing, China, 20–24 August 2018; pp. 3616–3621.
42. Su, T.R.; Lee, H.Y. Learning chinese word representations from glyphs of characters. *arXiv* **2017**, arXiv:1708.04755.
43. Shaukat, K.; Luo, S.; Chen, S.; Liu, D. Cyber threat detection using machine learning techniques: A performance evaluation perspective. In Proceedings of the 2020 International Conference on Cyber Warfare and Security (ICCWS), Islamabad, Pakistan, 20–21 October 2020; pp. 1–6.

electronics

MDPI

Article

Optimising Health Emergency Resource Management from Multi-Model Databases

Juan C. Arias [1], Juan J. Cubillas [2,*] and Maria I. Ramos [3]

1 Grupo TIC-144 del Plan Andaluz de Investigación, Universidad de Jaén, 23071 Jaén, Spain
2 Department Tecnologías de la Información y Comunicación aplicadas a la Educación, Universidad Internacional de La Rioja, 26006 Logroño, Spain
3 Department Ingeniería Cartográfica, Geodésica y Fotogrametría, Campus Las Lagunillas, Universidad de Jaén, Edif. A3, 23071 Jaén, Spain
* Correspondence: juanjose.cubillas@unir.net

Abstract: The health care sector is one of the most sensitive sectors in our society, and it is believed that the application of specific and detailed database creation and design techniques can improve the quality of patient care. In this sense, better management of emergency resources should be achieved. The development of a methodology to manage and integrate a set of data from multiple sources into a centralised database, which ensures a high quality emergency health service, is a challenge. The high level of interrelation between all of the variables related to patient care will allow one to analyse and make the right strategic decisions about the type of care that will be needed in the future, efficiently managing the resources involved in such care. An optimised database was designed that integrated and related all aspects that directly and indirectly affected the emergency care provided in the province of Jaén (city of Jaén, Andalusia, Spain) over the last eight years. Health, social, economic, environmental, and geographical information related to each of these emergency services was stored and related. Linear and nonlinear regression algorithms were used: support vector machine (SVM) with linear kernel and generated linear model (GLM), and the nonlinear SVM with Gaussian kernel. Predictive models of emergency demand were generated with a success rate of over 90%.

Keywords: healthcare; database design; geospatial data

Citation: Arias, J.C.; Cubillas, J.J.; Ramos, M.I. Optimising Health Emergency Resource Management from Multi-Model Databases. *Electronics* **2022**, *11*, 3602. https://doi.org/10.3390/electronics11213602

Academic Editors: Agnieszka Konys and Andrei Kelarev

Received: 27 September 2022
Accepted: 3 November 2022
Published: 4 November 2022

1. Introduction

The health sector is one of the most sensitive sectors in our society, and proof of this are the resources and efforts that are invested worldwide in trying to improve the management of the health system, especially in the optimisation of health resources [1]. Although there has been a huge change in the way diseases are diagnosed and treated, there has been little change in the way health services are managed in the 21st century. A number of academic studies have emerged in the field of service design, but not much of this research is available, especially in the field of health services [2]. An overview of the current state-of-the-art in this area shows that the vast majority of it is aimed at achieving greater economic efficiency in some aspects of the sector. There is scientific work in which different techniques have been tested in order to improve the management of health care resources. In this sense, Cubillas et al. [3] used tools of data mining to improve the appointment scheduling in primary health care centres. The results show that it is possible to predict, with a very acceptable level of precision, the number of patients who will attend the health centre each day. For this purpose, a series of historical assistance data were used. In this type of work, the quantity and quality of available data are the keys to generate an adequate predictive model. Similarly, other research has used spatial analysis to improve the effectiveness of these predictive algorithms, and confirms that the use of spatial data extends the scope of predictive models [4]. Additionally, the use of statistical methods to anticipate patient arrival rates in health care organisations allows one to schedule the internal staff in order to meet the demand for service driven by the patient arrival rate [5]. Nevertheless, this research had the limitation of the use of a few months of data to draw inferences about the patient arrival. This issue generates insights that are less reliable and more subject to short-term idiosyncrasies in data. In short, all types of research based on models that provide advance information on the behaviour of a phenomenon such as the demand for health resources are highly dependent on aa large volume of available data with a high spatio-temporal quality.

There is no relevant history of the implementation of systems that provide a daily and sufficiently early forecast of the demand for resources in emergency health services (i.e., that provide direct location and management of the resources that attend to patients on a daily basis) [6]. There are some studies that have highlighted an important increase in the need to optimise the structure of databases in order to face the demand for new necessary data in health care management [7,8]. An example of this is the implementation of telemedicine systems and the adaptation to the new information laws, thanks to the new broadband communications that are beginning to become generalised [9].

In the second decade of the 21st century, there has been an increase in publications aimed to improve the structure of databases adapted to daily management as a way to obtain more detailed health information. New channels of communication between the patient and health care services are proliferating by means of portable devices such as tablets and smartphones, or using a PC [10,11]. Subsequently, research in this sector has maintained this line of work and new challenges appear on the horizon. Moreover, the global pandemic initiated in 2020 by the novel severe acute respiratory syndrome (SARS)-CoV-2 virus (coronavirus disease 2019 [COVID-19]), has led to drastic changes in health priorities. Biomedical priorities have come to dominate the agenda, highlighting the multi-sectoral knowledge gaps and the challenges to be addressed for health management of the pandemic. In contrast, information management and decision support systems took a back seat [12]. However, once the process of immunisation of the population has begun, it is time to take stock and analyse how health resources have been managed and whether, in some way, correct decision making could have prevented the saturation of health services. There are many and varied data to be assessed in order to carry out an adequate management of health care resources. Despite the current pandemic, the demand for health care continues to be motivated by different pathologies and for different reasons.

In short, nowadays, more and more data are available, all from different sources, with different formats and different temporality and resolution. It is therefore necessary to properly manage and integrate a variety of data from new input devices used in health into centralised databases. It is also important that these databases are able to integrate any new variables resulting from the progress of health research. In this way, the usefulness of the database, in addition to assuming an effective resource management tool and providing quality to the service, would also have an important role in disease monitoring [13,14]. This scenario requires the development of specific tools and methodologies aimed at achieving these health management goals such as Hamami et al. (2019) [15], who highlighted that achieving the best model is a complex task due to the interaction of many components and the variability of parameter values that lead to radically different dynamics. It therefore points out that the modelling process can be improved through the use of data mining techniques [16]. Another example of the use of data mining techniques in health care management for decision making has already concluded that they can influence the costs, revenue, and operational efficiency while maintaining a high level of patient care [17].

There are, therefore, many aspects to consider and, above all, the large amount of data that is generated every day around a health service must be managed. Thus, in addition to data mining tools, it is an important area of application for big data [18], which is known as medical big data [19]. Medical big data comes from a variety of sources such as administrative records, clinical records, biometric data, data from patient reports, etc. They also are large in scale, extremely fast in update, polymorphic, incomplete, and time sensitive [20]. In addition, whether or not the data are used appropriately remains an open question. The data warehouse (DW) is the answer to data processing, but the applications of traditional DW methods in the health care domain require considerable attention due to the unique business nature of this industry [21]. Muji et al. (2010) [22] proposed a data-driven approach to the development of health information systems, which involved a database-centric system where different applications share the same integrated data source. The database design provides the necessary scalability to cover other specialised applications without the need for structural changes at the database level. The achievement of this objective requires databases with administrative and health care information data from several consecutive years [6,23,24] as well as an efficient model for storing and retrieving big health data to achieve valid estimates for optimal and quality management [25]. In short, in terms of offering an improvement in the quality of health care, it is essential to adapt database systems for use in DW and big data technologies and in their exploitation techniques.

In the field of health emergencies, we can cite the work of Graham et al. 2018 [26] in which a predictive study was carried out on the flow of patients to the emergency department from hospitals by using records from two large hospitals in the city of Northern Ireland. This work achieved a reliability of more than 80%.

Other more recent studies such as those by Gurazada et al. (2022) [27] have conducted predictive work on the length of stay of patients in the emergency department. Sixteen potentially relevant factors impacting on waiting times were identified through a literature review.

All of this work contributes to improving patient care by providing health care resource managers with advance information. These studies handle a large volume of patient data. However, in an emergency department, a large amount of data is recorded. Not only patient data, but also data about the service provided, the resources used, and external factors at the time of the emergency. The correct organisation and storage of this heterogeneous information in a database multiplies the possibilities of extracting hidden knowledge as well as predictive capabilities from the data.

This work focused on the management aspect of emergency health resources. It presents the design of a database that is complex enough, that is, with multiple variables extracted from each health emergency demand, to integrate all types of health information to allow for advanced knowledge and better management of these resources, thus providing quality patient care. The aim of this work was the design and implementation of a multidisciplinary database containing all of the information of the complete process of management and resolution of emergencies in the city of Jaén, in Andalusia, southern Spain. This database will serve as a source to apply and analyse regression algorithms of data mining in order to predict the demand for emergency resources that will occur in the coming days.

2. Methodology/Methods

2.1. Methodology of Work in Emergencies in Andalusia

An emergency is defined as a situation in which a person's life is in danger, otherwise, it is identified as an urgency. Currently, there is a free emergency telephone number available to citizens in Andalusia (061), Spain, and an urgency number (902505061). This service is provided by EPES (Public Company of Sanitary Emergencies) [28], which has eight provincial services in Andalusia, one per province. The provincial services are the headquarters from which all of the urgencies and emergencies of each province are managed. The most important nucleus of each provincial service is the coordination room, where all calls made by citizens to the emergency number of each province are received. Each coordination room is formed by one or several coordinating doctors and by a set of telephone managers that manage the citizen's health demand. These telephone managers attend to the citizen by gathering as much information as possible about the current request, the coordinating doctor participates in this management and finally decides, depending on the seriousness of the patient, which resource is mobilised to resolve the request.

All of the necessary health resources for all of the urgencies and emergencies of the province are mobilised from the coordination room. EPES has its own resources such as the terrestrial emergency teams (mobile UVI) and the air emergency teams (Sanitary Helicopters) as well as coordinating and mobilising all of the emergency resources of the Andalusian Health Service (SAS) and all ambulances from the province's urgent transport network (RTU). There are approximately 60 units in the Provincial Service of Jaen.

Once the resources have been activated from the coordination room, they are directed to the place of assistance, with all movements of said units being recorded in the computer system, knowing in real-time the geolocation of all of them and the exact moment of resolution of the assistance where it is. All emergency and emergency units have electronic devices (tablets) in which they register the patient's medical history of the care provided (HCDM Digital Clinical History in Mobility), which is computerised at the same time as the resolution. The most relevant information of this history is sent to the coordinating centre for storage, along with the demand created by the telephone manager at the beginning of the process, ending this demand and giving the cycle a new start (Figure 1).

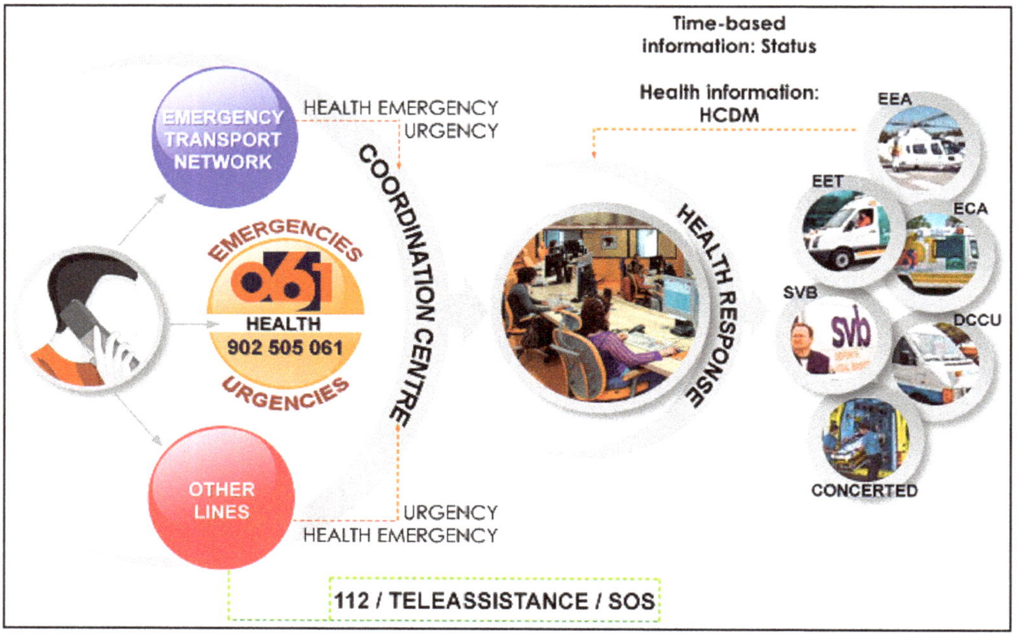

Figure 1. Complete cycle of urgencies and emergencies management.

The previously indicated units are located in pre-set and static locations. A significant improvement in this system would be to change the position of the available resources dynamically, according to the type of assistance they provide, the number of them, and the time of year they are available. For this purpose, the database presented in this work can be an important step forward, as the design, structure, and level of detail of the information stored allow for this objective to be achieved.

2.2. Dataset

The first phase of the work carried out consisted of information gathering, which involved data collection from various agencies involved in the health emergency service. The idea was to identify the idiosyncrasies of the emergency event at each point in time, together with the factors that may have influenced its occurrence, thus enriching the information currently stored in the system. This also involves downloading data from different websites with official statistical data, meteorological, social, and economic data. In the case of this research, data from the last eight years (2013–2020) of health activity in urgencies and emergencies were used. The main characteristic of most of the information integrated in the system is its spatial component. The geolocated data and their descriptions are as follows:

- Health data: All information related to the health care that the patient has received is stored, since the telephone call is received in the coordinating centre, until the medical team states the case as finished. EPES provided us with all of the data corresponding to the user's requests for assistance in urgency and emergency situations including diagnosis, clinical trial, treatment, antecedents, resources mobilised and detailed action times as well as the geolocation and their resolution.

  In order to enrich the data stored in the system, the following data were included:

- Atmospheric data: Data collected by the environmental information network of Andalusia (REDIAM) [29]. Data on temperatures, rainfall, humidity, and air quality were downloaded.
- Sociological data: Data related to the personal information of users were divided into:
  - Economic level of the patients in each area: Analysing, on one hand, the cadastral value of real estate, extracted from the Directorate General for Cadastre of Spain website [30]. We also added an analysis of the current price of housing in each district of the city through real estate web portals.
  - Level of unemployment in patients and their family units: This variable was obtained from data provided by the Spanish National Statistics Institute [31]. This public institution provided us with the type of population in each census tract. In Spain, a census tract composes a small region of the city, 1000 and 2500 residents.
  - Level of study, age of citizens, and members of the family unit: These data were obtained from the website of the Institute of Statistics and Cartography of Andalusia [32].

2.3. Data Mining Algorithms

As indicated in the introduction, in this work, data mining techniques were used, based on the attributes stored in the designed database. Prior to the development of the models, the following techniques were revised:

The predictive study carried out in this work was based on the use of regression algorithms. These included linear regression, logistic regression, the generalised regression model, one-class support vector machine (SVM), etc. In this study, the nature of the variables a priori was unknown, in fact, as discussed above, they are heterogeneous in nature. Linear and nonlinear regression algorithms were used as follows:

- Linear: Purely linear algorithms have great strength due to their characteristics and simplicity as they are calculated with a simple weighted sum of the variables:

$$y = \beta 0 + \beta 1 \times 1 + \ldots + \beta pxp + \epsilon \quad (1)$$

The first algorithm selected in this study was the generalised linear models (GLM) algorithm, which works mathematically as the weighted sum of the features with the mean value of the distribution assumed by the link function g, which can be chosen flexibly depending on the type of result.

$$g(EY(y \mid x)) = \beta 0 + \beta 1 \times 1 + \ldots \beta pxp \quad (2)$$

In other words, this algorithm is an extension of linear algorithms that allows linear or normal distributions and non-constant variances to be modelled. Linear models make a set of restrictive assumptions, in which the target is normally distributed conditional on the value of the predictors with a constant variance, regardless of the value of the predicted response. In this sense, GLM relaxes these restrictions, and for a binary response example, the response is a probability in the range [0, 1] [33,34].

Another linear algorithm selected was SVM, which has the advantage of being able to be used with different kernels. Kernels allow the data to be distributed on a hyperplane according to a function, which facilitates the adaptation of the algorithm to the nature of the data, allowing for infinite transformations.

In this study, we worked with the SVM with linear kernel. When the linear kernel is used, the following transformation is performed

$$K(x,x') = x \cdot x' \quad (3)$$

This algorithm has the advantage that it fits very well if the nature of the data is linear, and if there are many predictor variables (as in the case study). Note that in this algorithm, there is no upper limit on the number of predictor attributes, and the only limitations are those imposed by the hardware.

- Nonlinear: In this case, the SVM algorithm is applied with a Gaussian kernel. This kernel applies the following transformation to the data:

$$K(x,x') = \exp(-\gamma \mid \mid x - x' \mid \mid 2) \quad (4)$$

The value of γ controls the behaviour of the kernel. When it is very small, the final model is equivalent to that obtained with a linear kernel, as its value increases, the data move away, forming a Gaussian bell in the hyperplane, fitting very well when the nature of the data does not have a linear distribution.

In summary, these are the advantages of these three algorithms in this study, starting from the hypothesis of a priori ignorance of the relationship between the variables with the target attribute and also considering that the training data we had were limited and the predictor variables were numerous. Moreover, the complexity of these algorithms means that the relationship between the attributes used cannot be described by a specific equation.

In short, the following algorithms were applied in this study:

- Minimum description length (MDL) algorithm [35]: For attribute importance detection, all attributes that do not relate to the target attribute are discarded.
- Regression algorithms: Once the valid attributes are known, several algorithms are tested in order to determine which of them has better accuracy. Generalised linear models (GLM) [33] and support vector machines (SVM) (linear and Gaussian kernel) [36] algorithms are used. The GLM algorithm is a pure linear model. On the other hand, support vector machines (SVM) is a powerful algorithm based on statistical learning theory. The main advantage of the SVM algorithm is that it can be configured with different kernels, in this case, we used a linear and Gaussian kernel.

3. Results and Discussion

3.1. Database

The workflow of the database design was divided into three phases. The first consisted of data collection from all sources described above. In the second, a data cleaning process was developed to facilitate data management and analysis. In this phase, a process of cleaning, normalisation, and grouping was carried out. We started with the original table, demands, which had 84 attributes, in which all the details of the assistance demands made by the emergency teams can be found. In order to prepare the structure of the database for different types of exploitation, this information was restructured into specific blocks that include several tables. These blocks are as follows:

- Resources mobilised in emergency assistance

This is the most important block in the database as they are tables that contain the health data. These tables contain all the information corresponding to the assistance provided by the emergency teams during the last eight years in the province of Jaen.

- Patient personal data

This block includes the patient information fields that contain the personal information of each patient. Most significant fields are the age, sex, date of birth, and address (Table 1).

Table 1. Personal patient information.

Attribute	Description
Age	Patient's years old
Sex	Man/Woman
Birth	Date of birth
Id_Province	ZIP code
Id_District	District code
Id_Town	Town code
Id_Street	Street code
Id_Number	Number
Id_Door	Door

- Patient health information fields

These contain the health information of all patients attended. The most significant fields are shown in Table 2.

Table 2. Patient health information.

Attribute	Description
IdTypeofdemand	Classification of demand Type of demand: A Attendance/T Transport/ I Informative
IdTypeofdemand1	First level of detail of the type of demand. E.g: 01 Transport accident 02 Alteration of vital signs 04 Dyspnoea …
IdTypeofdemand2	Second level of detail of the type of demand
IdTypeofdemand3	Third level of detail of the type of demand
IdSResource	Resource Code
IdAssistance	Attendance Code
ZipCode	Zipcode
ClinicalJudgment1	First Clinical Trial
ClinicalJudgment2	Second Clinical Trial
ClinicalJudgment3	Third Clinical Trial
IdResolveCode	Resolution Code: 1* do not arrive to see patient 2* Arrive but do not act 3* Attend to patient
IdFinance	Financing code. (State funding, Private companies …)
AdrDTDestination Situation	Destination code of the resource. Team attending: U Urban P Peripheral, if the assistance is covered by either of the two teams.
AdrDTSituation	Equipment Coverage Zone: Urban/Peripheral
IdAdmissionCenter	Hospital Admission Centre

- Chronological information fields of the assistance

In relation to the information on the ambulance mobilised for each assistance, the start and end time of each assistance interval is recorded. This information includes the time at which the mobile resource is activated by the coordinating centre, the time at which it arrives at the site of medical assistance, the time elapsed during the action on the patient, the time of transport of the patient to the hospital, and the time at which the mobile resource is available again for the next assistance (Table 3).

Table 3. Chronological information of the assistance.

Attribute	Description
Year_D	Year of date of attendance.
Month_D	Month of date of attendance.
Day_D	Day of date of attendance.
IdProvince	Province code.
Requestdate	Request date
IdRequest	Request code
IdCall	Call code
IdLine	Line code
IdLineType	Line type code
IdAlertant	Alert source code (User, General emergency service . . .)
InLetTime	Incoming call time
RingTime	Time at which the system rings
AnswerTime	Time the call is answered
ResourceCreationTime	Time of resource creation
ActivationTime	Resource activation time
ExitTime	Time of departure of the resource
ArrivalTime	Time of arrival of the resource
LoadTime	Time of patient loading in the ambulance (resource)
DestinationTime	Time of arrival at destination
OperationTime	Time at which the resource becomes operational
AvailableTime	Time when you are fully available for other assistance
CoordinationTime	Demand coordination time
ActivationTiming	Time taken to activate the resource
AttentionTiming	Patient care time
AnswerTiming	Response time (from the time the call comes in until the resource reaches the patient)
IdExclusionGround	Reason for exclusion in case of failure to send the appropriate resource for that request
IdResourceUnit	Resource unit sent.
IdResourceType	Resource unit sent.

- Weather information

Numerous studies indicate that environmental and weather factors directly influence conditions such as allergies and directly influence the onset of certain diseases such as allergies or certain chronic illnesses. Thus, it is important to bear in mind that Jaén is the largest oil producer in the world and, therefore, the flowering of the olive tree in spring, when temperatures are high, means that many people allergic to pollen demand emergency services. Adverse weather conditions also lead to a proliferation of accidents requiring emergency services. Therefore, it is clear that for a better management of emergency resources, these meteorological factors have to be considered as they can influence a sudden increase in the demand for emergency assistance. In this study, some external factors have been considered that can influence the number of required assistance such as meteorological factors (e.g., minimum, maximum, and average temperature, precipitation, humidity, and daily air quality data). The data downloaded from the website of the Environmental Information Network of Andalusia REDIAM [29] has a field ESTACION_ID, indicating the number of meteorological stations, with a total of 20 meteorological stations monitored throughout the province of Jaén. All atmospheric data collected in the meteorological stations of the urban core of the city of Jaén corresponded to the same period of assistance considered in this research (Figure 2).

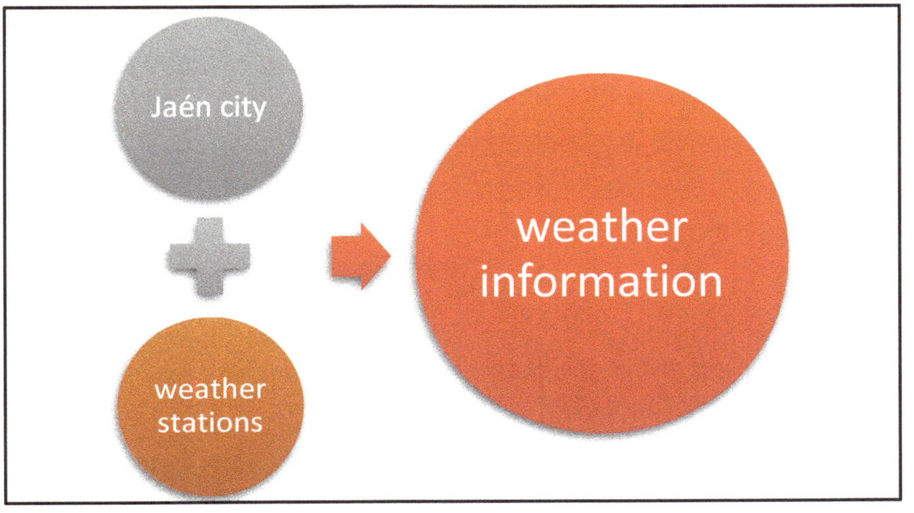

Figure 2. Definitive table composition of meteorological information.

- Sociological information

The urban core is divided into nine districts called postcodes. Because the location of the assistance provided is given by postcode, as much information as possible was collected for each postcode. The source of information used was the Institute of Statistics and Cartography of Andalusia [32], distributed in 100 fields with these generic areas: (1) total number of inhabitants, by sex and age group; (2) marital status of the population, by age group; (3) level of studies, by age group; (4) types of housing, use, regime, size; and (5) households, and number of people that compose it.

- Geolocated quadrants

One of the factors that enriches the database is the incorporation of geolocated information. This allows the exploitation of the database to take into account the variability of the information depending on its location. The minimum geographical unit considered is a geolocated quadrant of 250 m, which is the one used by the Institute of Statistics and Cartography of Andalusia [32]. This was not considered in the fragmentation of the urban core map of Jaén in 128 quadrants with cells of 250 m (Figure 3). The information stored for each quadrant was:

- Quadrant identification;
- X, Y (UTM Zone 30, ETRS89) coordinates of the four corners;
- Streets and numbers of them included;
- Total population by age groups;
- Employment information: affiliates, for others, in their own and pensioners;
- Link with zip code.

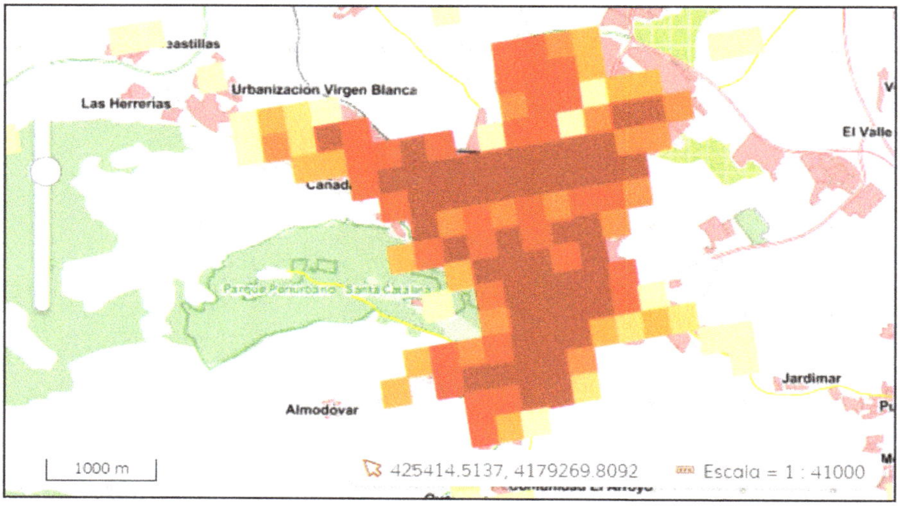

Figure 3. Geolocated quadrants that divide the town of Jaén.

Finally, in the third phase, the database was designed and created. The structure of the entity relationship diagram (ERD) is presented in Figure 4.

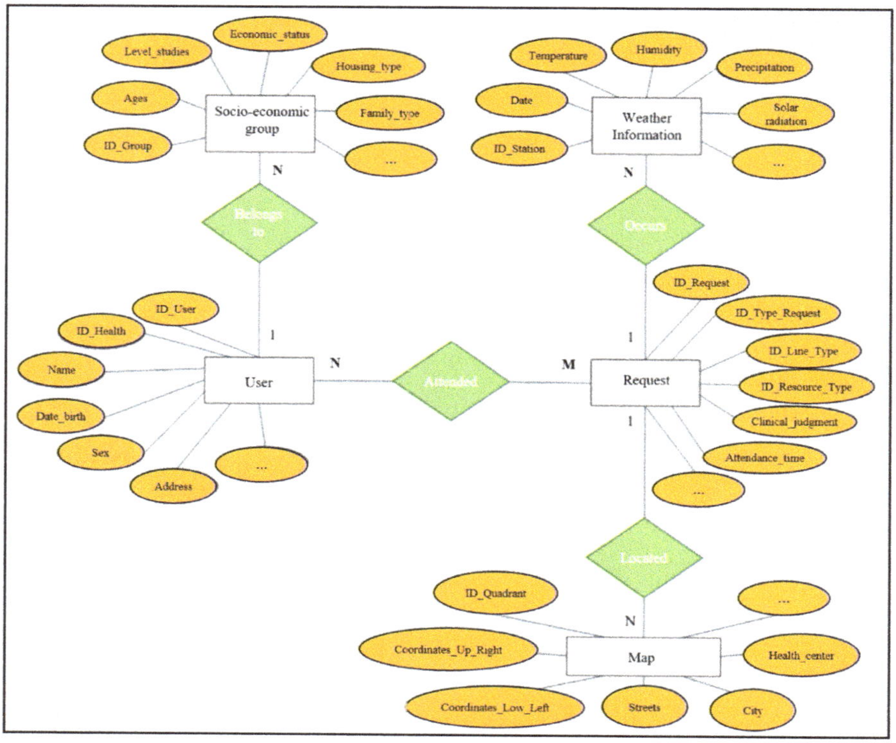

Figure 4. Entity relationship diagram (ERD).

The entity relationship model conceptually represents the organisation and relationship of the data in the designed database. In this case, it is a simplified representation as the database has multiple tables (more than 50 tables). The purpose of Figure 4 is to show the type of information stored and its relationship. Each entity was grouped as a block of information, and the data blocks represented were as follows: socio-economic data of the users that can potentially be attended, meteorological and environmental information, clinical data of the users, geographical information of the users, registers of, and finally, the emergency resources mobilised.

The database management system used was the Oracle Database [37], which is a system of object-relational type (ORD). The development environment used was Oracle SQL Developer [38], an integrated development environment that allows working with SQL in Oracle databases. This environment allowed us to create and execute SQL queries and procedures for the integration of different types of information.

- Debugging tables and preparing data

The structure explained above has many applications, one of the most important is to study the number of attendances that are expected by demarcation, taking into account the different variables that affect the result. Deepening this assumption, new tables in which the information grouped by days, zones, and resources will be stored can be generated. A forecast of the number and type of expected attendances will be obtained as well as valuable information on the mobilisation of resources that is expected at different times of the year, also taking into account variables such as atmospheric data, day of the week, holiday, or work, economic values of the area, and all the variables that surround an assistance (Table 4).

Table 4. New generated table for resource management optimisation.

Attribute	Description
Date	Date of data registered
Year	Year of data registered
Month	Month of data registered
Weekday	This field records whether it is a weekend or not.
Holiday	This field records whether it is a holiday or not.
Day_Month	Number between 1 and 12
Day_Year	Number between 1 and 365
Week_Month	Number between 1 and 5
Week_Year	Number between 1 and 52
Max_Temperature	Highest temperature recorded on that date
Min_Temperature	Lowest temperature recorded on that date
Avg_Temperature	Average temperature recorded on that date
Max_Humidity	Highest humidity recorded on that date
Min_Humidity	Lowest humidity recorded on that date
Avg_Humidity	Average humidity recorded on that date
Wind_Speed	Average wind speed value recorded on that date
Radiation	Average solar radiation value recorded on that day
Precipitation	Average precipitation value recorded on that day
Demarcation	Delimitation
Resource_Type	Type of resource
Number_Resources	Total number of resources mobilised on that date

As indicated above, the urban core of Jaén was divided into 128 quadrants, which allowed us to exploit the information in a georeferenced manner. Each of the quadrants were stored using the coordinates that defined them, which allowed us to study the information of the demands and the population in a detailed and geographical manner.

The way the database was designed and the inclusion of geographic information and other external factors such as meteorological factors allowed the data to be exploited for predictive analysis. The growth of the database and the increase in the volume of information stored means that valuable historical data are now available. Future exploitation and analysis of these data such as detecting patterns of behaviour and relationships between variables will allow advance planning of the emergency resources available at each time of the year and in each part of the city.

3.2. Predictive Model

The results of the first phase of the study corresponded to the application of the MDL algorithm to identify which attributes had the most influence on the target attributes. In this case, the number of resources mobilised (Table 5) shows the attributes that are related to the target attribute and are therefore used by the algorithms. The attributes in the database were checked by the minimum description length (MDL) algorithm, which returned a value between 0 and 1, with 0 indicating that the attribute has no relationship with the target attribute, and 1 indicating that the attribute has the maximum relationship. The attributes that were related to the target and that were used to train the model are shown below. The model was trained with real demand data from 2011 to 2019, and 2020 was left out to produce the result of the predictive algorithms with real demand data from 2020.

Table 5. Attributes used in the model.

Name	Description
Demarcation	Location of the demand requested
Type of holiday	Working day or public holiday
Month	Month of the request
Day week	Day of the week (Monday, Tuesday, ...)
Maximum humidity	Predicted maximum relative humidity
Average humidity	Expected average relative humidity
Minimum humidity	Predicted minimum relative humidity
Precipitation	Precipitation forecast
Solar radiation	Solar radiation
Max temperature	Predicted maximum temperature
Average_temperature	Predicted average temperature
Minimum_temperature	Predicted minimum temperature
Type of resource mobilised	Type of resource mobilised (ambulance, intensive care unit, doctor, nurse, etc.)
Wind speed	Wind speed

In this work, as mentioned in Section 2.3, two types of regression algorithms were used: linear and nonlinear. In the case of the linear algorithms, support vector machine (SVM) with linear kernel and the generated linear model (GLM) were used; on the part of the nonlinear models, SVM with Gaussian kernel was used. Each model has its advantages and disadvantages, in the case of SVM, it provides great performance when there is little training data available, however, GLM is an extension of a linear regression model, which is very useful when the conditional distribution of the target attribute is not normal, introducing a link function g (2). Its adjustment in practice is conducted using the maximum likelihood method, therefore, this model was based on calculating the weighted sum of the predictors. These models were formulated by John Nelder and Robert Wedderburn as a way of unifying statistical models such as linear regression, logistic regression, and Poisson regression

The prediction focuses on determining the number of emergency resource activations that will be required to meet the demand for emergency health care, for which the three regression models were tested, and to measure their efficiency, a model was generated with data for the years 2011–2019 and the prediction was made for the year 2020, comparing the absolute error of the prediction with the real data for the year 2021. The results obtained were as follows: GLM had an error of 9%, SVM with linear kernel 16%, and SVM with Gaussian kernel 21%. The efficiency of the model can be seen in the form of a graph. Figure 5 shows the actual number of emergency resource mobilizations each day during 2020. For this purpose, a predictive model was generated with the training data of the actual realised demands in the year 2011 until 2019. Then, from the three models tested, it the prediction of resources to be mobilized in 2020 was generated, and finally, it calculated the absolute error by comparing the prediction with the actual value of the mobilised resources. The graph showed the prediction of the GLM model for each day (red line) and the blue line represents the actual number of resources mobilised in this year, so the accuracy of the model could be seen graphically. The absolute error of the GLM regression algorithm was 9%; this value was very good since the variation in the mobilisation of the demands can vary from 50 on the day when the most were mobilised and nine on the day when the least were mobilised (i.e., the variation was higher than 555%).

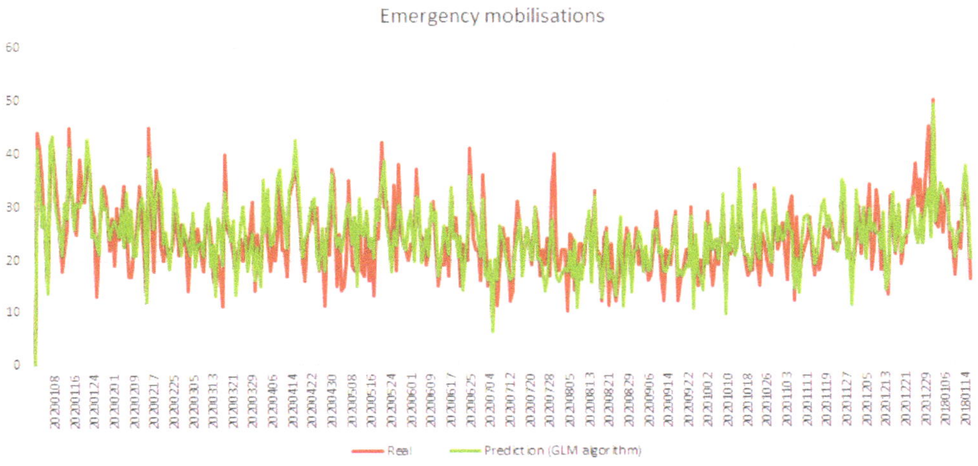

Figure 5. Prediction on the emergency resource activation. Comparison of the actual data and predictions for the year 2020.

Considering that the number of activations varies greatly, ranging from 20 to 45 per day, it is very important to be able to have a temporary forecast in advance, as each ambulance is equipped with a doctor, nurse, and driver. Therefore, it involves a significant expenditure of health care resources.

4. Conclusions

The general objective of the project was to create a database as complete as possible and with a great diversity of information, which would represent in detail all possible aspects of the emergency health activity. We did not just want to store data, but to obtain the maximum details of the entire process of attending to an emergency, that is, from the moment the call is received in the coordination room until the end of the assistance received by the patient, thus closing the health claim that said patient originated.

An additional objective that we addressed was to study and store all the non-health aspects that surround an emergency and that may affect that emergency. As previously mentioned, the economic, social, environmental, and geographical aspects of each of the emergencies have been studied. The next step was to analyse all of this information and study the percentage of relationship that each variable had with the appearance or alteration of said emergencies. In this sense, it has been concluded that there is a direct relationship between the environmental factors and the activation of emergency services in Jaén. This relationship was statistically quantified with the MDL algorithm, which quantifies the relationship of each attribute with the target attribute.

Another important achievement is that a model was designed using the multi-model database where not only clinical data, but also other very basic environmental and air quality factors are stored, these attributes being precisely some of the input system data for the prediction. These data are available on several websites with up to a 10-day forecast.

The main conclusion of this work is that we managed to develop models that are able to predict the number of activations of the emergency services with an absolute error of 6%, considering the large variation in the number of activations from one day to another, with variations of more than 110%. Other predictive studies in the health sector have achieved a reliability of around 80% [26]. This study achieved better accuracy. It is also important to note that this study worked with health data captured at the time of care by the doctor or nurse. These data are stored in the optimised database, which allows these data to form part of the training data of the predictive model by recalculating the predictions and readjusting the model each day as the database grows. This is disruptive to other work [39], where public or non-clinical data sources are used.

There are predictive works that use machine learning to address the evolution of patients in the emergency department, more specifically, the level of mortality [40], and others have focused on predicting the population groups that are more likely to use health services [41]. In this sense, what is innovative about the study presented here is that it focused on accounting for the resources that will be mobilised each day (i.e., being able to know in advance the emergency health demand that will be received on a given day). It is therefore a prediction that makes it possible to anticipate the resources available, improving the quality of patient care. This information, in advance, is an indicator that can be very important for emergency resource managers, being a useful tool, better than a naïve model based on the average of historic values. The use of this tool can also help to improve several aspects of health care management. The first is the economic plan, if the demand is known well in advance. Another important aspect is that the application of the model will increase efficiency, as we will be able to anticipate the demand for resources, a key aspect in health emergencies.

Finally, it can be concluded that this multi-model database allowed us to exploit the information with predictive models. Furthermore, it is a first step toward further work in the future to analyse the type of resources requested in the demands and the main pathologies of the activations, or even determine or predict the location where the emergency activation will take place.

Author Contributions: All authors whose names appear on the submission made substantial contributions to the conception or design of the work, nevertheless, here are the concrete contributions of each author: Writing—original draft: J.C.A.; Methodology and formal analysis: J.C.A. and J.J.C.; Writing—Reviewing and Editing, M.I.R. All authors have read and agreed to the published version of the manuscript.

Funding: This research received no external funding.

Informed Consent Statement: All methods were carried out in accordance with the relevant guidelines and regulations. All experimental protocols were approved by EPES (Empresa Pública de Emergencias Sanitarias) Consejería de Salud y Familias. Junta de Andalucía (Spain). No clinical data were used in this research. Informed consent was obtained from all subjects and/or their legal guardian(s).

Data Availability Statement: Data included in the database cannot be shared due to the data protection law in Spain. The design of the database is available and can be shared openly. To request the data, please contact the first author. In this work, clinical data were not used.

Acknowledgments: This work would not have been possible without the support of EPES (Public Company for Health Emergencies), a company belonging to the Andalusian Health System. This work was also partially supported by the Graphics and Geomatics Group of Jaén (TIC-144), and the research PREDIC_I-GOPO-JA-20-0006 which is co-financed with European agricultural fund for rural development and the Junta de Andalucía funds.

Conflicts of Interest: We confirm that there are no potential competing interests and all authors have approved the manuscript for submission.

References

1. Institute of Medicine (US). *Improving the Nation's Health Care System*; National Academies Press (US): Washington, DC, USA, 2009.
2. Vaz, N.; Venkatesh, R. Service Design in the Healthcare Space with a Special Focus on Non-Clinical Service Departments: A Synthesis and Future Directions. *Health Serv. Manag. Res.* **2022**, *35*, 83–91. [CrossRef] [PubMed]
3. Cubillas, J.J.; Ramos, M.I.; Feito, F.R.; Ureña, T. An Improvement in the Appointment Scheduling in Primary Health Care Centers Using Data Mining. *J. Med. Syst.* **2014**, *38*, 89. [CrossRef] [PubMed]
4. Ramos, I.; Cubillas, J.J.; Feito, F.R.; Ureña, T. Spatial Analysis and Prediction of the Flow of Patients to Public Health Centres in a Middle-Sized Spanish City. *Geospat. Health* **2016**, *11*, 452. [CrossRef] [PubMed]
5. Ganguly, A.; Nandi, S. Using Statistical Forecasting to Optimize Staff Scheduling in Healthcare Organizations. *J. Health Manag.* **2016**, *18*, 172–181. [CrossRef]
6. Wiréhn, A.-B.E.; Karlsson, H.M.; Carstensen, J.M. Estimating Disease Prevalence Using a Population-Based Administrative Healthcare Database. *Scand. J. Public Health* **2007**, *35*, 424–431. [CrossRef]
7. Kerr, K.; Norris, T.; Stockdale, R. Data Quality Information and Decision Making: A Healthcare Case Study. In Proceedings of the 18th Australasian Conference on Information Systems, Toowoomba, Australia, 5–7 December 2007.
8. Salman, O.H.; Rasid, M.F.A.; Saripan, M.I.; Subramaniam, S.K. Multi-Sources Data Fusion Framework for Remote Triage Prioritization in Telehealth. *J. Med. Syst.* **2014**, *38*, 103. [CrossRef]
9. Pérez, J.; Iturbide, E.; Olivares, V.; Hidalgo, M.; Martínez, A.; Almanza, N. A Data Preparation Methodology in Data Mining Applied to Mortality Population Databases. *J. Med. Syst.* **2015**, *39*, 152. [CrossRef]
10. Trifirò, G.; Coloma, P.M.; Rijnbeek, P.R.; Romio, S.; Mosseveld, B.; Weibel, D.; Bonhoeffer, J.; Schuemie, M.; van der Lei, J.; Sturkenboom, M. Combining Multiple Healthcare Databases for Postmarketing Drug and Vaccine Safety Surveillance: Why and How? *J. Intern. Med.* **2014**, *275*, 551–561. [CrossRef]
11. Ramos, M.I.; Cubillas, J.J.; Feito, F.R. Improvement of the Prediction of Drugs Demand Using Spatial Data Mining Tools. *J. Med. Syst.* **2016**, *40*, 6. [CrossRef]
12. Burkle, F.M.; Bradt, D.A.; Ryan, B.J. Global Public Health Database Support to Population-Based Management of Pandemics and Global Public Health Crises, Part I: The Concept. *Prehospital Disaster Med.* **2021**, *36*, 95–104. [CrossRef]
13. Mezghani, E.; Exposito, E.; Drira, K.; Da Silveira, M.; Pruski, C. A Semantic Big Data Platform for Integrating Heterogeneous Wearable Data in Healthcare. *J. Med. Syst.* **2015**, *39*, 185. [CrossRef] [PubMed]
14. Wang, Y.; Kung, L.; Byrd, T.A. Big Data Analytics: Understanding Its Capabilities and Potential Benefits for Healthcare Organizations. *Technol. Forecast. Soc. Chang.* **2018**, *126*, 3–13. [CrossRef]
15. Hamami, D.; Atmani, B.; Cameron, R.; Pollock, K.G.; Shankland, C. Improving Process Algebra Model Structure and Parameters in Infectious Disease Epidemiology through Data Mining. *J. Intell. Inf. Syst.* **2019**, *52*, 477–499. [CrossRef]
16. Benhar, H.; Idri, A.; Fernández-Alemán, J.L. A Systematic Mapping Study of Data Preparation in Heart Disease Knowledge Discovery. *J. Med. Syst.* **2018**, *43*, 17. [CrossRef] [PubMed]
17. Silver, M.; Sakata, T.; Su, H.C.; Herman, C.; Dolins, S.B.; O'Shea, M.J. Case Study: How to Apply Data Mining Techniques in a Healthcare Data Warehouse. *J. Healthc. Inf. Manag. JHIM* **2001**, *15*, 155–164. [PubMed]
18. Oussous, A.; Benjelloun, F.-Z.; Ait Lahcen, A.; Belfkih, S. Big Data Technologies: A Survey. *J. King Saud Univ. Comput. Inf. Sci.* **2018**, *30*, 431–448. [CrossRef]
19. Lee, C.H.; Yoon, H.-J. Medical Big Data: Promise and Challenges. *Kidney Res. Clin. Pract.* **2017**, *36*, 3–11. [CrossRef] [PubMed]
20. UNDP. *Human Development Report 2015*; UNDP: New York, NY, USA, 2015.
21. George, J.; Kumar, B.V.; Kumar, V.S. Data Warehouse Design Considerations for a Healthcare Business Intelligence System. In Proceedings of the WCE 2015, London, UK, 1–3 July 2015; Available online: http://www.iaeng.org/publication/WCE2015/ (accessed on 17 March 2022).
22. Muji, M.; Ciupa, R.; Dobru, D.; Bică, C.; Olah, P.; Bacarea, V.; Marusteri, M. Database Design Patterns for Healthcare Information Systems. In Proceedings of the International Conference on Advancements of Medicine and Health Care through Technology, Cluj-Napoca, Romania, 23–26 September 2009; pp. 63–66, ISBN 978-3-642-04291-1.
23. Brookhart, M.A.; Stürmer, T.; Glynn, R.J.; Rassen, J.; Schneeweiss, S. Confounding Control in Healthcare Database Research: Challenges and Potential Approaches. *Med. Care* **2010**, *48*, S114–S120. [CrossRef]
24. Yue, X.; Wang, H.; Jin, D.; Li, M.; Jiang, W. Healthcare Data Gateways: Found Healthcare Intelligence on Blockchain with Novel Privacy Risk Control. *J. Med. Syst.* **2016**, *40*, 218. [CrossRef]
25. Goli-Malekabadi, Z.; Sargolzaei-Javan, M.; Akbari, M.K. An Effective Model for Store and Retrieve Big Health Data in Cloud Computing. *Comput. Methods Programs Biomed.* **2016**, *132*, 75–82. [CrossRef]
26. Graham, B.; Bond, R.; Quinn, M.; Mulvenna, M. Using Data Mining to Predict Hospital Admissions From the Emergency Department. *IEEE Access* **2018**, *6*, 10458–10469. [CrossRef]
27. Gurazada, S.G.; Gao, S. (Caddie); Burstein, F.; Buntine, P. Predicting Patient Length of Stay in Australian Emergency Departments Using Data Mining. *Sensors* **2022**, *22*, 4968. [CrossRef] [PubMed]
28. Empresa Pública de Emergencias Sanitarias. *EPES—061 | Gestión de las Emergencias y Urgencias Sanitarias en Andalucía*; Empresa Pública de Emergencias Sanitarias: Malaga, Spain, 2021.
29. Red de Información Ambiental de Andalucía—Portal Ambiental de Andalucía. Available online: https://www.juntadeandalucia.es/medioambiente/portal/acceso-rediam (accessed on 21 February 2020).

30. Sede Electrónica Del Catastro—Inicio. Available online: http://www.sedecatastro.gob.es/ (accessed on 18 January 2020).
31. INE. Instituto Nacional de Estadística. Available online: https://www.ine.es/ (accessed on 10 January 2020).
32. Instituto de Estadística y Cartografía de Andalucía. Available online: https://www.juntadeandalucia.es/institutodeestadisticaycartografia (accessed on 13 January 2020).
33. Dobson, A.J. *An Introduction to Generalized Linear Models*, 2nd ed.; Chapman & Hall/CRC texts in statistical science series; Chapman & Hall/CRC: Boca Raton, FL, USA, 2002; ISBN 978-1-58488-165-0.
34. Bolker, B.M.; Brooks, M.E.; Clark, C.J.; Geange, S.W.; Poulsen, J.R.; Stevens, M.H.H.; White, J.-S.S. Generalized Linear Mixed Models: A Practical Guide for Ecology and Evolution. *Trends Ecol. Evol.* **2009**, *24*, 127–135. [CrossRef]
35. Grünwald, P.D.; Myung, J.I.; Pitt, M.A. (Eds.) *Advances in Minimum Description Length: Theory and Applications*; Neural Information Processing series; Bradford Books: Cambridge, MA, USA, 2005; ISBN 978-0-262-07262-5.
36. Cortes, C.; Vapnik, V. Support-Vector Networks. *Mach. Learn.* **1995**, *20*, 273–297. [CrossRef]
37. Gestión de Datos Autónoma. Available online: https://www.oracle.com/es/autonomous-database/ (accessed on 23 February 2020).
38. SQL Developer. Available online: https://www.oracle.com/database/technologies/appdev/sqldeveloper-landing.html (accessed on 23 February 2020).
39. Bobashev, G.; Warren, L.; Wu, L.-T. Predictive Model of Multiple Emergency Department Visits among Adults: Analysis of the Data from the National Survey of Drug Use and Health (NSDUH). *BMC Health Serv. Res.* **2021**, *21*, 280. [CrossRef] [PubMed]
40. Sánchez-Salmerón, R.; Gómez-Urquiza, J.L.; Albendín-García, L.; Correa-Rodríguez, M.; Martos-Cabrera, M.B.; Velando-Soriano, A.; Suleiman-Martos, N. Machine Learning Methods Applied to Triage in Emergency Services: A Systematic Review. *Int. Emerg. Nurs.* **2022**, *60*, 101109. [CrossRef]
41. Garcia-Canton, C.; Rodenas, A.; Lopez-Aperador, C.; Rivero, Y.; Anton, G.; Monzon, T.; Diaz, N.; Vega, N.; Loro, J.F.; Santana, A.; et al. Frailty in Hemodialysis and Prediction of Poor Short-Term Outcome: Mortality, Hospitalization and Visits to Hospital Emergency Services. *Ren. Fail.* **2019**, *41*, 567–575. [CrossRef]

electronics MDPI

Article

Acquiring, Analyzing and Interpreting Knowledge Data for Sustainable Engineering Education: An Experimental Study Using YouTube

Zoe Kanetaki [1], Constantinos Stergiou [1], Georgios Bekas [1], Sébastien Jacques [2,*], Christos Troussas [3], Cleo Sgouropoulou [3] and Abdeldjalil Ouahabi [4]

1 Laboratory of Mechanical Design, Department of Mechanical Engineering, University of West Attica, 12241 Athens, Greece; zoekanet@uniwa.gr (Z.K.); stergiou@uniwa.gr (C.S.); gb331984@googlemail.com (G.B.)
2 Research Group on Materials, Microelectronics, Acoustics and Nanotechnology (GREMAN), University of Tours, UMR 7347, CNRS, INSA Centre Val-de-Loire, 37100 Tours, France
3 Educational Technology and eLearning Systems Laboratory, Department of Informatics and Computer Engineering, University of West Attica, 12243 Athens, Greece; ctrouss@uniwa.gr (C.T.); csgouro@uniwa.gr (C.S.)
4 UMR 1253, iBrain, Université de Tours, INSERM, 37000 Tours, France; abdeldjalil.ouahabi@univ-tours.fr
* Correspondence: sebastien.jacques@univ-tours.fr; Tel.: +33-6-667-639-46

Abstract: With the immersion of a plethora of technological tools in the early post-COVID-19 era in university education, instructors around the world have been at the forefront of implementing hybrid learning spaces for knowledge delivery. The purpose of this experimental study is not only to divert the primary use of a YouTube channel into a tool to support asynchronous teaching; it also aims to provide feedback to instructors and suggest steps and actions to implement in their teaching modules to ensure students' access to new knowledge while promoting their engagement and satisfaction, regardless of the learning environment, i.e., face-to-face, distance and hybrid. Learners' viewing habits were analyzed in depth from the channel's 37 instructional videos, all of which were related to the completion of a computer-aided mechanical design course. By analyzing and interpreting data directly from YouTube channel reports, six variables were identified and tested to quantify the lack of statistically significant changes in learners' viewing habits. Two time periods were specifically studied: 2020–2021, when instruction was delivered exclusively via distance education, and 2021–2022, in a hybrid learning mode. The results of both parametric and non-parametric statistical tests showed that "Number of views" and "Number of unique viewers" are the two variables that behave the same regardless of the two time periods studied, demonstrating the relevance of the proposed concept for asynchronous instructional support regardless of the learning environment. Finally, a forthcoming instructor's manual for learning CAD has been developed, integrating the proposed methodology into a sustainable academic educational process.

Keywords: computer-aided design (CAD); educational data mining; engineering education; online and hybrid learning environments; social media analytics

Citation: Kanetaki, Z.; Stergiou, C.; Bekas, G.; Jacques, S.; Troussas, C.; Sgouropoulou, C.; Ouahabi, A. Acquiring, Analyzing and Interpreting Knowledge Data for Sustainable Engineering Education: An Experimental Study Using YouTube. *Electronics* **2022**, *11*, 2210. https://doi.org/10.3390/electronics11142210

Academic Editors: Agnieszka Konys and Agnieszka Nowak-Brzezińska

Received: 24 May 2022
Accepted: 13 July 2022
Published: 14 July 2022

1. Introduction

Today, many higher education institutions are integrating learning activity data analytics into their operations [1]. Universities are recognizing the benefits of information solutions that not only better students at all stages of their education, even the most challenging, but also implement ever more effective educational resources to enhance the learning experience for students and their instructors [2]. Teaching module organizers and instructors focus on analytics results and the use of algorithms to improve their content and flexibility to identify students at risk of academic failure as early as possible and then provide them with more targeted learning solutions [3].

Long before the global health crisis due to the SARS-CoV-2 virus, nearly a decade of student education data could be used to understand key aspects of learner characteristics that would differentiate those who are capable of graduating from those at risk of dropping out. With all educational procedures going online, the acquisition of electronic data from a variety of online sources has led to an increase in all analytical procedures, and their management is a concern for the academic community [4].

It has been more than two years since the learning experience in universities was totally disrupted by the consequences of the COVID-19 pandemic, resulting in a significant decline in student retention and academic performance. Sustainable measures were then gradually put in place to ensure students' virtual presence and support them in their asynchronous tasks. Today, after almost a year and a half of exclusive and mandatory distance learning imposed by the pandemic, the learning processes of higher education have been tested in terms of applicability, feasibility and long-term sustainability. Although the health situation is not yet fully stabilized, the educational models deployed during the most critical period must be evaluated [5–8].

The first semester of the 2021–2022 academic year is undoubtedly a defining moment in the educational community. Although the post-COVID-19 period has not yet arrived, it is now appropriate to speak of the beginning of the "meta-COVID-19 period," certainly defining a period of disruption, but offering many exciting opportunities in the academic world. Specifically, in the era of digital transformation of the educational procedure [9], the authors of [10] applied the Greek term "meta", meaning beyond, to describe a promising future for academia, followed by a global health phenomenon. With the increasing implementation of digital learning systems during the health crisis, it is important to ensure that the system design can motivate and support active student engagement to achieve the required educational goals. Therefore, socially oriented technology tools can be incorporated into the design of online learning systems to increase student engagement and improve student performance during the learning process [11]. By adjusting learning tactics and introducing technological features applied during the epidemic into meta-COVID-19 instruction, academic institutions will be able to progress and thrive in sustainable educational models.

The work presented here consists of an experimental study that focuses on instructor monitoring of learner behaviors and engagement throughout the teaching module to assess the effectiveness and sustainability of the applied learning tactic. The methodology applied is based on video analysis and, more precisely, on the processing and interpretation of data coming from the consultation by students of online digital content directly from the reports of a YouTube channel made available as part of a computer-aided mechanical design (CAD) module [12]. The analysis of the experimental data collected, analyzed and interpreted should make it possible to evaluate the benefit of this YouTube channel dedicated to asynchronous pedagogical support in online or hybrid teaching environments. To convince the university community of the sustainable integration of the YouTube channel into the educational process, learners' viewing habits were analyzed in depth from the channel's 37 instructional videos, all in conjunction with the execution of the CAD course. To develop the results of this work, two distinct learning periods are considered: the first refers to exclusively distance learning spaces (i.e., during the most critical period of the health crisis), and the second reflects both face-to-face and mixed learning environments (i.e., mixing face-to-face and distance).

The study proposed here covers a wide range of skills: from the creation of a complete asynchronous educational and social environment based on a YouTube channel where digital observations can be acquired, to the extraction and exploitation of visualization data. The novelty of this study lies in the fact that the use of social media channels in online and hybrid learning spaces has not yet been analyzed in depth, for its sustainable integration in the academic educational procedure.

The development of this work will be articulated as follows: Section 2 will first present a review of the literature related to the objectives of this work. The methodological and organizational aspects will be presented in Section 3, starting with the creation of the educational YouTube channel, the sources of data exploration, and proceeding to the statistical analysis of the acquired data where the results will be presented. The main results obtained will be presented in Section 4, and a discussion, based on these results, will be conducted in Section 5. Finally, the conclusions and research perspectives will be analyzed in Section 6.

2. Related Work

Long before the emergence of distance learning imposed by the COVID-19 pandemic, institutions around the world incorporated learning management systems (LMS) into their instructional schemes, whether in online or blended learning environments [13]. During the health crisis, several learning tools were used, individually or in combination. At the University of West Attica (Greece), as well as at the University of Tours (France), the Microsoft (MS) Teams learning platform was, for example, widely used for synchronous transmissions, in combination with the E-Class (or Moodle) LMS for asynchronous support. In [14], researchers proved that using a single learning platform (MS Teams), supported by a social media channel such as YouTube for asynchronous support, limited the dropout rate of learners, compared to MS Teams supported by the LMS Moodle.

Just as considering customer preferences is critical to the development of a business, in education, analyzing students' preferences and taking steps to provide them with learning materials in innovative learning spaces could be the key to improving their academic performance [15,16].

Educational data mining (EDM) seems to have a major effect in the field of education [17]. EDM and data analytics promise a better understanding of student learning, as well as new insights into the hidden aspects that influence learner performance [18]. Learning analytics (LA) is an emerging area of learning management systems that tracks and records student activities in online and virtual learning environments [19]. The role of LA is to use the data generated by students as they interact with new technology features to improve the teaching–learning process (TLP) and enable instructors to make better decisions in terms of structuring teaching modules [20]. It is generally accepted that the more data you collect, the more information you get. Therefore, the more information one acquires, the more accurate predictions, forecasts and estimates can be obtained. Data quality is very important, as it is affected by the number of variables and the amount of data acquired, which can lead to information sparsity, especially in cases where the quality of the data appears to be poor [21]. In addition, process analysis allows for the observation of unusual activities and behaviors, which can lead to the detection of "outliers", alarm objects, and calls for intervention [22]. Given the power of the method, LA can therefore be a major feedback tool for educators and instructional designers to improve the learning experience [23].

Researchers in [24] developed a technology acceptance model (TAM) to examine which factors of social networking sites such as YouTube and TikTok can support and facilitate online knowledge acquisition. In this study, data collection was conducted using an online questionnaire on four external factors: content richness, innovativeness, satisfaction, and enjoyment. The results showed that both social networking sites contribute to knowledge sharing and acquisition. Although reports from YouTube channels were not retrieved and considered in this research, the authors concluded that to increase acceptance, the focus should be on uploaded video content.

Although the authors of [25] conducted a systematic review of the literature describing the sources and use of educational data, data analyses from the YouTube channel were unfortunately not considered. Additionally, in [26], the researchers examined the influence of instructor-generated video content on student engagement and participation in a course using the number of posts per week and the number of characters per post as parameters. They conducted an independent-sample *t*-test to compare student evaluation of the course by dividing the population into two groups: those who had been exposed to instructor-generated videos and those who had not. The test showed statistical significance between the two groups.

In [27], the authors discussed the use of YouTube analytics both to assess student attendance during lectures and to measure the impact of lectures on the student learning experience. Going further, the authors in [28] planned campaigns by processing analytics data from the tools offered by social media channels.

The remarkable popularity of social media applications can be attributed to the encryption technology used, which ensures user privacy and limits access to personal information [28]. Social media is thought to offer benefits such as enhancing human interaction through the use of electronic media, increasing creativity, creating a sense of affiliation and acceptance, encouraging engagement and cooperative learning, reducing restrictions in terms of space and social or economic position within a community, increasing interaction and communication among members, and improving users' technological expertise [29]. With technological tools now widely available to people under the age of 20, it is possible to access social media sites with one hand, thanks to mobile technology and the use of inexpensive electronic devices such as tablets and smartphones [30,31].

Sharing videos or their URL links has become easier for instructors with the adoption of virtual learning environments (VLE). Content in teaching modules, discussion forums, and targeted sequences in online courses can be shared via YouTube video links embedded in the assessment features of learning platforms [32]. The analytics of YouTube channels can provide valuable information about learners' viewing habits, as well as measures of how students engage in the learning process with videos [33]. With YouTube analytics, instructors can thus track video viewing behaviors on supportive tasks to better understand their usefulness [34]. Previous studies in this area have shown that students do not follow a video in its entirety. They play, stop, rewind, and replay the educational content in order to review segments of the video recordings. This specific learner behavior is intended to recall the part of the newly acquired knowledge that was not clearly defined [33,35]. In [35], the authors investigated the use of pausing and searching in videos of course recordings provided by the channel interface, and how these two features relate to students' learning tactics and performance in a specific curriculum. Before the COVID-19 pandemic imposed restrictions on academia, researchers studied how learners in traditional learning environments and flipped classrooms interacted with the videos, as well as the nature of those interactions through analysis of processing data [32]. Instructors recorded short or long videos of portions of their lectures and uploaded them to VLE [32].

In transforming the traditional learning space into a virtual one, one of the most significant problems has been the loss of contact with the engineering environment itself, which is a key aspect of engineering education [36,37]. In addition, the authors of [10] associated educational data mining, processing, and analysis with the term "sustainability", which is present and promoted in most aspects of everyday life, from business operations to manufacturing and the environment [38]. Data mining, the processing of data and eventual interpretation of the results, is a fundamental process that allows researchers to establish the relationship between raw digital data and the assessment of real world conditions [39]. With digital data now available by tracking activities, analysts can get lost in unnecessary information. To facilitate the process, researchers should be able to set their boundaries, creating controlled environments for data production, i.e., targeted to the goals of their research area.

In this study, the data collected and analyzed provides insight into students' video content viewing behaviors from a dedicated YouTube channel. These behaviors were analyzed over two distinct time periods: the first in 2020–2021, i.e., during the most critical period of restrictions due to the COVID-19 pandemic, and the second in 2021–2022, during the early hours of the meta-COVID-19 period. The objective is to study the similarities between the delivery of engineering training modules in exclusively online mode and in mixed mode. This objective had already been set well before the implementation of the new pedagogical environments, whether online or hybrid; the social communication channel was then used as an asynchronous pedagogical support allowing the collection of information directly related to students' behaviors, while avoiding the "noise" due to irrelevant details.

Given the results already available and the gaps identified by the literature review, this work is guided by the following five research questions:

- Can a YouTube channel provide adequate and quality asynchronous support on student tasks in online learning environments?
- Can we provide a guide for future instructors and organizers of CAD modules, willing to implement pedagogical methods supported by technology and social media sites such as YouTube, which would benefit the learning process in online and hybrid spaces?
- Can learners' viewing habits be revealed in online and hybrid learning spaces through information provided by YouTube?
- Do learners' behaviors and visualization patterns follow the flow of the module?
- Can educational data mining (EDM) from social media sites like YouTube provide a solid foundation for addressing student needs?

The answers to the above research questions should help demonstrate the relevance of educational data from social media channels such as YouTube to the academic community, and help institutions implement their strategies for sustainable digital transformation in higher education.

The research objective of this study is not limited to evaluating a specific YouTube channel as a tool to support asynchronous teaching. It also aims to provide feedback to instructors and suggest steps and actions to implement in their teaching modules to ensure students' access to new knowledge while promoting their engagement and satisfaction, regardless of the learning environment, i.e., face-to-face, fully remote, and hybrid.

3. Methodological and Organizational Aspects

3.1. Foreword

The methodology described in this section was deployed in a 12-week "computer-aided mechanical design (CAD I)" module, a teaching module within the Department of Mechanical Engineering of the University of West Attica (Greece). Prior to the health crisis, this module was divided between traditional mechanical design in a room equipped with drawing boards and computer-aided design with Autodesk Inventor software in a computer lab.

During the most critical hours of the COVID-19 pandemic, videos (directly downloadable into the MS Teams environment available to students) were integrated into flipped classrooms to provide asynchronous instructional support to complement the online courses delivered synchronously via the MS Teams platform. All of this work was done, including the integration of the following three tasks:

- Provide asynchronous support to students and incorporate it into assessments of individual tasks.
- Guarantee students the robustness of the learning process, including avoiding any disruption caused by multiple platforms or LMS.
- Create a virtual lab to connect distance learning spaces with real engineering environments related to students' future work [36,37,40].

During the period when pandemic restrictions were relaxed, the 12-week CAD module was divided into two main stages. For the first 4 weeks, all classes were conducted face-to-face in the classroom equipped with drawing boards. The objective of this phase was to teach students to represent views of a three-dimensional object by freehand drawings (sketches). In the second stage, conducted in the computer lab, students enrolled in activities to create three-dimensional mechanical objects in different views (top, side, and cross-section) using CAD tools. To complete these activities, learners had the option of participating face-to-face or online (the class was streamed live by the instructor via

MS Teams). In both stages, asynchronous support videos were attached to the students' assigned task. The YouTube channel "MCAD I UNIWA" was created to provide learners with all the video support needed to complete their learning independently. This YouTube channel, and all of its video content (see Appendix A), is managed by a CAD instructor with over twenty years of CAD experience. The administrator of the YouTube channel was also responsible for coordinating all activities of the various instructors in the teaching module. The MCAD I UNIWA's YouTube channel policy is public. In order to target specific tasks and associate them with their asynchronous support video links, the tasks were signed as "assignments" on the MS Teams communication platform. Each task was announced in the students' MS Teams dashboard, where instructions were provided in text form [36]. The YouTube channel links for each task were uploaded as reference material targeting the specific task to avoid confusion when searching the 37 videos to find the one relevant to the task, as shown in Figure 1. The instructor was motivated to attach the URLs of the videos to each task after considering that most viewers of today's social media channels do not easily subscribe to the channels. In this way, we were able to reach both non-subscribing and subscribing students, with the latter being immediately notified of new videos.

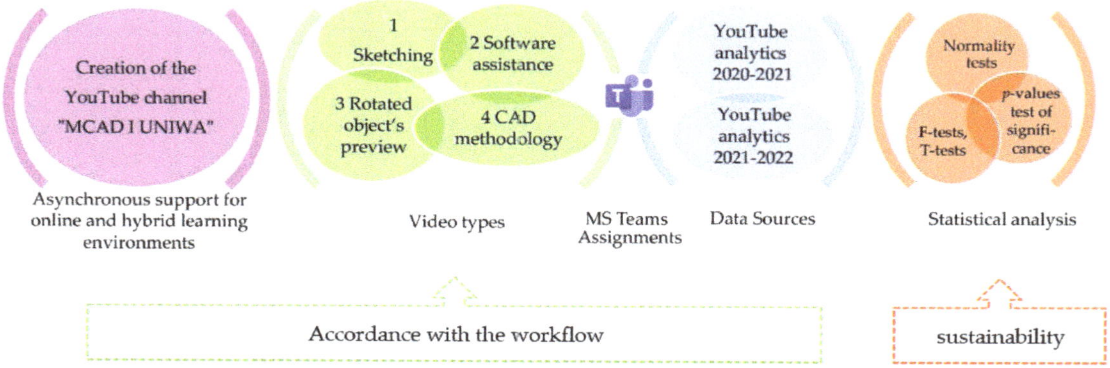

Figure 1. Flow chart illustrating the data mining and processing methodology applied.

What we seek to highlight in this manuscript is the presence of a significant relationship between learners' listening habits and their behaviors, particularly in the acquisition of knowledge and skills necessary for graduation. To do so, we draw on feedback from the CAD I module, looking in depth at data collected from the "MCAD I UNIWA" YouTube channel. To achieve this objective, we implemented the methodology described in Figure 1. The first step was the creation of a YouTube channel in which most of the instructional video content was created from screen recordings and audio recordings intended exclusively for student use. We did not use the raw recording of the full laboratory lecture because we wanted to target specific tasks, i.e., fundamental to the mechanical engineering profession, to help students perform them outside of class [41].

The uploaded videos were divided into four categories, based on the learning objectives to be achieved at each stage of the teaching process. Video categories 1 and 3 were devoid of audio, primarily to invite the instructor to explain the content presented at their own pace and in their own style. Specifically, the sketch videos showed the instructor's sketchbook, complete with pencil, as he or she drew freehand views of the object. The videos showing the model of the object being studied in three dimensions were intended to help students design the geometric shapes in all views around the object. The third and fourth categories were generated by a combination of screen recordings and audio recordings from the software modeling environment.

The YouTube channel administrator examined viewing patterns since the first video was uploaded to progressively interpret learners' needs and determine if these screen and audio recordings met the demand for asynchronous support. During learning phases conducted exclusively online, visualization patterns could be identified by the end of the first semester. When the health regulations were relaxed, i.e., when university educational spaces were allowed to transform into hybrid learning environments (i.e., combining online and face-to-face learning), a challenge arose: to analyze the visualization patterns of learners in hybrid educational spaces and correlate the results with those of spaces conducted exclusively at a distance. Once all the data was collected, the method then consisted of filtering the data from the social media channel reports corresponding to the two distinct time periods: the first corresponding to the first semester of the 2020–2021 academic year in the exclusively distance learning environments and the second, the first semester of the 2021–2022 year in the face-to-face, virtual, and mixed learning environments. Thus, two years were necessary to collect sufficient data and to ensure the results that will be discussed in the rest of the manuscript.

The final step in the methodology is to perform statistical tests on the defined variables. First, normality tests were performed to assess the normality of the distribution. In the case where the distribution is Gaussian, parametric tests were chosen. In the opposite case, non-parametric tests were performed [42]. For determining if the variances analyzed for the two separate academic years' time periods are equal or unequal, a series of *F*-tests were performed. These tests can eventually be completed by *t*-tests depending on the equality or inequality of the variances generated by the *F*-tests.

3.2. Participant Demographics

Although not such an easy task, analysis of YouTube channel reports provides detailed information about the demographics of the participants and gives a "typical user profile" while revealing repetitive viewing behaviors. In our case, the term "user" refers to engineering students attending the computer-aided mechanical design module. Since one of the research questions focuses on developing a manual for future instructors, it is essential that module coordinators observe student activities outside the classroom and ultimately develop a profile of the typical student, which each instructor must consider, before taking steps to improve module delivery. For this reason, information about the status of the YouTube channel subscription, the type of viewing device used, as well as the preferred operating system, is collected, in addition to the standard demographic data. Note that in the YouTube studio, this type of data is available in separate tabs for each of the variables, after filtering for the two observation periods (i.e., 2020–2021 and 2021–2022). The goal of this strategy is to provide a clear picture with enough numerical data to allow for the most accurate comparisons and conclusions possible.

Accumulating demographic characteristics directly from the source, as opposed to self-reported responses in questionnaires, can increase the validity of the acquired data. It should be noted that many studies, including [43], have relied exclusively on demographic attributes to assess and even predict students' academic performance with high accuracy.

Table 1 provides a summary of the key demographic characteristics of the mechanical engineering students who participated in the study during the two time periods noted above. This summary was compiled from data directly extracted from YouTube channel reports, content available in YouTube Studio's advanced analysis mode, and by specifying a custom time period. Each metric (gender, age, geography, etc.) was exported from the YouTube Studio view to Google Sheets and converted to an MS Excel file [44].

The number of students who took the online module was 212 in the winter semester of the 2020–2021 year. In 2021–2022, Table 1 shows a slightly higher number of students (i.e., 230) who took the module in a blended learning mode. Regardless of the two time periods considered, 90% of the learners were male. This percentage reflects the true majority of male students enrolled in the Department of Mechanical Engineering at the University of West Attica, as verified by the student registry. In the hybrid learning environments, 97.6% of the students were between the ages of 18 and 24, and the total number of participants was located in Greece. Considering that, in the exclusive online learning environments,

60.9% of the students were between the ages of 18 and 24 and 37.7% were between the ages of 25 and 34, it can be inferred that the online learning spaces offered a unique opportunity for older students to attend their courses without their physical presence. In addition, the exclusive online learning spaces allowed a small number of international resident students to take the module from another country.

Table 1. Demographic information of mechanical engineering students who participated in the study.

Demographics	2020–2021 (n = 212)	Viewing Time (Hours)	2021–2022 (n = 230)	Viewing Time (h)
		Gender of the participant and age distribution		
Female	10.0%		10.0%	
Male	90.0%		90.0%	
Age of the participant				
18–24	60.9%		97.6%	
25–34	37.7%		2.4%	
35–44	0.9%		0.0%	
45–54	0.5%		0.0%	
		Geography		
National	97.8%	261.6	100%	174.6
International	2.2%	0.1	0%	0.0
		Subscription status		
Not subscribed	75.9%	193.6	82.1%	151.4
Subscriber	24.1%	70.9	17.9%	30.2
		Type of viewing device used		
Computer	75.0%	210.39	70.7%	137.4
Mobile phone	23.6%	52.91	27.7%	38.0
Tablet	0.9%	0.81	1.3%	5.6
TV	0.5%	0.32	0.3%	0.2
		Operating system		
Windows	78.1%	209.4	67.2%	133.4
Android	17.7%	46.0	22.2%	31.1
iOS	3.7%	7.8	6.4%	12.5
Macintosh	0.4%	1.0	2.6%	4.4
Smart TV	0.1%	0.02	0.0%	0.0
PlayStation	0.0%	0.0	0.4%	0.2
Linux	0.0%	0.0	0.4%	0.1
WebOS	0.0%	0.0	0.4%	0.02
Xbox	0.0%	0.0	0.4%	0.01

The registration status of viewers is already an interesting factor: Only 24.1% of viewers are registered on the educational channel, indicating an initial trend that most students watch the videos repeatedly, without finding a reason to subscribe to the YouTube channel. This suggests that students have a similar attitude toward the educational channel, as probably with other social media sites.

Although personal computers were widely used in both periods studied, they are losing ground to mobile devices each year. Tablet use increased in the second period. Finally, television as a viewing device decreased in 2021–2022. Finally, the most used operating system was Microsoft Windows, but there is a 10.9% decline between 2020–2021 and 2021–2022, which is explained by the increasing use of tablets and the doubling of iOS android devices.

3.3. Reports on Students' Views, Comparative Analysis and Discussion

A total of thirty-seven videos were uploaded to the YouTube channel. Since there was no preparation time available before the universities closed, the timing of the posting of each video was scheduled in parallel with the running of the laboratory module. The name of each video corresponds to the title of the assigned task. All videos can be distinguished by content and learning objective into the following four categories, as shown in Table 2:

1. Sketch and make freehand drawings in two dimensions of views of given objects.
2. Assist the CAD software with the basic commands needed to accomplish the assigned tasks.
3. Preview of the object to be studied that was modeled in the three-dimensional CAD modeler. This specific type of video was created not only to help students perceive shapes and geometric entities, generate multiple representations of them (views), but also to develop their optimized design principle [45].
4. Support on tasks by screen and sound recordings of the drawing procedure in the CAD environment.

Table 2. Analysis of data from the "MCAD I UNIWA" YouTube channel.

	Number of Videos	Number of Views	Average Viewing Time (s)	Average Percentage of Viewing
Video category no. 1	11	3758	727	37.4
Video category no. 2	5	2107	444	28.5
Video category no. 3	9	4777	383	67.1
Video category no. 4	12	5576	1644	29.9
Total	37	16,218		

In order to understand the actual size of the learners, the unique users' metric was filtered from the reports provided by the YouTube channel. This specific metric provides a clearer picture of the estimated number of views during the two different time periods (i.e., during the pandemic period and during the start-up period of meta-COVID-19). The metrics for specific aspects related to viewing by video categories are presented in Figures 2 and 3. Specifically, the fourth category, relating to the methodology of carrying out computer-aided design tasks, is the most viewed and, as expected, has the longest viewing time. The videos with the highest average viewing percentage also belong to this category. The third category of videos, showing an overview of filmed objects, comes next in the students' preferences. It should be noted that this type of video has the shortest duration, limited to ten to fifty seconds. Furthermore, three of the nine videos of this type were generated in the second period, i.e., in hybrid learning environments, after taking into account the students' requests. The second category of videos contains the smallest number of videos due to the fact that most of the software support tools were integrated in the fourth category. This type of video has a considerable number of views in relation to the number of views. This suggests that these videos are aimed at a specific audience, who will persist in watching the content to better understand the use of the software. Finally, the first category, drawing without sound, has a considerable number of views, duration of viewing and average percentage of viewing that mainly reflect the first period, when the educational process was carried out exclusively online.

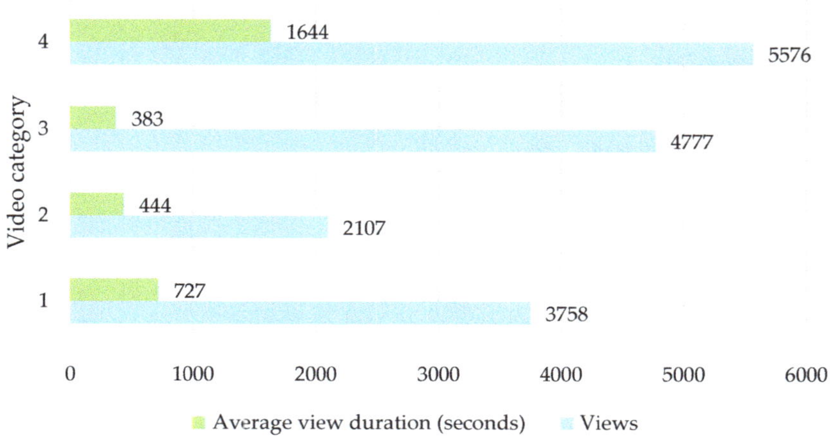

Figure 2. Average viewing time and number of views by video category.

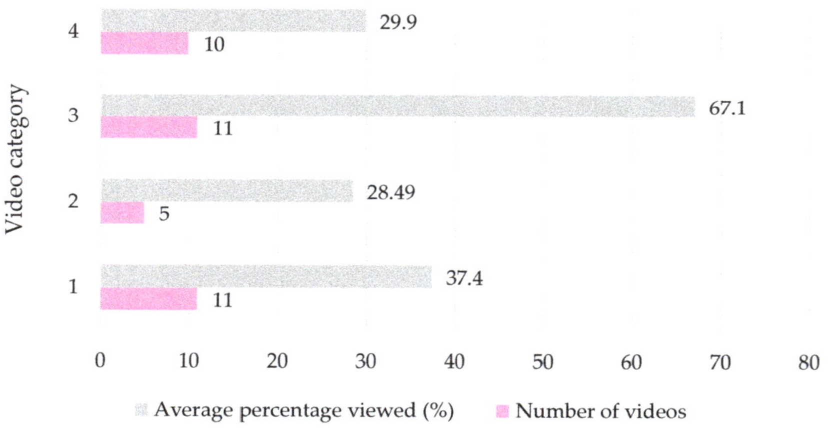

Figure 3. Average percentage of views and number of videos by video category.

4. Main Results

4.1. The Number of Views and Unique Viewers: Two Major Variables in the Foreground

Table 3, whose data is best depicted in Figure 4, summarizes the number of views per week for the two time periods studied (in 2020–2021, when distance learning was exclusive, and in 2021–2022, when hybrid learning was the norm) based on the sequence of course modules. The interest of Table 3 and Figure 4 is also to examine the variations between the two periods studied. It is important to note that the number of views decreased in 2021–2022, which may be due to the fact that students were attending classes in face-to-face mode and asynchronous support was not as necessary as in distance learning environments.

Table 3. Variation of the number of views per week for two time periods per module step.

	Number of Weeks												
		Sketching (First Stage)					CAD Drawings (Second Stage)					Final Exam (Third Stage)	
Year	1	2	3	4	5	6	7	8	9	10	11	12	13
2020–2021	54	137	137	84	**154**	184	**224**	186	116	72	110	**190**	202
2021–2022	27	72	109	134	**120**	154	**177**	161	143	96	47	**97**	143
Variation	27	65	28	−50	**34**	30	**47**	25	−27	−24	63	**93**	59

Separate colors were applied in the two periods according to Figure 4.

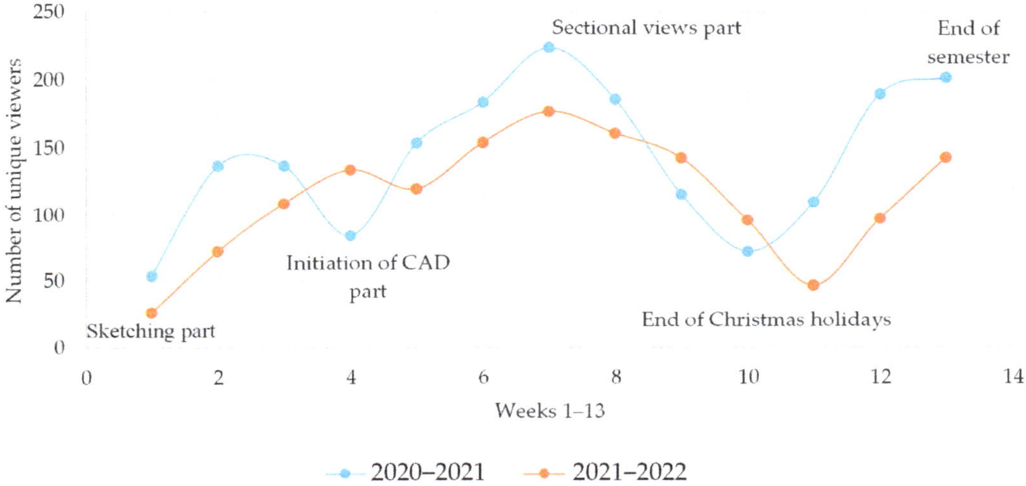

Figure 4. Chart of unique viewers per week. Separate colors were applied in the two periods according to Table 3.

Figure 4 shows two trend curves for the number of unique viewers as a function of CAD module length (in number of weeks) for the two time periods studied (2020–2021 and 2021–2022). Both curves show similar trend dynamics, but only from the fourth course onward. From the beginning of the teaching until the fourth laboratory lecture, the trends are slightly different and can be explained by the fact that the module was delivered exclusively online in 2020–2021 and face-to-face in the first semester of 2021–2022. Higher values of unique viewers in both time periods, can be noticed in week seven. At this point in the educational process, students enroll in the third stage of the CAD module from the second stage of the module workflow. The seventh laboratory lecture is where the most difficult part of the new knowledge has been delivered, referring to the sectional views. It should be noted that at this stage, students who have not yet assimilated the layout of the plans, have difficulties in completing their weekly tasks. Finally, new knowledge was accumulated in the first seven lessons, combining the use of software, the rules of mechanical drawing and the perception of object views, therefore asynchronous task support was necessary in this specific period. From the 8th to the 10th week, the Christmas vacations took place and a clear downward trend can be observed in both lines of the time series. Upward trends are again observed from week 10 onwards, when students resume module attendance after the Christmas break. High values of unique viewers can also be distinguished in week 13, which can be interpreted as the fact that the laboratory lectures have reached their final point and the exams are only one week away.

These first graphically illustrated results must now be rigorously demonstrated by statistical data, which will be analyzed in the following sections.

4.2. Selected Statistical Variables

The statistical study proposed in this section is conducted on the following six variables: "Impressions", "Unique viewers", "Viewing time", "Impressions Click-Through Rate (CTR)", "Number of views" and "Average viewing time". In particular, we will use YouTube Analytics terminology to describe these variables. The term "Impressions" is used to express the number of times one of the video thumbnails from the YouTube channel appears on the participant's screen. Therefore, the "Impressions Click-Through Rate" (CTR) indicates how many impressions were converted into views. Its calculation (expressed as a percentage) is based on the ratio between the number of clicks and the number of impressions. This metric aims to reveal how many people saw the thumbnail, found it interesting and clicked on it. The "Views" metric indicates the number of times a video has been viewed. The term "viewers" refers to the number of individuals who watch a certain piece of content. A viewer may click more than once on a specific piece of content. This is why the "Unique viewers" metric is used, which is a more accurate and relevant variable since it only counts one person, even if they click on the same video content multiple times or use multiple devices or browsers. "Viewing time" is expressed in YouTube terminology as "watch time". This audience retention metric, expressed here as an average value, counts the hours that a specific video is watched by individual viewers. For each of the six variables defined above, two pairs of datasets were compared:

- Dataset 1: data from the academic period 2020–2021;
- Dataset 2: data from the academic period 2021–2022.

4.3. Normality Test Results

First, for each variable, it was necessary to verify the normality of the data distribution. This normality test is very important insofar as it determines the choice to carry out a test, either parametric or non-parametric. In the remainder of this subsection, we will only discuss the main results of the procedure. The basis and steps of the Shapiro–Wilk test, along with detailed results, are presented in Appendix B.

The results in Appendix B show that, regardless of the two time periods studied (i.e., 2020–2021 and 2021–2022), only the two variables "Impressions" and "Number of views" have p-values greater than the 5% threshold, which does not allow us to reject the null hypothesis and thus to consider that each sample in question is normally distributed. For these variables, an F-test was performed to evaluate whether the variances of the variables (whose distributions can be considered as normal according to the Shapiro–Wilk test) are equal or not. The final objective is to reject or not the null hypothesis of the existence of statistically significant differences between the academic years 2020–2021 and 2021–2022 [36]. These F-tests will be complemented by Student's t-tests to compare the means of the two data distributions considered.

For the other four variables (i.e., "Unique viewers"; "Watch time"; "Impressions CTR"; and "Average viewing time") with non-normal distributions, a Mann–Whitney–Wilcoxon test (which is a non-parametric method), was performed to assess whether there is a statistically significant difference between the two periods of 2020–2021 and 2021–2022.

4.4. Hypothesis Testing Results

For each of the two normally distributed variables mentioned above (i.e., "Impressions" and "Number of views"), the Fisher–Snedecor test or F-test was performed. As with the results of the normality tests presented earlier, here we analyze only the results obtained (see Table 4). However, the foundations and main steps of the method are recalled in Appendix C. The results in Table 4 show that for the "Impressions" variable, there is no equality of variances between the 2020–2021 and 2021–2022 data, since the p-value is below the 5% threshold. As for the variable "Number of views", there is equality of variances as long as the p-value is greater than the 5% threshold. For this variable in particular, this allows us to conclude that for the two periods considered, since the variances are equal, learners' need for assistance in completing their tasks from viewing asynchronous educational content does not depend on the learning environment.

Table 4. Results of the F-test for the two normally distributed variables "Impressions" and "Number of views".

	2020–2021 "Impressions"	2021–2022 "Impressions"	2020–2021 "Number of Views"	2021–2022 "Number of Views"
p-value	0.01	0.01	0.12	0.12

The F-tests were supplemented with t-tests, as shown in Table 5. The results of these tests reflect the comparison of the means of the data sets of the variables "Number of Views" and "Impressions". As the p-values in Table 5 are above the 5% threshold, the null hypothesis cannot be rejected. Therefore, no statistically significant difference was observed between 2020–2021 and 2021–2022.

Table 5. Results of the t-test for the two normally distributed variables "Impressions" and "Number of views".

	"Impressions"	"Number of Views"
Assumption	Unequal variance	Equal variance
p-value	0.055	0.08

For each of the four remaining variables (i.e., "Unique viewers"; "Viewing time or Watch time"; "Impressions CTR"; and "Average viewing time") whose distributions are not Gaussian for the two periods studied (i.e., 2020–2021 and 2021–2022), a non-parametric Mann–Whitney–Wilcoxon test was performed.

The results in Table 6 show that there is no statistically significant difference between the two periods studied for the variables "Viewing time" and "Unique viewers" because their respective p-values are above the defined risk of 5%. However, this is not the case for the other variables (i.e., "Impressions CTR" and "Average viewing time"). Indeed, their respective p-values are below the 5% risk, confirming the statistically significant difference between the 2020–2021 and 2021–2022 data.

Table 6. Mann–Whitney–Wilcoxon test results for the four variables "Unique viewers"; "Watch time"; "Impressions CTR"; and "Average viewing time".

	Unique Viewers (10 Weeks)	Viewing Time or Watch Time (h)	Impressions CTR (%)	Average Viewing Time (s)
p-value	0.443	0.122	0.000	0.000
Null hypothesis rejected or not rejected	Null hypothesis not rejected	Null hypothesis not rejected	Null hypothesis rejected	Null hypothesis rejected

All the statistical results described above, whether parametric or non-parametric, confirm the preliminary results established in Section 4.1. in that the two variables "Number of views" and "Unique viewers" do not depend on the learning environment (i.e., exclusively remote or hybrid environment).

5. Discussion

Beyond the analysis of the statistical tests proposed above, Figure 4 allows us to draw some major conclusions that we will review and discuss in this section.

The total number of unique viewers in the first semester of the 2020–2021 academic year was 719, and in the same period of the 2021–2022 year was 570. By calculating the percentage of unique viewers for each of the two periods compared to the total number, a percentage difference of 11.55% was calculated. The two curves in Figure 4 representing the data series for both time periods show similar trends in student viewing habits for both time periods, leading us to conclude that learners' need for assistance in completing their tasks is not dependent on the learning environment. This observation is made when comparing online and hybrid modes of instruction. When comparing the visual behavior of students during the first four laboratory lectures taught in the face-to-face mode in 2021–2022 with the same period in 2020–2021 in the online spaces, the curve does not show similar trends. This specific observation can be explained by the fact that when teaching the online module, all twelve lectures were delivered exclusively at a distance, whereas in the hybrid learning spaces, the first four courses were delivered exclusively in face-to-face mode. In addition, the theme of the first four lectures was "sketching", which involves freehand drawings. For the non-computer tasks, videos were only used for object representation, and their contribution was limited to categories 1 and 3, with tasks related to the first category being repeated face-to-face.

The variable "Impressions CTR" was one of the viewing measures that showed statistically significant differences between the two time periods in the non-parametric tests conducted. By comparing the proportion of reduction between the "Number of views" (11.86%) and "Impressions CTR" (1.76%), we can conclude that even though the "Number of views" decreased in the hybrid learning spaces, the "Impressions CTR" variable showed a very low percentage of reduction, which proves the positive attitude of students clicking on the thumbnails of the videos to watch them.

Although the results of this study revealed viewing patterns for both time periods examined, there are still some limitations, based on the circumstances in which the module was delivered during each time period. The learning experience at universities was affected by the consequences of the pandemic, resulting in a significant decline in student retention and academic performance. Sustainable measures, such as those implemented in this study, had to be taken to first ensure students' virtual presence, as well as support for asynchronous tasks.

In the exclusive online instruction, the lack of physical contact between learners and their educators allowed the former to engage in the asynchronous support channel, which allowed instructors to track their viewing activities and analyze their viewing behaviors through data analysis by retrieving high-precision information. In the hybrid learning modes, specifically in the face-to-face delivered modules, learners had the opportunity to physically communicate with their instructors and get support for their tasks.

In synthesis of the above, the YouTube channel created and used in this work as an asynchronous tutoring tool has been integrated into the educational process. It provides quality asynchronous support when needed and is part of a long-term viability and sustainability approach. These new tools for supporting individual tasks outside of the classroom can benefit pedagogical practices and ultimately the learning process in very implicit ways and primarily by being "masked" by popular social media sites like YouTube. The channel analysis provided valuable information about learners' visualization habits that can serve as guidelines for future instructors and developers of instructional module structure. The log of visualization measures revealed viewing patterns indicating that students' visualization behaviors follow the flow of the module, and especially regardless of their mode of attendance and teaching space. The ability to follow the content of a module through educational videos at one's own pace and preference contributes to the development of senses of quality and equity [19]. The EDM and its statistical analysis showed that the foundation of the YouTube channel met the needs of students regardless of the learning environment on which this tool was applied.

Moving forward, the methodology applied in this study provides direct feedback for future CAD instructors and instructional module developers. Our proposed recommendations and action plan are as follows:

- Recognize module gaps throughout the course. Recognize the needs of learners, for the promotion of sustainable engineering education. Reorganize the flow of tasks according to the learning objectives of the modules and determine the critical points of difficulty according to the knowledge to be acquired.
- Do not focus on a specific task, but on the unit by creating categories of units. Each unit may include several tasks, but the focus of the knowledge introduced is not the technical instruction itself, but the concept of the learning methodology. At the stage where the new knowledge is accumulated, call it the "pick of the curve". Determine the points at which the new knowledge needs to be performed, as well as its nature in terms of asynchronous support. Do not be afraid to combine new knowledge with entertainment by using user-friendly digital tools, such as YouTube channels.
- Create direct access to certain parts of the course to redirect learners if necessary. Make access clear to avoid confusion.
- Analyze, evaluate and reconstruct the course based on the results. This means questioning the actions taken, modifying them if they fail, and adapting them according to the nature of the learning environments.
- Never neglect the social aspects: learners must be prepared for their professional future. Technical aspects can be learned through training, but methodology is the expertise of higher education instructors.

6. Conclusions and Future Work

In 2020–2021, during the most critical hours of the COVID-19 pandemic, higher education instructors, specifically those at the University of West Attica (Greece), created a social media channel (in this study, using YouTube), as part of a mechanical engineering CAD module, to provide students with asynchronous support for their teaching tasks. To provide learners with the most direct access, links to the 37 videos were attached to each assessed task. One year later, at the beginning of the meta-COVID-19 period, the same asynchronous task support technique was applied, but this time in face-to-face and blended learning spaces.

The experimental analysis proposed in this manuscript is based on the process of processing and interpreting acquired knowledge data extracted directly from the reports of a YouTube channel; this YouTube channel having been created and used by an instructor and containing educational videos of different categories based on the learning objectives of a CAD module. The main challenge here was to investigate the potential of the data as acquired from the YouTube channel and whether the raw material downloaded in the form of previews could be processed and reveal valuable information about how students use this form of asynchronous digital educational material; YouTube having been hijacked from its primary function, i.e., entertainment.

The shift from exclusively online to hybrid learning environments first showed that the use of asynchronous task support decreased. YouTube analytics were the most appropriate tool in terms of efficiency and accuracy for expressing student retention beyond physical, online, or hybrid engineering labs, as they not only recorded the number of times a video was viewed, but also differentiated between users who watched the same video multiple times. Specifically, we defined and used variables and measures that are very common on social media sites, expressing the level of audience acceptance, to examine the contribution of video to the learning process. The following six statistical variables were selected as primary measures expressing audience retention in YouTube channels: "Impressions", "Unique viewers", "Viewing time or watch time", "Impressions Click-Through Rate (CTR)", "Number of views" and "Average viewing time".

The comparative analysis of the YouTube reports showed similar trends in student viewing habits over the two time periods studied (i.e., at the most critical time of the global health crisis and at the beginning of the meta-COVID-19 period), with a slight decrease of nearly 12% in viewing due to a return to traditional learning environments, where most students solved their tasks in class and did not require asynchronous support. In contrast to face-to-face and 100% distance learning, the analysis showed that trends in learner viewing habits are similar during online and hybrid learning spaces. In particular, reports from the social media channel showed that educational videos followed the weekly stream trend, resulting in an increase in active viewers, which was directly related to the increase in workload and workload accumulation.

After testing the normality of each of the above six variables, a series of hypothesis tests (i.e., parametric and non-parametric based on normality tests) were performed to accept or reject the null hypothesis of this study, which concerns the absence of statistically significant changes in learners' listening habits over the two periods studied (i.e., 2020–2021 and 2021–2022). Of the six variables analyzed, only two—"Impressions CTR" and "Average viewing time"—show display metrics with statistically significant differences between the two periods studied. For both of these statistically significant variables, it is appropriate to focus on impressions CTR, which expresses the number of times viewers click to watch a video after seeing its thumbnail. Although there was a decrease in views and impressions CTR between the two time periods, the percentage reduction was not proportional for either variable.

Although we have answered the five research questions listed in Section 2, the current study has some limitations. Although many universities have begun to use data and analytics, there is still a long way to go before these tools can fully prove their potential in terms of improving the learning experience. This is especially true today, due to the unstable health conditions caused by the COVID-19 outbreak, although overall they seem to be gradually normalizing.

Future work will involve processing analyses of a second semester CAD module Computer Aided Mechanical Design (CAD II) and performing similar statistical tests to improve the reliability of the results. Due to the increase in the number of students and institutions participating in online learning and using digital tools over the past two years, there is now a plethora of data available that may not have been available before. Institutions of higher education may want to start using this data with an eye toward serving students ever better in the years to come.

Author Contributions: Conceptualization, Z.K. and C.S. (Constantinos Stergiou); methodology, Z.K., C.S. (Constantinos Stergiou), G.B. and S.J.; software, G.B.; validation, G.B.; formal analysis, Z.K. and G.B.; investigation, Z.K. and C.T.; resources, Z.K. and C.S. (Constantinos Stergiou); data curation: Z.K.; writing—original draft preparation: Z.K.; writing—review and editing: Z.K., C.S. (Constantinos Stergiou), G.B., S.J. and A.O.; visualization: Z.K., C.S. (Constantinos Stergiou), G.B. and S.J.; supervision, C.S. (Constantinos Stergiou) and C.S. (Cleo Sgouropoulou); project administration, C.S (Constantinos Stergiou) and S.J.; funding acquisition, S.J. and A.O. All authors have read and agreed to the published version of the manuscript.

Funding: This research received no external funding.

Institutional Review Board Statement: This study involves the analysis of data sets obtained from reports on a YouTube channel created and administered by a member of the authors' team. Therefore, all rights to the reports are reserved to the authors.

Informed Consent Statement: Informed consent was obtained from all study participants at the time of initial data collection.

Data Availability Statement: Not applicable.

Acknowledgments: These research activities are currently supported by the University of West Attica and more particularly by its Department of Mechanical Engineering, as well as by the University of Tours. The authors of this manuscript would like to thank their colleagues at the following institutions, as well as the students, who contributed greatly to the success of this work.

Conflicts of Interest: The authors declare no conflict of interest.

Abbreviations

The following abbreviations are used in this paper:

CAD	Computer-aided design
CTR	Impressions Click-Through Rate
EDM	Educational data mining
LA	Learning analytics
LMS	Learning management system
MS	Microsoft
TAM	Technology acceptance model
TLP	Teaching—learning process
VLE	Virtual learning environments

Appendix A. URL of Each Educational Video Content of the YouTube Channel "MCAD I UNIWA"

Video	Video Title	URL
Total		
1	CAD 06A	https://www.youtube.com/watch?v=Pp2KWt28haM (accessed on 23 May 2022).
2	CAD 07 PART 13B	https://www.youtube.com/watch?v=N01BuI3DCAU&t=426s (accessed on 23 May 2022).
3	CAD 07 PART 13A	https://www.youtube.com/watch?v=ezeYLUihbn4 (accessed on 23 May 2022).
4	CAD 06B ASK 11B	https://www.youtube.com/watch?v=-VrzgqM-x5Q (accessed on 23 May 2022).
5	CAD 08 ASK 15	https://www.youtube.com/watch?v=HT-oJB2JZ9k (accessed on 23 May 2022).
6	CAD 06B ASK 11A	https://www.youtube.com/watch?v=il4Nyba1dn8 (accessed on 23 May 2022).
7	CAD 05D	https://www.youtube.com/watch?v=rARN13OfGig (accessed on 23 May 2022).
8	CAD 10 PART 20	https://www.youtube.com/watch?v=v40Dg4ftOXw (accessed on 23 May 2022).
9	CAD 07 DIMENSION	https://www.youtube.com/watch?v=xBQ_qv-szIQ (accessed on 23 May 2022).
10	CAD 09 PART 17	https://www.youtube.com/watch?v=MxraNn92Wfo (accessed on 23 May 2022).
11	CAD 07 PART 13	https://www.youtube.com/watch?v=AIeUy0Q_1v8 (accessed on 23 May 2022).
12	ASK 03A	https://www.youtube.com/watch?v=A7UeOz1wluE (accessed on 23 May 2022).
13	CAD 08 PART 14	https://www.youtube.com/watch?v=g2eXTlYJm0Y (accessed on 23 May 2022).
14	CAD 05A	https://www.youtube.com/watch?v=HtUdjt8w89Y&t=8s (accessed on 23 May 2022).
15	CAD 07 PART 12	https://www.youtube.com/watch?v=qSENFhmbZfw (accessed on 23 May 2022).
16	CAD 11	https://www.youtube.com/watch?v=Kp-vX2eBpLQ (accessed on 23 May 2022).
17	ASK 03B	https://www.youtube.com/watch?v=c7CaOlUXO0E (accessed on 23 May 2022).
18	ASK 06C	https://www.youtube.com/watch?v=l8LVvr5uIz0&t=17s (accessed on 23 May 2022).
19	ASK 06A	https://www.youtube.com/watch?v=3BP7jxdpt7c (accessed on 23 May 2022).
20	CAD 04B	https://www.youtube.com/watch?v=Dlc8CvUG5wk (accessed on 23 May 2022).
21	CAD 09 PART 18	https://www.youtube.com/watch?v=vopGLakGO_g (accessed on 23 May 2022).
22	CAD 08 LIBRARY	https://www.youtube.com/watch?v=RWY5EuLB9nQ (accessed on 23 May 2022).
23	ASK 03A TOP	https://www.youtube.com/watch?v=knAEEWcHqLA (accessed on 23 May 2022).
24	CAD 04A	https://www.youtube.com/watch?v=9y2bkS7T3Wc (accessed on 23 May 2022).
25	ASK 06B	https://www.youtube.com/watch?v=FjmdZa2ix2o&t=11s (accessed on 23 May 2022).
26	ASK 04A	https://www.youtube.com/watch?v=W5FzH5-RF3w (accessed on 23 May 2022).
27	ASK 04B	https://www.youtube.com/watch?v=4ft1ZpLzcj8 (accessed on 23 May 2022).
28	ASK 03A FRONT LEFT	https://www.youtube.com/watch?v=cRUM-BOB8jM&t=105s (accessed on 23 May 2022).
29	CAD 05B	https://www.youtube.com/watch?v=0m-9ofdEgJs (accessed on 23 May 2022).
30	CAD 05C	https://www.youtube.com/watch?v=gWf-1wpQnAA (accessed on 23 May 2022).
31	CAD 11 PART 22	https://www.youtube.com/watch?v=TPNZibk1CJs (accessed on 23 May 2022).
32	CAD 07 PART 006	https://www.youtube.com/watch?v=p5YSKekgDmw (accessed on 23 May 2022).
33	CAD 07 PART 007	https://www.youtube.com/watch?v=5hUMxgh3AzE (accessed on 23 May 2022).
34	ASK 03A RIGHT	https://www.youtube.com/watch?v=gDi0qexf1KI (accessed on 23 May 2022).
35	CAD 10 TEXT	https://www.youtube.com/watch?v=tG39K1aFc8I (accessed on 23 May 2022).
36	01 SKETCH	https://www.youtube.com/watch?v=v2BTZmS6YIk (accessed on 23 May 2022).
37	AUTOCAD DESIGN CENTER	https://www.youtube.com/watch?v=3g3CqBEB7MQ (accessed on 23 May 2022).

Appendix B. Summary of the Basics and Main Steps of the Shapiro–Wilk Normality Test

The Shapiro–Wilk normality test is designed to detect all deviations from normality. In particular, this test rejects the normality hypothesis when the *p*-value is less than or equal to a threshold value (usually 5%). The null hypothesis is that there is no difference between the distribution studied and a normal distribution. The alternative hypothesis is that there is a difference. If the *p*-value at the end of the test is less than the set threshold (usually 5%), then the null hypothesis can be rejected and the data are not considered normal. In this case, a series of non-parametric tests can be applied. Conversely, if the null hypothesis cannot be rejected, the data are considered normal and parametric tests can be implemented [46,47]. Note that if only one of the data sets does not meet the set threshold and the other data set does, the variable of interest is considered a non-parametrically valued variable.

In addition to the *p*-value and to decide the normality of each distribution, we will focus on two metrics in particular: skewness and kurtosis of each of the variables examined for the two datasets (i.e., variables referring to a sample of 37 YouTube videos, with Dataset 1 containing observations from the 2020–2021 academic year and Dataset 2 from the 2021–2022 academic year). While skewness focuses on the overall shape, kurtosis focuses on the tail shape. The normal distribution is characterized by a zero skewness coefficient and a zero kurtosis coefficient. Concerning the skewness, a positive coefficient indicates a left asymmetry, while a negative coefficient indicates a right asymmetry. With respect to kurtosis, a negative value indicates that the distribution is "platykurtic", i.e., more flattened than a normal density. A positive kurtosis coefficient indicates that the distribution is "leptokurtic", i.e., less flattened.

The following tables summarize the *p*-values, skewness and kurtosis of the six variables defined in this study for the two periods studied (i.e., 2020–2021 and 2021–2022).

	2020–2021 "Impressions"	2021–2022 "Impressions"	2020–2021 "Unique Viewers" (10 Weeks)	2021–2022 "Unique Viewers" (10 Weeks)
p-value (Shapiro–Wilk test)	0.675	0.120	0.561	0.038
Skewness	−0.102	0.150	0.138	−0.029
Kurtosis	−0.688	−0.917	−0.884	−0.946
Type of test allowed	Parametric		Non-parametric	
	2020–2021 "Watch Time (h)"	2021–2022 "Watch Time (h)"	2020–2021 "Impressions CTR (%)"	2021–2022 "Impressions CTR (%)"
p-value (Shapiro–Wilk test)	0.000	0.000	0.310	0.000
Skewness	1.280	1.717	0.550	1.772
Kurtosis	0.289	3.320	0.358	3.591
Type of test allowed	Non-parametric		Non-parametric	
	2020–2021 "Number of Views"	2021–2022 "Number of Views"	2020–2021 "Average Viewing Time (s)"	2021–2022 "Average Viewing Time (s)"
p-value (Shapiro–Wilk test)	0.479	0.086	0.000	0.000
Skewness	0.177	0.447	−0.323	1.086
Kurtosis	−0.730	0.844	0.924	1.369
Type of test allowed	Parametric		Non-parametric	

Appendix C. Summary of the Basics and Main Steps of and *t*-Tests for Parametric variables, and the Mann–Whitney–Wilcoxon Test for Non-Parametric Variables

As stated in [48,49], the hypothesis is an interpretation of the reasons for a certain phenomenon. Two types of hypotheses can be defined in a scientific approach:

- The first is the research hypothesis, which aims to state the subject of the research. If it is well defined, it will include the factors being studied and their expected relationship.
- The second is the statistical hypothesis, which converts the research hypothesis into a mathematical complex and a statistically testable statement about the presumed value of the variable being studied in the population.

Therefore, the null hypothesis must be tested. In our case, it concerns the lack of significant difference between the variances of the six variables examined, when comparing the samples of variables from the two academic semesters mentioned above [50].

Parametric statistical tests (based on the hypothesis that the sample under consideration is drawn from a population following a distribution belonging to a given family, i.e., the normal distribution), when their use is well justified, they generally have greater statistical power than non-parametric tests (i.e., without a distribution). More precisely, they are likely to detect a significant effect when it actually exists. Normality is tested here by the Shapiro–Wilk test mainly because, for a given significance level, the probability of rejecting the null hypothesis (i.e., a sample is drawn from a normally distributed population) if it is false is higher than for other tests of normality [42].

Appendix C.1. Fisher-Snedecor or F*-Test for Parametric Variables*

For the two variables examined (i.e., "Impressions" and "Number of views"), a sample size of 37 was set, referring to the number of videos uploaded to the YouTube channel. The degrees of freedom are determined by subtracting one from the sample size. In statistical calculations, degrees of freedom measure the mathematical complexity of a calculated parameter. As mentioned earlier, the probability that the tested parameters have statistical significance is tested by setting a threshold for their statistical significance.

The Fisher–Snedecor test or F-test consists in comparing the resultant value of the test with the critical value (F_{critical}) of the Fisher–Snedecor distribution for the risk sought (the risk is equal to 5% here); this critical value is determined from a table. If the resulting value of the test (p-value) is higher than the critical value (5%), then the null hypothesis (i.e., the two variances are equal) is rejected. Otherwise, it is not rejected [51]. Note that the more dissimilar the variances are, the more the p-value tends to zero.

Appendix C.2. Student's t-Test for Parametric Variables

The Student's t-test, or t-test, is a popular statistical test used to measure the differences between the means of two groups or a group compared to a standard value. It is based on a probability distribution called Student's T distribution. Performing this test is used to understand whether the differences are statistically significant.

The 2-sample t-test, which is the most standard and classical analysis technique, aims at comparing the means of two independent populations to identify a significant difference. To run a Student's T distribution, the following must be available: the difference between the mean values of the data sets; the variance of each sample; the number of data in each group; and the acceptable error threshold (usually 5%). In this study, for a variable under consideration, the null hypothesis is that the two populations (of 2020–2021 and 2021–2022) are identical and that there is no significant difference between them. At the end of the test, if the p-value is lower than the set threshold (usually 5%), the null hypothesis can be rejected. Otherwise, it cannot be rejected.

Appendix C.3. Mann–Whitney–Wilcoxon Test for Non-Parametric Variables

The Mann–Whitney–Wilcoxon test is used to test the hypothesis that the distributions of each of the two groups of data are close. Like any statistical test, it consists in highlighting an event whose probability distribution is known (at least its asymptotic form) from what is observed. The p-value obtained, if it is unlikely according to this law, will suggest rejecting the null hypothesis. More precisely, if the p-value is greater than the fixed risk (here 5%), then the null hypothesis cannot be rejected. Otherwise, it can be rejected.

References

1. Chango, W.; Lara, J.A.; Cerezo, R.; Romero, C. A review on data fusion in multimodal learning analytics and educational data mining. *WIREs Data Min. Knowl. Discov.* **2022**, *e1458*, 1–19. [CrossRef]
2. Wang, Q.; Mousavi, A.; Lu, C. A scoping review of empirical studies on theory-driven learning analytics. *Distance Educ.* **2022**, *43*, 6–29. [CrossRef]
3. Hantoobi, S.; Wahdan, A.; Al-Emran, M.; Shaalan, K. A Review of Learning Analytics Studies. In *Recent Advances in Technology Acceptance Models and Theories*; Al-Emran, M., Shaalan, K., Eds.; Studies in Systems, Decision and Control; Springer International Publishing: Cham, Switzerland, 2021; pp. 119–134, ISBN 978-3-030-64987-6.
4. Gutierrez-Bucheli, L.; Kidman, G.; Reid, A. Sustainability in engineering education: A review of learning outcomes. *J. Clean. Prod.* **2022**, *330*, 129734. [CrossRef]
5. Hodges, C.; McCullough, H. The Adjacent Possible for Higher Education: The Digital Transformation of Faculty. Available online: https://er.educause.edu/articles/2021/9/the-adjacent-possible-for-higher-education-the-digital-transformation-of-faculty (accessed on 21 November 2021).
6. Jacques, S.; Ouahabi, A.; Lequeu, T. Remote Knowledge Acquisition and Assessment during the COVID-19 Pandemic. *Int. J. Eng. Pedagog. IJEP* **2020**, *10*, 120–138. [CrossRef]
7. Jacques, S.; Ouahabi, A.; Lequeu, T. Synchronous E-learning in Higher Education during the COVID-19 Pandemic. In Proceedings of the 2021 IEEE Global Engineering Education Conference (EDUCON), Vienna, Austria, 21–23 April 2021; pp. 1102–1109.
8. Shloul, T.; Javeed, M.; Gochoo, M.; Alsuhibany, S.; Ghadi, Y.; Jalal, A.; Park, J. Student's Health Exercise Recognition Tool for E-Learning Education. *Intell. Autom. Soft Comput.* **2022**, *35*, 149–161. [CrossRef]
9. Lutfi, A.; Alsyouf, A.; Almaiah, M.A.; Alrawad, M.; Abdo, A.A.K.; Al-Khasawneh, A.L.; Ibrahim, N.; Saad, M. Factors Influencing the Adoption of Big Data Analytics in the Digital Transformation Era: Case Study of Jordanian SMEs. *Sustainability* **2022**, *14*, 1802. [CrossRef]
10. Kanetaki, Z.; Stergiou, C.; Bekas, G.; Jacques, S.; Troussas, C.; Sgouropoulou, C.; Ouahabi, A. Grade Prediction Modeling in Hybrid Learning Environments for Sustainable Engineering Education. *Sustainability* **2022**, *14*, 5205. [CrossRef]
11. Orji, F.A.; Vassileva, J. A Comparative Evaluation of the Effect of Social Comparison, Competition, and Social Learning in Persuasive Technology on Learning. In Proceedings of the Intelligent Tutoring Systems, Virtual Event, 7–11 June2021; Cristea, A.I., Troussas, C., Eds.; Springer International Publishing: Cham, Switzerland, 2021; pp. 369–375.
12. Jackman, W.M. YouTube Usage in the University Classroom: An Argument for its Pedagogical Benefits. *Int. J. Emerg. Technol. Learn. IJET* **2019**, *14*, 157–166. [CrossRef]
13. Cabrera, I.; Villalon, J.; Chavez, J. Blending Communities and Team-Based Learning in a Programming Course. *IEEE Trans. Educ.* **2017**, *60*, 288–295. [CrossRef]
14. Kanetaki, Z.; Stergiou, C.; Bekas, G.; Troussas, C.; Sgouropoulou, C. The impact of different learning approaches based on MS Teams and Moodle on students' performance in an on-line mechanical CAD module. *Glob. J. Eng. Educ.* **2021**, *23*, 185–190. [CrossRef]
15. Kanetaki, Z.; Stergiou, C.; Bekas, G.; Troussas, C.; Sgouropoulou, C. Creating a Metamodel for Predicting Learners' Satisfaction by Utilizing an Educational Information System During COVID-19 Pandemic. *Nov. Intell. Digit. Syst.* **2021**, *338*, 127–136. [CrossRef]
16. Baker, R.S.J.D. International Encyclopedia of Education. In *Data Mining*, 3rd ed.; McGaw, B., Peterson, P., Baker, E., Eds.; 2010; pp. 112–118.
17. Govindarajan, M.; Govindarajan, M. Educational Data Mining Techniques and Applications. Available online: https://www.igi-global.com/gateway/chapter/www.igi-global.com/gateway/chapter/272957 (accessed on 8 May 2022).
18. Ihantola, P.; Vihavainen, A.; Ahadi, A.; Butler, M.; Börstler, J.; Edwards, S.H.; Isohanni, E.; Korhonen, A.; Petersen, A.; Rivers, K.; et al. Educational Data Mining and Learning Analytics in Programming: Literature Review and Case Studies. In Proceedings of the 2015 ITiCSE on Working Group Reports, Vilnius, Lithuania, 4–8 July 2015; Association for Computing Machinery: New York, NY, USA, 2015; pp. 41–63.
19. Krishnan, R.; Nair, S.; Saamuel, B.S.; Justin, S.; Iwendi, C.; Biamba, C.; Ibeke, E. Smart Analysis of Learners Performance Using Learning Analytics for Improving Academic Progression: A Case Study Model. *Sustainability* **2022**, *14*, 3378. [CrossRef]
20. Siemens, G.; Baker, R.S.J.d. Learning analytics and educational data mining: Towards communication and collaboration. In Proceedings of the 2nd International Conference on Learning Analytics and Knowledge, Vancouver, BC, Canada, 29 April 2012–2 May 2012; Association for Computing Machinery: New York, NY, USA, 2012; pp. 252–254.
21. Muttakin, F.; Wang, J.-T.; Mulyanto, M.; Leu, J.-S. Evaluation of Feature Selection Methods on Psychosocial Education Data Using Additive Ratio Assessment. *Electronics* **2022**, *11*, 114. [CrossRef]
22. Nowak-Brzezińska, A.; Łazarz, W. Qualitative Data Clustering to Detect Outliers. *Entropy* **2021**, *23*, 869. [CrossRef] [PubMed]
23. Dooley, L.; Makasis, N. Understanding Student Behavior in a Flipped Classroom: Interpreting Learning Analytics Data in the Veterinary Pre-Clinical Sciences. *Educ. Sci.* **2020**, *10*, 260. [CrossRef]
24. Al-Maroof, R.; Ayoubi, K.; Alhumaid, K.; Aburayya, A.; Alshurideh, M.; Alfaisal, R.; Salloum, S. The acceptance of social media video for knowledge acquisition, sharing and application: A com-parative study among YouTube users and TikTok Users' for medical purposes. *Int. J. Data Netw. Sci.* **2021**, *5*, 197–214. [CrossRef]
25. de Oliveira, C.F.; Sobral, S.R.; Ferreira, M.J.; Moreira, F. How Does Learning Analytics Contribute to Prevent Students' Dropout in Higher Education: A Systematic Literature Review. *Big Data Cogn. Comput.* **2021**, *5*, 64. [CrossRef]

26. Draus, P.J.; Curran, M.J.; Trempus, M.S. The Influence of Instructor-Generated Video Content on Student Satisfaction with and Engagement in Asynchronous Online Classes. *J. Online Learn. Teach.* **2014**, *10*, 240–254.
27. Rodrigo, M.M.T.; Ladrido, E.M.M. Promoting Equity and Assuring Teaching and Learning Quality: Magisterial Lectures in a Philippine University during the COVID-19 Pandemic. *Educ. Sci.* **2022**, *12*, 146. [CrossRef]
28. Isaenko, E.; Makrinova, E.; Rozdolskaya, I.; Matuzenko, E.; Bozhuk, S. Research of social media channels as a digital analytical and planning technology of advertising campaigns. In Proceedings of the IOP Conference Series: Materials Science and Engineering, St. Petersburg, Russian, 27–29 August 2020; Volume 986, p. 012014. [CrossRef]
29. Al-rahmi, W.M.; Othman, M.S.; Yusuf, L.M. Using Social Media for Research: The Role of Interactivity, Collaborative Learning, and Engagement on the Performance of Students in Malaysian Post-Secondary Institutes. *Mediterr. J. Soc. Sci.* **2015**, *6*, 536. [CrossRef]
30. Dahlstrom, E. *ECAR Study of Undergraduate Students and Information Technology, 2012*; EDUCAUSE Center for Applied Research: Louisville, CO, USA, 2012.
31. Kanaki, K.; Kalogiannakis, M.; Poulakis, E.; Politis, P. Employing Mobile Technologies to Investigate the Association Between Abstraction Skills and Performance in Environmental Studies in Early Primary School. *Int. J. Interact. Mob. Technol. IJIM* **2022**, *16*, 241–249. [CrossRef]
32. Walsh, J.N.; O'Brien, M.P.; Slattery, D.M. Video Viewing Patterns Using Different Teaching Treatments: A Case Study Using YouTube Analytics. *Res. Educ. Learn. Innov. Arch.* **2019**, *22*, 77–95. [CrossRef]
33. McGowan, A.; Hanna, P.; Anderson, N. Teaching Programming: Understanding Lecture Capture YouTube Analytics. In Proceedings of the 2016 ACM Conference on Innovation and Technology in Computer Science Education, Arequipa, Peru, 9–13 July 2016; Association for Computing Machinery: New York, NY, USA, 2016; pp. 35–40.
34. Schwarzenberg, P.; Navon, J.; Nussbaum, M.; Pérez-Sanagustín, M.; Caballero, D. Learning experience assessment of flipped courses. *J. Comput. High. Educ.* **2018**, *30*, 237–258. [CrossRef]
35. Le, A.; Joordens, S.; Chrysostomou, S.; Grinnell, R. Online lecture accessibility and its influence on performance in skills-based courses. *Comput. Educ.* **2010**, *55*, 313–319. [CrossRef]
36. Kanetaki, Z.; Stergiou, C.; Bekas, G.; Troussas, C.; Sgouropoulou, C. Analysis of Engineering Student Data in Online Higher Education During the COVID-19 Pandemic. *Int. J. Eng. Pedagogy IJEP* **2021**, *11*, 27–49. [CrossRef]
37. Binkley, M.; Erstad, O.; Herman, J.; Raizen, S.; Ripley, M.; Miller-Ricci, M.; Rumble, M. Defining Twenty-First Century Skills. In *Assessment and Teaching of 21st Century Skills*; Griffin, P., McGaw, B., Care, E., Eds.; Springer Netherlands: Dordrecht, The Netherlands, 2012; pp. 17–66, ISBN 978-94-007-2324-5.
38. Konys, A. An Ontology-Based Knowledge Modelling for Sustainable Entrepreneurship Domain. In Proceedings of the 55th Hawaii International Conference on System Sciences, Hawaii, HI, USA, 4–7 January 2022; pp. 5316–5325.
39. Petrila, L.; Goudenhooft, G.; Gyarmati, B.F.; Popescu, F.-A.; Simuț, C.; Brihan, A.-C. Effective Teaching during the COVID-19 Pandemic? Distance Learning and Sustainable Communication in Romania. *Sustainability* **2022**, *14*, 7269. [CrossRef]
40. Zabidi, N.; Wang, W. The Use of Social Media Platforms as a Collaborative Supporting Tool: A Preliminary Assessment. *Int. J. Interact. Mob. Technol. IJIM* **2021**, *15*, 138–148. [CrossRef]
41. Kanetaki, Z.; Stergiou, C.; Bekas, G.; Troussas, C.; Sgouropoulou, C. Evaluating Remote Task Assignment of an Online Engineering Module through Data Mining in a Virtual Communication Platform Environment. *Electronics* **2022**, *11*, 158. [CrossRef]
42. Rus-Arias, E.; Palos-Sanchez, P.R.; Reyes-Menendez, A. The Influence of Sociological Variables on Users' Feelings about Programmatic Advertising and the Use of Ad-Blockers. *Informatics* **2021**, *8*, 5. [CrossRef]
43. Batool, S.; Rashid, J.; Nisar, M.W.; Kim, J.; Mahmood, T.; Hussain, A. A Random Forest Students' Performance Prediction (RFSPP) Model Based on Students' Demographic Features. In Proceedings of the 2021 Mohammad Ali Jinnah University International Conference on Computing (MAJICC), Karachi, Pakistan, 15–17 July 2021; pp. 1–4.
44. Anantharamaiah, K.B. YouTube Analytics Using Google Data Studio 2020. Available online: https://doi.org/10.2139/ssrn.3655551 (accessed on 23 May 2022).
45. Chatzopoulos, A.; Kalogiannakis, M.; Papadakis, S.; Papoutsidakis, M. A Novel, Modular Robot for Educational Robotics Developed Using Action Research Evaluated on Technology Acceptance Model. *Educ. Sci.* **2022**, *12*, 274. [CrossRef]
46. Camuffo, A.; Gerli, F. Modeling management behaviors in lean production environments. *Int. J. Oper. Prod. Manag.* **2018**, *38*, 403–423. [CrossRef]
47. Zhang, L.; Lu, W.; Liu, X.; Pedrycz, W.; Zhong, C. Fuzzy C-Means clustering of incomplete data based on probabilistic information granules of missing values. *Knowl.-Based Syst.* **2016**, *99*, 51–70. [CrossRef]
48. Sakai, T. *t*-Tests. In *Laboratory Experiments in Information Retrieval: Sample Sizes, Effect Sizes, and Statistical Power*; Sakai, T., Ed.; The Information Retrieval Series; Springer: Gateway East, Singapore, 2018; pp. 27–41, ISBN 9789811311994.
49. Booth, T.; Doumas, A.; Murray, A.L. Null Hypothesis. In *Encyclopedia of Personality and Individual Differences*; Zeigler-Hill, V., Shackelford, T.K., Eds.; Springer International Publishing: Cham, Switzerland, 2020; pp. 3267–3270, ISBN 978-3-319-24612-3.
50. Saxon, E. Defining the null hypothesis. *BMC Biol.* **2015**, *13*, 68. [CrossRef] [PubMed]
51. F Distribution—An Overview | ScienceDirect Topics. Available online: https://www.sciencedirect.com/topics/mathematics/f-distribution (accessed on 23 May 2022).

electronics

MDPI

Article

A Computational Tool for Detection of Soft Tissue Landmarks and Cephalometric Analysis

Mohammad Azad [1,*], Said Elaiwat [1] and Mohammad Khursheed Alam [2]

1 Department of Computer Science, College of Computer and Information Sciences, Jouf University, Sakaka 72441, Saudi Arabia; smelaiwat@ju.edu.sa
2 Orthodontics, Department of Preventive Dental Science, College of Dentistry, Jouf University, Sakaka 72345, Saudi Arabia; mkalam@ju.edu.sa
* Correspondence: mmazad@ju.edu.sa

Citation: Azad, M.; Elaiwat, S.; Alam, M.K. A Computational Tool for Detection of Soft Tissue Landmarks and Cephalometric Analysis. *Electronics* **2022**, *11*, 2408. https://doi.org/10.3390/electronics11152408

Academic Editors: Agnieszka Konys, Agnieszka Nowak-Brzezińska and Hyunjin Park

Received: 31 May 2022
Accepted: 17 July 2022
Published: 2 August 2022

Abstract: In facial aesthetics, soft tissue landmark recognition and linear and angular measurement play a critical role in treatment planning. Visual identification and judgment by hand are time-consuming and prone to errors. As a result, user-friendly software solutions are required to assist healthcare practitioners in improving treatment planning. Our first goal in this paper is to create a computational tool that may be used to identify and save critical landmarks from patient X-ray pictures. The second goal is to create automated software that can assess the soft tissue facial profiles of patients in both linear and angular directions using the landmarks that have been identified. To boost the contrast, we employ gamma correction and a client-server web-based model to display the input images. Furthermore, we use the client-side to record landmarks in pictures and save the annotated landmarks to the database. The linear and angular measurements from the recorded landmarks are then calculated computationally and displayed to the user. Annotation and validation of 13 soft tissue landmarks were completed. The results reveal that our software accurately locates landmarks with a maximum deviation of 1.5 mm to 5 mm for the majority of landmarks. Furthermore, the linear and angular measurement variances across users are not large, indicating that the procedure is reliable.

Keywords: soft tissue; gamma correction; landmark detection; X-ray images; facial profile

1. Introduction

A pleasant-looking facial aesthetic is one of the purposes of healthcare treatment. Nowadays, many young males and females are looking for orthognathic surgery for a better and more attractive appearance in society. Therefore, it is necessary to study the facial skeleton and corresponding soft tissue. Bones and teeth are examples of hard tissue. Ligaments, tendons, and muscles are examples of soft tissue that connects and supports the body's surrounding structures and organs [1]. For successful orthognathic surgery, hard and soft tissue facial profile analysis should be included [2–4]. The hard tissue analysis for orthognathic surgery has been discussed by Burstone et al. [5] and the soft tissue analysis by Legan and Burstone [4]. However, several researchers found that the soft tissue (covering the teeth and face) can behave differently from patient to patient because of the thickness.

Structure inconsistencies have historically been considered the main treatment restrictions by orthodontists. In actuality, the therapeutic modifiability is more closely related to the soft tissues. As a result, the crucial step in orthodontic decision-making is the study of soft tissue. The extent to which the orthodontist can change the size of dental arches and the positioning of the mandible is determined by these soft tissues. Therefore, the cephalometric analysis of the soft tissue should be taken into account in a successful surgical treatment plan [6,7]. Furthermore, it is necessary to find the standard soft tissue profile analysis based on age, sex, ethnic group, etc. [8–11]. Similarly, Sahar [11] reported that an increasing number of Saudis are looking for orthognathic surgery. Therefore, this study should be carried out to find its own characteristics for Saudi Arabia.

There are two ways to analyze the soft tissue profile. The first way is by hand without any computer-aided solution; a transparent acetate sheet is superimposed above the printed X-ray image and then manually calculates the linear distances and angles. The second way is the computer-aided solution. There are many tools available which are used for research and teaching, but may cost a lot of money. Besides, the tool may not be customizable to suit the local community's needs. Nevertheless, most of the research regarding soft tissue profile analysis did not consider X-ray images, which are inevitable for the Saudi population since only X-ray images are acceptable in all Saudi hospitals, clinics, dental colleges, etc. Therefore, it is really necessary and of utmost importance to consider an inexpensive approach for the accurate investigation of the soft tissue profile analysis from the X-ray images.

Furthermore, other approaches are not created to study the linear and angular measurements automatically and need some type of manual intervention to obtain an accurate result. The novelty of our approach is that we implemented this process automatically without any type of manual intervention from the user, i.e., after successful annotations, the user can obtain the linear and angular measurements automatically.

The contribution of our study is:

1. The creation of a computational tool for identifying and saving important soft tissue landmarks in X-ray images;
2. The creation of a computational tool to automatically calculate linear and angular measurement of the patients' soft tissue facial profiles using the above identified landmarks.

We performed experiments with 14 male and 14 female subjects' X-ray images. We preprocess the images to improve the contrast by applying the gamma correction. After that, we annotated all these images by four examiners. As a result, we have a total of 112 annotated X-ray images. In the end, we calculated the variations among all annotators and found the variations are negligible, and hence, our approach is reliable. Nevertheless, we also calculated the linear and angular measurements and calculated the variations. We found that the variation is very small and our results are reliable.

An image must be in grayscale X-ray images with a minimum size of 1024 × 1024 to be included in the annotations. Since X-ray images are the main type of images for clinical practices in Saudi Arabia, we chose these types of images, and the mentioned size is the minimum best size for good X-ray images. In addition, the input images must include the areas of the forehead, nasal, labial, and chin. The reason is that these areas contain all our soft tissue landmarks. Furthermore, the intensity of the edge line, which contains such landmarks, and the background of the image should be distinguishable for the success of the detection of the landmarks. If these conditions are not met for any image sample, then it will not be processed further.

The rest of the paper is organized as follows. In Section 2, we consider the previous studies related to our paper. In Section 3, we explain the preprocessing steps before proceeding with our approach. Section 4 contains the detailed methods of the software architecture as well as methods related to the capturing of the annotations and measurements. Section 5 contains the results of the computer experiments and a discussion of the findings. At the end, Section 6 contains short conclusions and future work.

2. Previous Studies

A number of studies [2,12–20] have been carried out on the different aspects of soft tissue profile analysis. Some research focuses on linear measurement of the soft tissue profile [2] and some other research focuses on angular measurement [4,21]. The aim of this section is to review only the important and relevant studies related to the present research of soft tissue profile analysis.

One of the initial significant research attempts on the linear measurement of the soft tissue profile analysis was conducted by Paulo et al. [2], who considered 15 landmark points from the four major regions (facial, labial, chin, and nasal) and then calculated linear measurements based on the vertical, horizontal, and Canut's lines. Unfortunately, the work is based only on the photographic records (no X-ray images) and it is only for the European white population. They also did a similar analysis for angular measurement [21]. Sahar et al. [11,16] and Nasser [18] focus on the soft tissue profile analysis for the Saudi population in the Riyadh region. Their solution uses a market tool that is very expensive to buy and does not have the flexibility for custom usages. They did not consider major landmarks as in [2] and did not consider all the linear and angular measurements under consideration. Nevertheless, they did not consider any preprocessing steps such as gamma correction to improve the contrast and visibility of the X-ray images.

Furthermore, other researchers consider this problem based on the population of different regions in the world, e.g., Alcaledi et al. [19] consider the Japanese population, Hamdan et al. [20] consider the Jordanian population, Al-Azemi et al. [17] consider the Kuwaiti population, Filipović et al. [15] consider the Serbian population, Celebi et al. [14] consider the Turkish population, Akter et al. [13] consider the Bangladeshi population, Pandian et al. [12] consider the Indian population, etc.

For many years, manual cephalometric tracing was the only method where a transparent acetate sheet was superimposed above the printed X-ray images. The researchers used a pencil to locate the major landmarks and draw the lines and angles for the soft and hard tissue analysis [22].

Ricketts [23] illustrated computerized cephalometric tracing in 1969. After that, many tools are available on the market. Unfortunately, such tools are very expensive and non-customizable for use in the analysis of the cephalometric study. Therefore, it is necessary to look for a solution that will be used for the study of the soft tissue profile analysis, especially for the Saudi population. Another disadvantage of other tools is that they do not automatically provide linear and angular measurements; rather, a ruler must be used to obtain measurements for each subject of study, which is time-consuming and tedious work. In our work, our method automatically calculates the distance once the dental practitioners finalize the landmarks on the X-ray images. Furthermore, we use gamma correction to improve the contrast of the image, which was not performed by other tools.

3. Data Preprocessing

The borders of the soft tissue in the initial X-ray images are not clearly visible due to poor contrast, and hence we need to apply some preprocessing steps, i.e., to use the gamma correction method to sharpen and increase the contrast of the borders of the soft tissue.

According to [24], gamma correction, or gamma, is a nonlinear process which is performed for encoding and decoding luminance values in still images or video systems. In the simplest instances, gamma correction is specified by the power-law expression:

$$Y = aX^{\gamma} \tag{1}$$

where γ is a user defined parameter. The user can change the parameters and the brightness of the image will change based on the value of γ. The output value Y is obtained by raising the non-negative real input value X to the power γ and multiplying it by the constant a. In the case of $a = 1$, inputs and outputs are usually in the range of 0–1.

In Figure 1, we show a sample X-ray image before and after applying the gamma correction. It is clear that after applying the gamma correction, the soft tissue borders are now visible for further usage. In our tool, we first preprocess all X-ray images using the gamma correction before performing actual annotations.

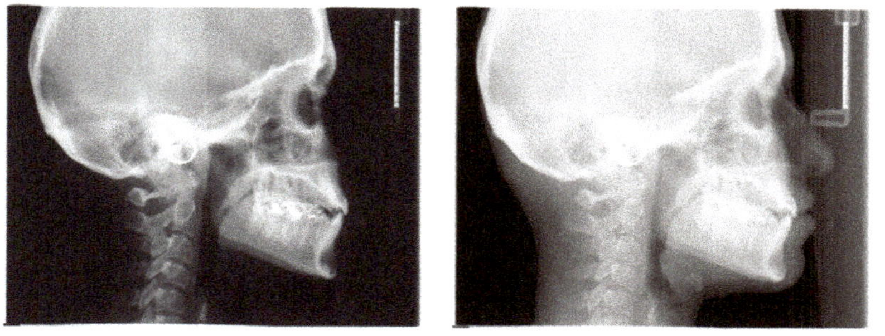

(a) Before (b) After

Figure 1. Effect of gamma correction of a sample X-ray image.

4. Method

The steps to developing the aforementioned tool for dental doctors' practices will be discussed in this section.

4.1. Architecture

We chose to use the client-server communication model paradigm. The reason behind this is that we have many users or clients (located separately) who will use this tool to annotate images. Such a distributive nature of clients necessitates the use of a client-server communication model rather than working as a single standalone program.

We have shown the client-server architecture in Figure 2. For the client-side, we used JavaScript (jQuery) along with CSS/HTML, and for the server-side, we used PHP (Laravel) with Apache server and MySQL database.

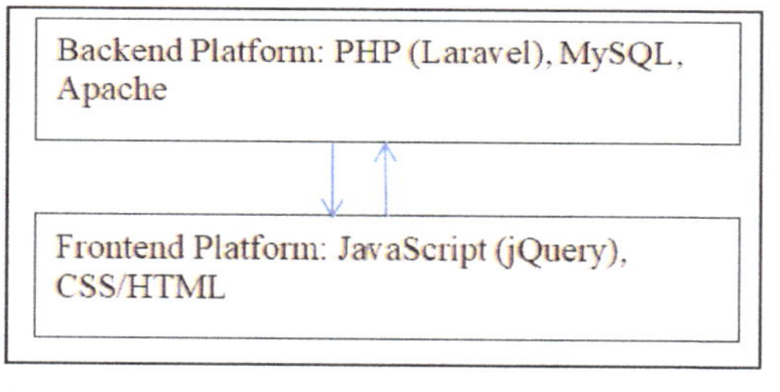

Figure 2. Software architecture.

The "admin" user controls the main functionality of this tool and can add other users to use this tool for annotations. The tool is divided into three main components. The first component is to add new users who will perform the job of annotation. The second component is to add the landmarks (tags in the tool) dynamically, i.e., we do not fix the number of landmarks; the admin can dynamically add new landmarks under consideration. Right now, the tool is using only thirteen soft tissue landmarks, as shown in Table 1.

The third component is to add new images for annotations, edit previously added images, and search for images by keywords, gender, etc. When an expert annotator first logs in then he will see the list of images that he has already uploaded and/or annotated and a button to add new images (see Figure 3).

At this stage, the annotator can add new images by clicking the button "Add Image". Furthermore, he can edit already added images by clicking the hyperlink "Edit" and can annotate the considered image by clicking on "Annotations".

Table 1. The set of soft tissue landmarks.

Landmark Name	Description
Tri	It is the sagittal middle point of the forehead that borders the hairline
G	It is the most anterior point of the forehead's central line
N	It is the point in the middle line that is placed at the root of the nose
Prn	It is the most noticeable part of the nose's tip
Cm	It is the nose's most inferior and anterior point
Sn	It's the junction of the upper lip and the columella
Ls	It's the place where the upper lip's mucocutaneous limit is indicated
Sts	It is the upper lip's most inferior point
Sti	It is the upper lip's most superior point
Li	It is the place where the lower lip's mucocutaneous limit is indicated
Sm	The inferior sublabial concavity's deepest point lies here
Pg	It is the chin's most anterior point
Me	It is the most inferior point of the chin's inferior edge

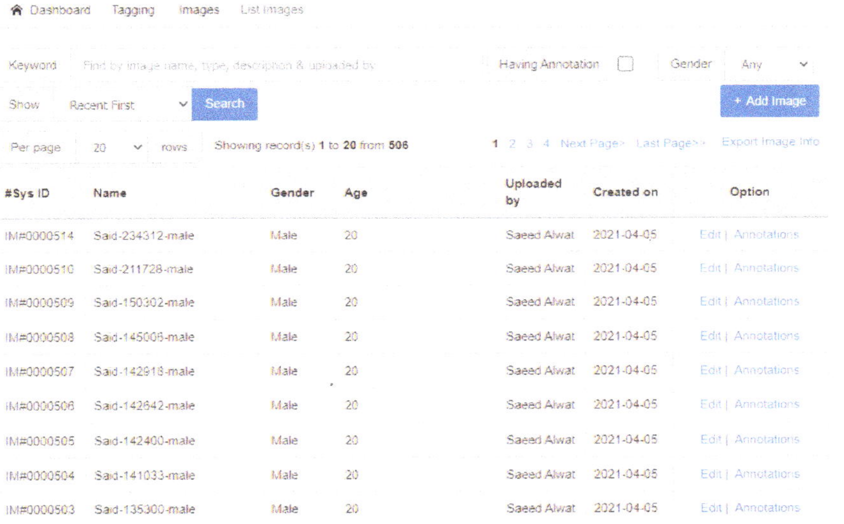

Figure 3. The first landing page after login to show the list of images.

4.2. Capturing Annotations

In this study, we use 13 landmarks from the soft tissue area as shown in Table 1.

The user should click "Annotations" to start annotating the above landmarks, or if he has already done so, he can update those annotations easily. For illustration purposes, we show a sample X-ray image with the expert annotations in Figure 4.

When an expert annotator uses this tool to annotate landmarks, our tool captures those landmark positions (2D coordinates) and stores them in the system. It is possible to zoom the X-ray images for better viewing and locating the position of the landmarks. The tool takes the position based on a percentage matrix, i.e., both x and y-axis are taken as 100% of its actual width and height, respectively. It then uses the captured landmark position to calculate the actual position from the actual width and height of that image. Therefore, the zoom does not affect the landmark's actual position. In addition, the position is taken with a long decimal fraction, which is up to 30 decimal points, to make the position more accurate.

It is always possible to change the landmarks manually and correct them. The "Edit" button is used for editing any images that have been previously annotated.

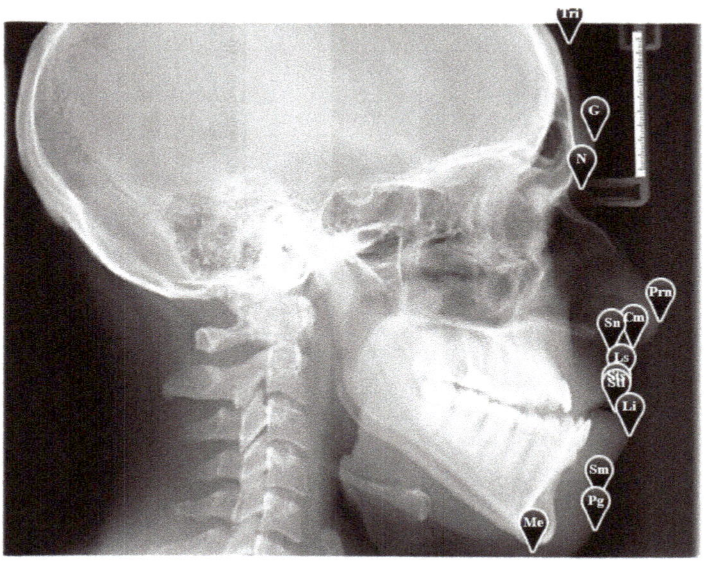

Figure 4. A sample X-ray image with annotation.

4.3. Linear and Angular Measurements

The tool automatically calculates the linear and angular measurements without any intervention from the user. Once the user finishes the annotation, then he needs to simply click the "View Distances" link and it will show both measurements immediately (see Figure 5). In addition, the user can save the measurement in an excel file.

4.3.1. Linear Measurements

The tool provides the horizontal (x-axis) distance and vertical (y-axis) distances for a fixed pair of landmarks. The horizontal and vertical distance calculations are trivial; that is the normal difference between the values of the corresponding axis. The distance output is given in pixels that can be converted to mm by Equation (4). Nevertheless, we can easily calculate the Euclidean distance (c) from the horizontal and vertical distances by the following formula:

$$c = \sqrt{a^2 + b^2} \tag{2}$$

where,
c = the Euclidean distance
a = the horizontal distance (Distance-X)
b = the vertical distance (Distance-Y)

4.3.2. Angular Measurements

The tool provides the angle for a given set of three landmarks in a degree unit. For example, G-N-Prn is a set of three landmarks where the middle landmark, N, is the vertex of the angle, and G-N and N-Prn are the sides of the angle. The angle has been calculated by the following formula:

$$\theta = \cos^{-1} \frac{b}{c} \tag{3}$$

where,
b = the distance G-N
c = the distance N-Prn

(**a**) Linear measurement

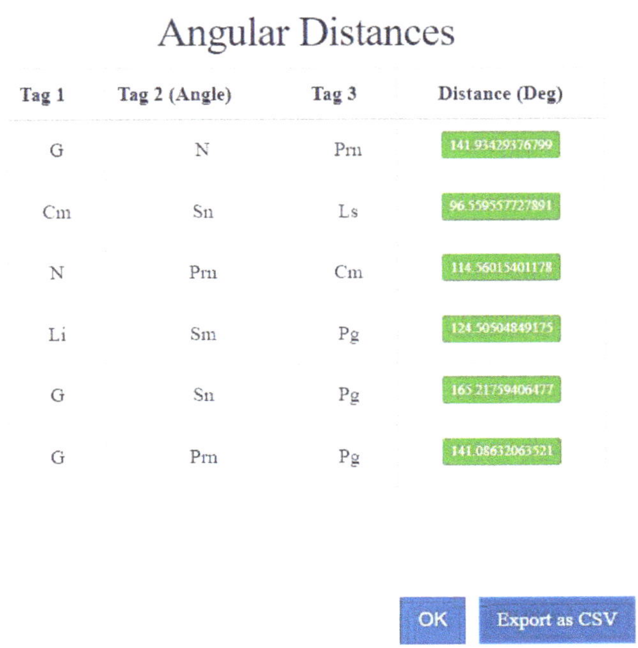

(**b**) Angular measurement

Figure 5. A sample window of linear and angular measurement.

5. Results & Discussion

For our experiments, we obtained 28 sample X-ray images (14 male and 14 female) from health centers in Saudi Arabia. Four examiners independently performed the annotation and validation of the concerned soft tissue landmarks. As a result, we have a total of 112 annotated X-ray images.

In the literature, there are many methods for the evaluation of the system for identifying the landmark position for the acceptance of clinical practices. The manual method of human visual judgement is prone to intrajudge and interjudge variations [25]. The second way is the mix of manual and computer systems recognition method that is also susceptible to human error [25]. The third way is to examine if the computer system's output is within the radius of 2 mm or not [25]. Our method is better where we obtained a radius for some landmarks even smaller than 2 mm.

5.1. Validation of Locating the Landmarks

We calculate the variation of landmarks by the computer system. For each landmark, we find the minimum variation that is the minimum distance between any two identified landmarks. Similarly, we calculate the average and maximum distance between any two landmarks. Note that, for each variation (minimum, average, and maximum), we take the average among the 28 samples and show them as our results. Table 2 displays the pixel and corresponding distance (mm) variation, with the first column displaying the landmark name, the second column displaying the minimum variation, the third column displaying the average variation and the fourth column displaying the maximum variation. We calculate the length in 'mm' from the pixel by the following formula (using 300 dpi):

$$Length\,[\mathrm{mm}] = pixel \times 25.4\,\mathrm{mm/dpi} \tag{4}$$

We can observe that the minimum variation is below 1 mm and the maximum variation for most landmarks is in the range of 1.5 mm to 5 mm except for the landmark G and Pg. It is due to the fact that they are the most difficult to identify. Similarly, the average variation for most landmarks is in the range of 1 mm to 2 mm except for the above-mentioned two landmarks. Figure 6 shows the variation in a sample X-ray image, and Figure 7 shows the variation of each landmark in the same X-ray image (red circle shows the variation area).

Table 2. Variation of different landmarks.

Landmark	Min		Avg		Max	
	Pixel	mm	Pixel	mm	Pixel	mm
Tri	3.6157	0.3061	13.6641	1.1569	23.7607	2.0117
G	11.0327	0.9341	69.8990	5.9181	118.3536	10.0206
N	5.5674	0.4714	24.1208	2.0422	40.3260	3.4143
Prn	4.0860	0.3460	15.2229	1.2889	26.4002	2.2352
Cm	4.0261	0.3409	15.5780	1.3189	28.5866	2.4203
Sn	5.3419	0.4523	18.6489	1.5789	31.4737	2.6648
Ls	3.4025	0.2881	12.6053	1.0672	21.8777	1.8523
Sts	4.9509	0.4192	20.0948	1.7014	36.3959	3.0815
Sti	6.7421	0.5708	20.8528	1.7655	37.0563	3.1374
Li	5.6839	0.4812	23.2916	1.9720	40.4615	3.4257
Sm	3.3831	0.2864	11.0309	0.9339	18.9436	1.6039
Pg	9.9248	0.8403	45.6980	3.8691	82.3182	6.9696
Me	8.6783	0.7348	32.0921	2.7171	58.5907	4.9607

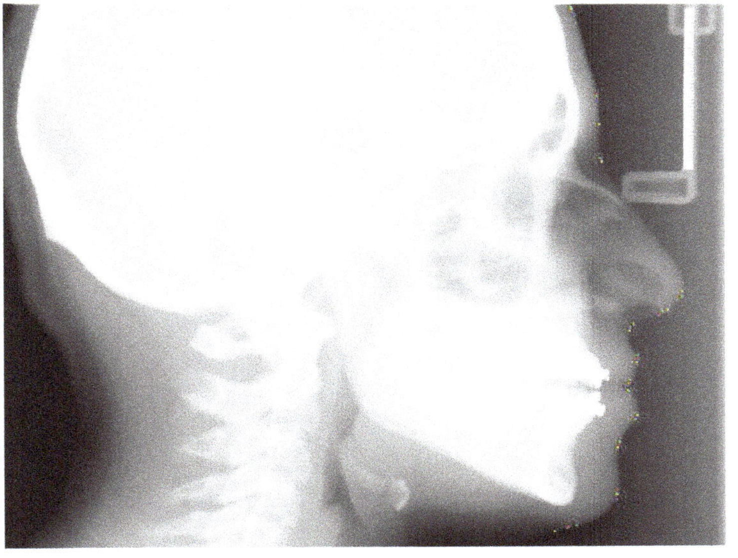

Figure 6. Graphical representation of variations of one sample image showing all the landmarks.

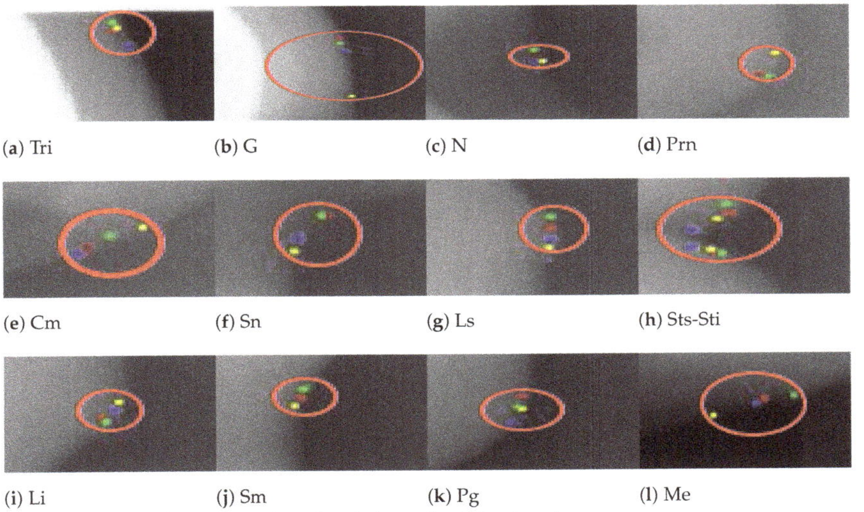

(**a**) Tri (**b**) G (**c**) N (**d**) Prn (**e**) Cm (**f**) Sn (**g**) Ls (**h**) Sts-Sti (**i**) Li (**j**) Sm (**k**) Pg (**l**) Me

Figure 7. Graphical representation of variations of each landmark.

5.2. Validation of Linear and Angular Measurements

We show the minimum (the column 'Min'), average (the column 'Avg'), and maximum (the column 'Max') value of linear measurement (Euclidean distance) in pixel and mm units in Table 3. Nevertheless, we compared statistically using a Student's paired t-test and we did not find any significant difference in variation, which shows that the results are stable using this tool.

Table 3. Variation of linear measurement.

Two Landmarks	Min		Avg		Max	
	Pixel	**mm**	**Pixel**	**mm**	**Pixel**	**mm**
Tri-G	326.7448	27.6644	376.6794	31.8922	430.2955	36.4317
G-Sn	784.4816	66.4194	837.6199	70.9185	889.4295	75.3050
Sn-Me	818.8464	69.3290	832.4976	70.4848	847.7765	71.7784
N-Sn	613.8476	51.9724	637.1901	53.9488	666.0193	56.3896
Sn-Sts	216.6776	18.3454	230.6743	19.5304	244.0921	20.6665
Sts-Sti	47.3594	4.0098	56.4537	4.7797	66.7857	5.6545
Sts-Sm	241.3957	20.4382	251.9805	21.3343	265.5227	22.4809
Ls-Sts	95.2157	8.0616	111.9558	9.4789	124.0368	10.5018
Li-Sti	104.0545	8.8099	121.4404	10.2820	135.8431	11.5014
Sm-Me	346.8453	29.3662	363.538	30.7796	381.9898	32.3418
Sn-Prn	180.9176	15.3177	202.0955	17.1108	227.1582	19.2327

Similarly, we show the minimum (the column 'Min'), average (the column 'Avg'), and maximum (the column 'Max') values of angular measurements (in degree units) in Table 4. Furthermore, we examined statistically using the Student's paired t-test and found no significant difference, indicating that the results produced by our tool are consistent.

Table 4. Variation of angular measurement (degree).

Three Landmarks	Min	Avg	Max
G-N-Prn	142.0364	145.784	149.6623
Cm-Sn-Ls	102.9063	106.5279	110.6642
N-Prn-Cm	105.6326	109.4912	115.0749
Li-Sm-Pg	127.4258	133.3732	139.8691
G-Sn-Pg	159.3171	160.547	162.3006
G-Prn-Pg	135.6283	137.3321	138.8915

5.3. Limitations and Special Cases

Even though this developed tool can detect soft tissue landmarks pretty accurately, it is not free from limitations. The success of landmark detection depends mainly on the quality, size, and resolution of images as well as the intensity of soft tissue edge lines compared to the background.

The most prevalent conditions affecting the facial region are cleft deformities [26] and craniofacial defects [27]. Soft tissue landmarks are difficult to identify in bilateral and unilateral complete cleft lip and palate cases because the alveolus and lip are not fused well. In such cases, a clear image and a zoom-in facility might help. As well, experienced orthodontists can follow an anatomical point of view if they feel difficulties.

6. Conclusions and Future Work

In this paper, we describe our tool to capture the soft tissue landmark positions. This tool is based on the paradigm of the client-server communication model. Any orthodontist can use this tool for his clinical practice, and it can accurately give the landmark positions up to 30 decimal points. It can also extract information on linear and angular measurements for orthodontic treatment, allowing for a more personalized healthcare experience. We conducted experiments on 28 human samples, which resulted in robust and accurate measurement of soft tissue landmarks within a 5 mm radius.

One of the limitations of this study is that this tool only annotates soft tissue landmarks. In the future, hard tissue landmarks will be explored.

In today's world, the smartphone is the most user-friendly technology available in the healthcare field. As a result, we will strive to integrate the proposed approach onto smartphones in the future, so that physicians may quickly recognize landmarks and complete the cephalometric analysis.

Author Contributions: Conceptualization, M.A., S.E. and M.K.A.; methodology, M.A., S.E. and M.K.A.; software, M.A.; validation, M.A.; formal analysis, M.A., S.E. and M.K.A.; investigation, M.A., S.E. and M.K.A.; resources, M.A., S.E. and M.K.A.; data curation, M.A., S.E. and M.K.A.; writing, M.A., S.E. and M.K.A.; visualization, M.A., S.E. and M.K.A.; supervision, M.A., S.E. and M.K.A.; project administration, M.A.; funding acquisition, M.A. All authors have read and agreed to the published version of the manuscript.

Funding: The authors extend their appreciation to the Deanship of Scientific Research at Jouf University for funding this work through research grant no. (DSR2020-04-2582).

Data Availability Statement: Available upon reasonable request.

Acknowledgments: The authors would like to express their gratitude to all of the volunteers who helped with this study.

Conflicts of Interest: The authors declare no conflict of interest.

References

1. Understanding Injuries: Soft Tissue versus Hard Tissue Injuries. Available online: https://www.mgmillerlaw.com/soft-versus-hard-tissue-injuries/ (accessed on 19 March 2022).
2. Fernández-Riveiro, P.; Suárez-Quintanilla, D.; Smyth, E.; Suarez-Cunqueiro, M. Linear photogrammetric analysis of the soft tissue facial profile. *Am. J. Orthod. Dentofac. Orthop.* **2002**, *122*, 59–66. [CrossRef] [PubMed]
3. Ricketts, R. Esthetics, environment, and the law of lip relation. *Am. J. Orthod.* **1968**, *54*, 272–289. [CrossRef]
4. Legan, H.; Burstone, C. Soft tissue cephalometric analysis for orthognathic surgery. *J. Oral Surg.* **1980**, *38*, 744–751. [PubMed]
5. Burstone, C.; James, R.; Legan, H.; Murphy, G.; Norton, L. Cephalometrics for orthognathic surgery. *J. Oral Surg.* **1978**, *36*, 269–277.
6. Sarver, D.; Pfoffit, W.; Ackerman, J. Evaluation of facial soft tissues. In *Contemporary Treatment of Dentofacial Deformity*; Proffit, W.R., White, R.P., Sarver, D.M., Eds.; Mosby: St. Louis, MO, USA, 2003; Volume 283.
7. Ackerman, J.L.; Proffit, W.R. Soft tissue limitations in orthodontics: Treatment planning guidelines. *Angle Orthod.* **1997**, *67*, 327–336.
8. Holdaway, R.A. A soft-tissue cephalometric analysis and its use in orthodontic treatment planning. Part I. *Am. J. Orthod.* **1983**, *84*, 1–28. [CrossRef]
9. Merrifield, L.L. The profile line as an aid in critically evaluating facial esthetics. *Am. J. Orthod.* **1966**, *52*, 804–822. [CrossRef]
10. Holdaway, R.A. A soft-tissue cephalometric analysis and its use in orthodontic treatment planning. Part II. *Am. J. Orthod.* **1984**, *85*, 279–293. [CrossRef]
11. Albarakati, S. Soft tissue facial profile of adult Saudis Lateral cephalometric analysis. *Saudi Med. J.* **2011**, *32*, 836.
12. Pandian, K.S.; Krishnan, S.; Kumar, S.A. Angular Photogrammetric Analysis of the Soft-tissue Facial Profile of Indian Adults. *Indian J. Dent. Res.* **2018**, *29*, 137–143.
13. Akter, L.; Hossain, M. Angular Photogrammetric Soft Tissue Facial Profile Analysis of Bangladeshi Young Adults. *APOS Trends Orthod.* **2017**, *7*, 279. [CrossRef]
14. Celebi, A.; Tan, E.; Gelgor, I.; Colak, T.; Ayyıldız, E. Comparison of Soft Tissue Cephalometric Norms between Turkish and European-American Adults. *Sci. World J.* **2013**, *2013*, 806203. [CrossRef]
15. Filipović, G.; Stojanovic, N.; Jovanovic, I.; Randjelovic, P.; Ilić, I.; Đorđević, N.; Radulović, N. Differences in Angular Photogrammetric Soft-Tissue Facial Characteristics among Parents and Their Offspring. *Medicina* **2019**, *55*, 197. [CrossRef]
16. ALBarakati, S.F.; Bindayel, N.A. Holdaway soft tissue cephalometric standards for Saudi adults. *King Saud Univ. J. Dent. Sci.* **2012**, *3*, 27–32. [CrossRef]
17. Al-Azemi, R.; Al-Jame, B.; Årtun, J. Lateral Cephalometric Norms for Adolescent Kuwaitis: Soft Tissue Measurements. *Med Princ. Pract.* **2008**, *17*, 215–220. [CrossRef]
18. Al-Jasser, N. Facial esthetics in a selected Saudi population. *Saudi Med. J.* **2003**, *24*, 1000–1005.
19. Alcalde, R.; Jinno, T.; Orsini, M.; Sasaki, A.; Sugiyama, R.M.; Matsumura, T. Soft tissue cephalometric norms in Japanese adults. *Am. J. Orthod. Dentofac. Orthop.* **2000**, *118*, 84–99. [CrossRef]
20. Hamdan, A. Soft Tissue Morphology of Jordanian Adolescents. *Angle Orthod.* **2010**, *80*, 80–85. [CrossRef]
21. Fernández-Riveiro, P.; Smyth, E.; Suárez-Quintanilla, D.; Suarez-Cunqueiro, M. Angular photogrammetric analysis of the soft tissue facial profile. *Eur. J. Orthod.* **2003**, *25*, 393–399. [CrossRef]
22. Leonardi, R.; Giordano, D.; Maiorana, F.; Spampinato, C. Automatic Cephalometric Analysis A Systematic Review. *Angle Orthod.* **2008**, *78*, 145–151. [CrossRef]
23. Ricketts, R.M. *Introducing Computerized Cephalometrics*; Rocky Mountain Data Systems, Inc.: Denver, CO, USA, 1969.
24. Poynton, C. *Digital Video and HDTV Algorithms and Interfaces*, 1st ed.; Morgan Kaufmann Publishers Inc.: San Francisco, CA, USA, 2003.

25. Tanikawa, C.; Yagi, M.; Takada, K. Automated Cephalometry: System Performance Reliability Using Landmark-Dependent Criteria. *Angle Orthod.* **2009**, *79*, 1037–1046. [CrossRef]
26. Paradowska-Stolarz, A.M.; Kawala, B. The Nasolabial Angle Among Patients with Total Cleft Lip and Palate. *Adv. Clin. Exp. Med.* **2015**, *24*, 481–485. [CrossRef]
27. Fatima, F.; Jeelani, W.; Ahmed, M. Current trends in craniofacial distraction: A literature review. *Dent. Med Probl.* **2020**, *57*, 441–448. [CrossRef]

electronics

MDPI

Review

A Systematic Review of the Applications of Multi-Criteria Decision Aid Methods (1977–2022)

Marcio Pereira Basílio [1,2,*], Valdecy Pereira [2], Helder Gomes Costa [2], Marcos Santos [3] and Amartya Ghosh [4]

1 Military Police of the Rio de Janeiro, Rio de Janeiro 21941-901, Brazil
2 Department of Production Engineering, Federal Fluminense University (UFF), Niteroi 24210-240, Brazil; valdecy.pereira@gmail.com (V.P.); heldergc@id.uff.br (H.G.C.)
3 Military Institute of Engineering (IME), Rio de Janeiro 21941-901, Brazil; marcosdossantos_doutorado_uff@yahoo.com.br
4 Symbiosis Institute of Business Management (SIBM), Hyderabad 509217, India; amartya.ghosh@sibmhyd.edu.in
* Correspondence: marcio_basilio@id.uff.br; Tel.: +55-21996379803

Abstract: Multicriteria methods have gained traction in academia and industry practices for effective decision-making. This systematic review investigates and presents an overview of multi-criteria approaches research conducted over forty-four years. The Web of Science (WoS) and Scopus databases were searched for papers on multi-criteria methods with titles, abstracts, keywords, and articles from January 1977 to 29 April 2022. Using the R Bibliometrix tool, the bibliographic data was evaluated. According to this bibliometric analysis, in 131 countries over the past forty-four years, 33,201 authors have written 23,494 documents on multi-criteria methods. This area's scientific output increases by 14.18 percent every year. China has the highest percentage of publications at 18.50 percent, followed by India at 10.62 percent and Iran at 7.75 percent. Islamic Azad University has the most publications with 504, followed by Vilnius Gediminas Technical University with 456 and the National Institute of Technology with 336. *Expert Systems with Applications*, *Sustainability*, and the *Journal of Cleaner Production* are the top journals, accounting for over 4.67 percent of all indexed works. In addition, E. Zavadskas and J. Wang have the most papers in the multi-criteria approaches sector. AHP, followed by TOPSIS, VIKOR, PROMETHEE, and ANP, is the most popular multi-criteria decision-making method among the ten nations with the most publications in this field. The bibliometric literature review method enables researchers to investigate the multi-criteria research area in greater depth than the conventional literature review method. It allows a vast dataset of bibliographic records to be statistically and systematically evaluated, producing insightful insights. This bibliometric study is helpful because it provides an overview of the issue of multi-criteria techniques from the past forty-four years, allowing other academics to use this research as a starting point for their studies.

Keywords: systematic review; multicriteria; MCDA; MCDM; MADM; MODM; AHP; TOPSIS; VIKOR; PROMETHEE; ANP

Citation: Basílio, M.P.; Pereira, V.; Costa, H.G.; Santos, M.; Ghosh, A. A Systematic Review of the Applications of Multi-Criteria Decision Aid Methods (1977–2022). *Electronics* **2022**, *11*, 1720. https://doi.org/10.3390/electronics11111720

Academic Editors: Agnieszka Konys and Agnieszka Nowak-Brzezińska

Received: 29 April 2022
Accepted: 25 May 2022
Published: 28 May 2022

1. Introduction

As the transmission of scientific knowledge in its most diverse fields of study expands, literature evaluation becomes a demanding work for the researcher [1]. The challenge is reflected in the volume of research published each month by thousands of academic publication outlets. According to [2]'s theory of limited rationality, a researcher's rationality is constrained by the knowledge available, the cognitive limitations of the individual mind, and the decision-making time availability.

Human activities require decision-making. All such decisions are based on an evaluation of individual decision options, typically based on the decision maker's preferences, experience, and other data [3]. Some decisions are simple, while others are complex [4].

According to Kahraman et al. [5] and Govindan and Jepsen [6], some decisions are relatively simple, especially if the consequences of making the wrong decision are minor, whereas others are highly complex and have significant effects. In most cases, real-life problem-solving involves several competing points of view that must be considered to reach a reasonable decision [7]. A decision can be defined formally as a choice made based on available information or a method of action aimed at solving a specific decision problem [8]. In practice, multiple-criteria decision analysis (MCDA) evaluates possible courses of action or options by selecting a preferred option or sorting the options from best to worst [9–12]. In everyday practice, the use of MCDA is critical in signaling the best rational alternative to the decision-maker so that he can allocate finite resources between competing and alternative interests. Whether in an organizational or domestic setting, the decision-maker is constantly confronted with multiple paths and limited resources. Researchers refer to multiple criteria methods in various ways. Some authors prefer the term multiple-criteria decision aid or aiding (MCDA), while others prefer to use the term multi-criteria decision-making or multiple-criteria decision-making (MCDM), multi-objective decision-making (MODM), or multi-attribute decision-making (MADM). Some authors prefer the term multiple-criteria decision aid or aiding (MCDA), while others prefer to use the term multiple-criteria decision analysis [13].

The most often used MCDA approaches, as opined by [3,14], are divided into two "schools": American and European. The American School of decision-support methods is based on a functional approach, namely the utilization of value or usability. These strategies typically do not account for data inconsistency, ambiguity, or decision-maker preferences. This collection of techniques is closely related to the operational approach based on a single synthesized criterion. MAUT, AHP, ANP, SMART, UTA, MACBETH, and TOPSIS are the critical methods used in the American School. The European School's techniques are based on a relational concept. As a result, they employ a synthesis of criteria based on outranking relations. Transgression between pairs of decision alternatives characterizes this relationship. Among the European School of decision support methods, the ELECTRE and PROMETHEE groups are the most prominent. NAIADE, ORESTE, REGIME, ARGUS, TACTIC, MELCHIOR, and PAMSSEM are other methodologies from the European MCDA sector. Many multi-criteria decision-making strategies integrate ideas from the American and European decision-making schools. EVAMIX, QUALIFLEX, PCCA, MAPPAC, PRAGMA, PACMAN, IDRA, COMET, and DRSA are a few examples.

Furthermore, as stated by [6,14–16], MCDA methods are used to solve decision-making problems in several areas, including the information and communication technology; business intelligence; environmental risk analysis; environmental impact assessment and environmental sciences; water-resource management; solid-waste management; remote sensing; flood-risk management; health-technology assessment; healthcare; transport; nanotechnology research; climate change; energy; international law and policy; human resources; financial management; performance and benchmarking; supplier selection; e-commerce and m-commerce; agriculture and horticulture; chemical and biochemical engineering; software evaluation; network selection; education and social policy; heating, ventilation, and air conditioning and small-scale energy management systems; and public security.

According to Sałabun et al. [3], despite the numerous MCDA approaches available, it is essential to note that no method is ideal and can be deemed acceptable for use in every decision-making context or for solving every choice problem [17]. As a result, different multi-criteria techniques may yield various choice suggestions [18]. However, if multiple multi-criteria methods produce inconsistent findings, the accuracy of each option is called into doubt [19]. In such a case, selecting a decision-support technique relevant to the given problem [20] becomes essential because only an appropriately chosen approach allows one to acquire the correct answer that reflects the decision maker's preferences [21].

Humans make decisions regularly, and decision-making is an inherent element of people's character. Some decisions are simple and have little impact on people's lives; others, on the other hand, directly impact people's lives, cities, and nations. In this regard, and given the importance of multi-criteria decision-making methods in assisting decision-makers in a variety of fields, the current study aims to answer the following research questions (RQ) and develop a reference framework on academic productivity regarding multi-criteria decision-making methods:

RQ1: Who are the most influential authors and researchers in their scientific productivity in multi-criteria decision-making methods?
RQ2:What is the annual scientific publication growth in multi-criteria decision-making methods?
RQ3: Which countries have the most significant production of articles on the multi-criteria methods of decision support?
RQ4: Which journals have the highest number of publications?
RQ5: What are the most used methods, and in which research areas?
RQ6: What are the conceptual structures of the multi-criteria decision-support methods?

Three hundred forty-two systematic literature studies on multi-criteria methods were discovered during the literature survey. The ten largest categories classified by Web of Science using multi-criteria methods were green sustainable science technology [22], energy fuels [23], environmental sciences [24], operations research and management science [25,26], computer science and artificial intelligence [27], management [28], economics [29], engineering environmental [30], computer science and interdisciplinary applications [31], and civil engineering [32].

This article is structured as follows: Section 2 briefly describes the methods and materials. Section 3 presents the preliminary bibliometric results and visualizes the collaborative relationships between countries and authors using R and the VOSviewer software. Keyword co-occurrences are analyzed, and strategic diagrams are constructed in the same section to reveal thematic trends on the multi-criteria decision support theme. The main discussions are summarized in Section 4.

2. Materials and Methods

This section presents the fundamental concepts that guided this study. The intention is not to cover all the subjects but rather to provide essential supporting information for understanding the research, the context, and the results.

The volume of academic publications is increasing at an accelerating rate. In this way, keeping up-to-date and knowing a given topic's state of the art is becoming increasingly difficult. As stated by Aria and Cuccurullo [33], the emphasis on empirical contributions has resulted in voluminous and fragmented research flows, which contributes to the heavy work of the researcher to keep up to date. Researchers affirm that literature reviews are prevalent in the state-of-the-art synthesis of various themes [33,34].

The structured literature review is a traditional way to analyze and review scientific literature. This type of review provides an in-depth analysis according to the content of the literature [35–39]. However, this method suffers from several limitations. For instance, it is very time-consuming, and the number of analyzed papers is limited. It is almost impossible to analyze hundreds of documents through the structured literature-review process. Although the authors carefully select the documents according to several criteria, it is challenging to eliminate subjective factors, and some essential studies may be omitted. With the digitization of scientific journals, the volume of published papers has increased dramatically. A bibliometric analysis effectively handles hundreds, even thousands, of documents and reviews the related literature from a macro perspective [37].

The term bibliometric refers to the quantitative study of bibliographic materials [40,41]. It can characterize the development in a research field or capture the changes in a specific journal. Various techniques have been developed to conduct bibliometric analysis, and the most-used methods are social network analysis and co-word analysis [37].

Social network analysis is based on the premise that the relationships between units can be interpreted as a graph [42]. It is an effective method to evaluate the importance of nodes and reveal the network structure. In the bibliometric networks, different types of networks, such as coauthorship networks [43,44], bibliographic coupling networks [45], and co-citation networks [46], are constructed by bibliometrics [47].

Co-word analysis is a content-analysis technique proposed by [48,49]. It is applied to map the strength of associations between information items in textual data [50]. It involves a co-occurrence analysis of keywords in a selected body of literature. Co-occurrence analysis, a central task of association analysis in data mining, is used to group keywords with high relevance in clusters [51]. Typically, each set corresponds to a search theme. Researchers use co-occurrence analysis to identify established and emerging research themes or tracking patterns [52–54].

Numerous software tools support bibliometric analysis; however, many do not assist scholars in a complete recommended workflow. The most relevant tools are CitNetExplorer [55], VOSviewer [56], SciMAT [50], BibExcel [57], Science of Science (Sci2) Tool [58], CiteSpace [59], HistCite, Pajek, Gephi, Bibliometrix [33], and VantagePoint (www.thevantagepoint.com (accessed on 24 April 2022)). In this study, VOSviewer and Bibliometrix were used to conduct a co-citation analysis.

In this study, a topical query on 29 April 2022, was conducted in the Web of Science (WoS) and Scopus database, using the following search query: ((“multi-attribute decision making” or “madm” or “mcda” or “modm” or “mcdm” or “multi-criteria” or “multi-criteria” or “multiplecriteria”) and (“ahp” or “todim” or “topsis” or “promethee” or “electre” or “vikor” or “maut” or “fitradeoff” or “dematel” or “copras” or “multimoora” or “swara” or “analytical network process” or “anp” or “simple multi-attribute rating technique” or “smart” or “goal programming” or “thor” or “cbr” or “saw” or “condorcet” or “drsa” or “macbeth” or “paprika” or “wpm” or “wsm” or “utadis” or “waspas”)). The search was only restricted to titles, abstracts, keywords, and articles published between 1977 and 2022. Additionally, the search in the WoS database was limited to the Core Collection. The search query yielded 35,643 entries from the WoS and Scopus databases. Following the download of the records, the RStudio bibliometrix package version 1.2.1335 was installed on a Win64 operating system. Bibliometric analysis was performed using the Bibliometrix R package. The Bibliometrix tool was used to build the descriptive and co-citation networks. The function convert2df embedded in the Bibliometrix package was used to extract and create a data frame corresponding to the unit of analysis within the exported files from WoS and Scopus databases. After making the data frames from the WoS and Scopus files, the mergeDbSources function merged the WoS and Scopus data frames and excluded duplicate records from both files. Twelve thousand one hundred forty-nine duplicate records were removed, resulting in a data frame with 23,494 records for the bibliometric analysis. The process of obtaining the bibliographic records file can be seen in Figure 1.

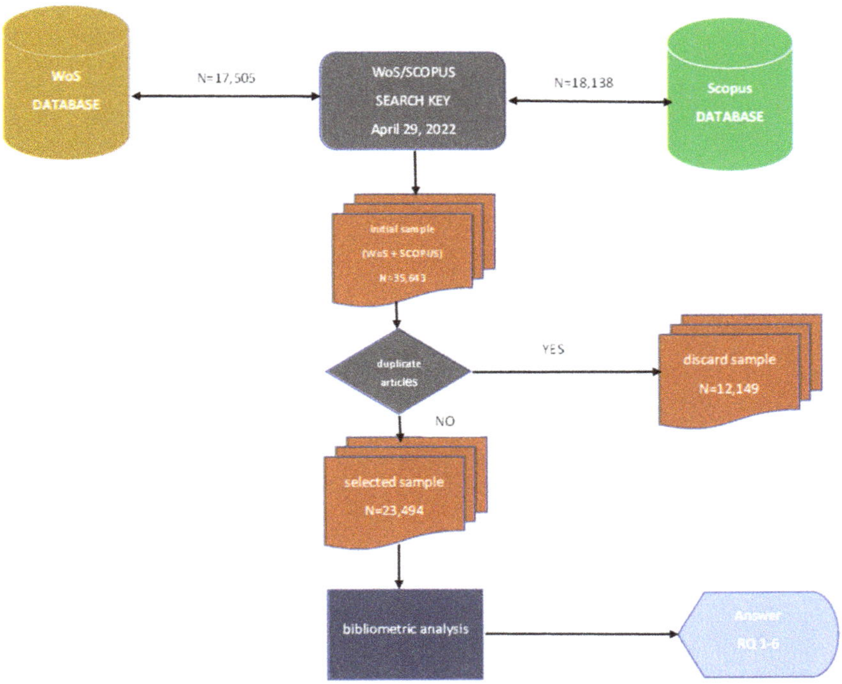

Figure 1. Search strategy and extraction of data. Source: Prepared by the authors based on Basilio et al. [60] and Ghosh and Prasad [61].

3. Results

The results from the bibliometric analysis show that 33,201 authors produced 23,494 documents in the period from 1 January 1977, to 29 April 2022. The types of documents identified in the sample, despite the limitations, are described in the methods and data section and further illustrated in Figure 2.

Regarding academic production, studies on multi-criteria decision-support methods had their genesis in 1977. Figure 3 depicts the publishing trajectory until April 2022. The graph shows that the upward trend began in 1986 with a modest inclination. During this time, the average number of publications each year was 7.3. From 1987 to 1996, the average number of papers per year climbed to 28.3 documents. This average increased to 123.2 records per year during the next ten years and finally reached 1265.73 from 2007 to 2021, indicating a strong level of interest in the topic among researchers. Taking the entire period into account, publications on multi-criteria decision-support methods grew at an annual percentage rate of 14.18. Figures 4 and 5 show the average total citations per year (16.06) and the average years from publication (6.36), respectively.

Five peaks are depicted in the graph shown in Figure 4. In 1983, the earliest and most important studies were conducted. In that year, six documents were published. The article by Van Laarhoven and Pedrycz [62], with a total citation count of 2158, had the most impact on citations in 1983. The authors presented a fuzzy variant of Saaty's pairwise comparison method for deciding between many options when there are competing choice criteria. Eleven publications were included in the sample in 1986. The article by Brans et al. [63] had a significant impact that year, increasing the yearly average of 1609 citations. Brans et al. [63] introduced the PROMETHEE approach in this study. Chen et al. [64] had the most-cited paper in 1994, with 967 citations. Chinese researchers provided novel methods for dealing with fuzzy multi-criteria decision-making based on the theory of fuzzy sets. There were

2454 citations to Chen's paper [64] in 2000, which affected the average of the 63 articles published that year. Chen [64] extended the TOPSIS model to the fuzzy environment. Furthermore, in 2004, two publications significantly impacted the average number of citations among the 128 papers published: Opricovic and Tzeng [65] had 2590 citations, while Pohekar and Ramachandran [66] had 1270 citations. The VIKOR and TOPSIS approaches were compared by Opricovic and Tzeng [65]. Pohekar and Ramachandran [66] conducted a systematic review of multi-criteria techniques for sustainable energy management. Table 1 provides a summary of the sample's most cited articles.

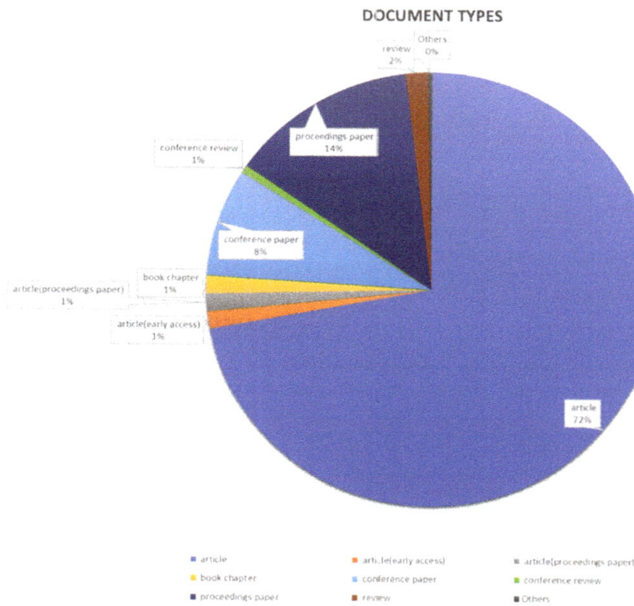

Figure 2. Graphical representation of the documents contained in the sample.

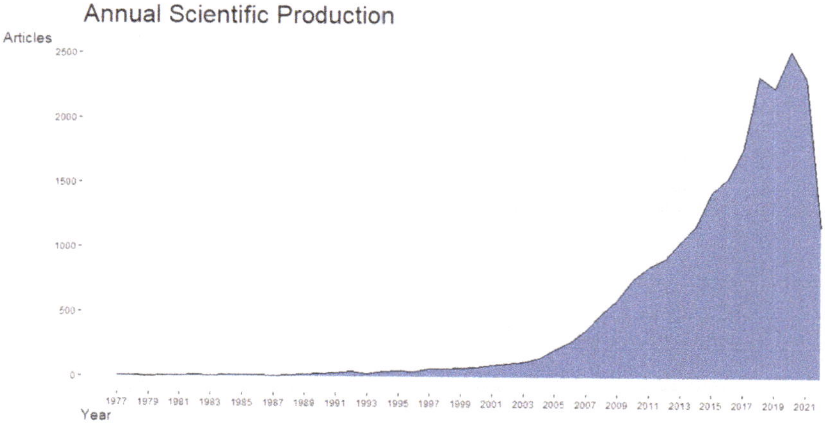

Figure 3. Graphical representation of the annual scientific production. Note: The data for 2022 corresponds to partial values quantified up to 29 April 2022.

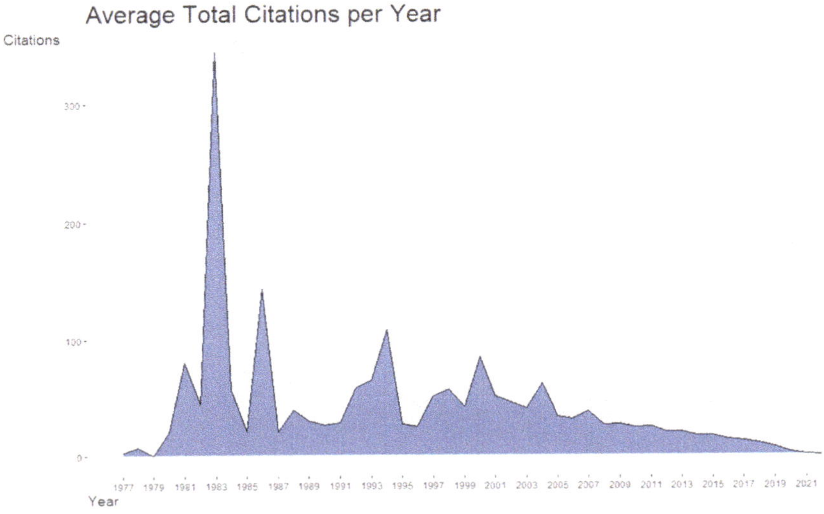

Figure 4. Graphical representation of the average total citations per year.

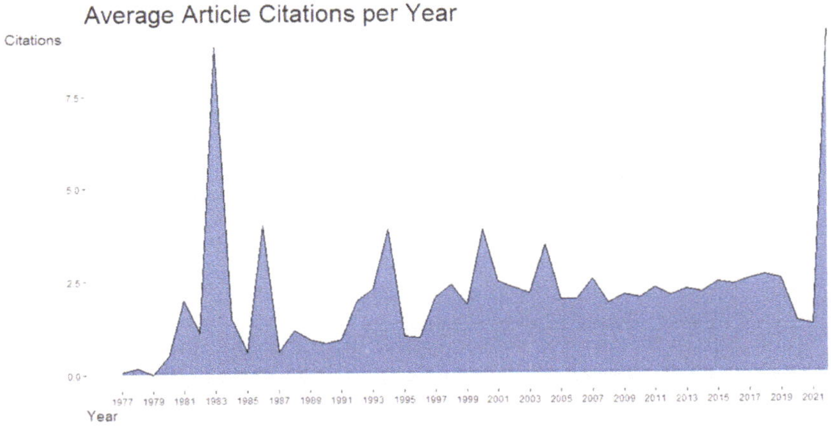

Figure 5. Graphical representation of the average article citations per year.

Table 1. Top 10 manuscripts per citations.

Rank	Title	Journal	First Author	Publication Year	Total Citations	TC per Year
1	A fuzzy extension of Saaty's priority theory	*Fuzzy Sets and Systems*	van Laarhoven, PJM	1983	1950	50.0
2	Compromise solution by MCDM methods: A comparative analysis of VIKOR and TOPSIS	*European Journal of Operational Research*	Opricovic S	2004	1834	101.9
3	Extensions of the TOPSIS for group decision-making under fuzzy environment	*Fuzzy Sets and Systems*	Chen CT	2000	1815	82.5

Table 1. *Cont.*

Rank	Title	Journal	First Author	Publication Year	Total Citations	TC per Year
4	How to select and how to rank projects: The Promethee method	*European Journal of Operational Research*	Brans JP	1986	1422	39.5
5	Application of multi-criteria decision making to sustainable energy planning—A review	*Renewable and Sustainable Energy Reviews*	Pohekar SD	2004	960	53.3
6	Handling multicriteria fuzzy decision-making problems based on vague set theory	*Fuzzy Sets and Systems*	Chen SM	1994	888	31.8
7	A fuzzy approach for supplier evaluation and selection in supply chain management	*International Journal of Production Economics*	Chen CT	2006	854	53.4
8	A state-of the-art survey of TOPSIS applications	*Expert Systems with Applications*	Behzadian M	2012	742	74.2
9	A multi-criteria intuitionistic fuzzy group decision making for supplier selection with TOPSIS method	*Expert Systems with Applications*	Boran FE	2009	732	56.3
10	Extended VIKOR method in comparison with outranking methods	*European Journal of Operational Research*	Opricovic S	2007	706	47.1

The year 2022 is shown as an outlier in Figure 5. The average number of papers cited every year was calculated using only the year of publication, which skews the results by overestimating this value. However, there are no distinguishing traits in this year's sample compared to earlier times. The volume of publications resulted in a total of 472,345 references.

3.1. Monitoring of Scientific Production around the World

Figure 6 shows that at least 120 countries or regions contributed to the research on multicriteria methods. China (n = 4327) is the largest contributor to multicriteria methods research, followed by India (n = 2485), Iran (n = 1812), Turkey (n = 1788), Taiwan (n = 1192), United States (n = 794), Brazil (n = 752), Spain (n = 608), Italy (n = 555), and Malaysia (n = 493). Regarding citations, Table 2 offers a slightly different order, but China continues to lead scientific production in terms of both knowledge generation and references to the scientific community: China (n = 82,615), Taiwan (n = 32,535), Turkey (n = 28,739), India (n = 23,643), Iran (n = 23,613), United States (n = 20,217), Lithuania (n = 12,292), United Kingdom (n = 10,917), Spain (n = 10,071), and Italy (n = 8601). As shown in Table 1, the top 10 research universities are Islamic Azad University (n = 504), Vilnius Gediminas Technical University (n = 456), National Institute of Technology (n = 336), University of Tehran (n = 334), Indian Institute of Technology (n = 265), and Istanbul Technical University (n = 243), as seen in Table 1.

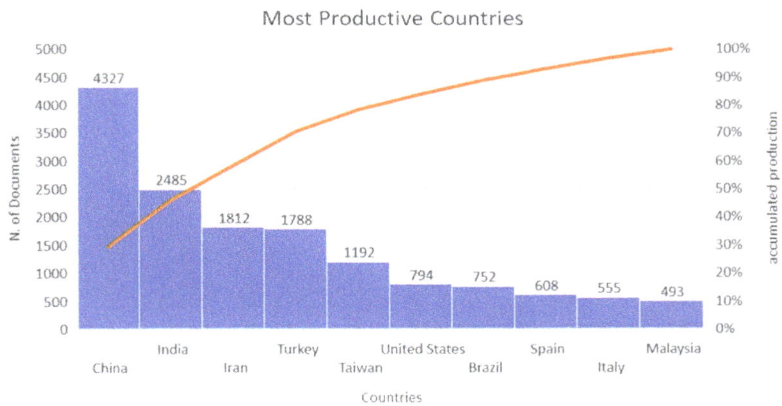

Figure 6. Graphical representation of the top 10 most productive countries.

Figure 7 illustrates the relationships between organizations through the coauthorship analysis, using universities as the unit of analysis. The research was based on the following criteria: (1) the minimum number of documents per organization ($n \geq 50$); (2) the minimum number of citations per organization ($n \geq 50$). With the established criteria, 50 organizations out of the 7619 analyzed were separated. The nodes represent the universities. The diameter of the nodes represents the number of citations, and the thickness of the connecting lines between the nodes represents the level of cooperation between the institutions. As a result, Islamic Azad University and Vilnius Gediminas Technical University stand out in this analysis.

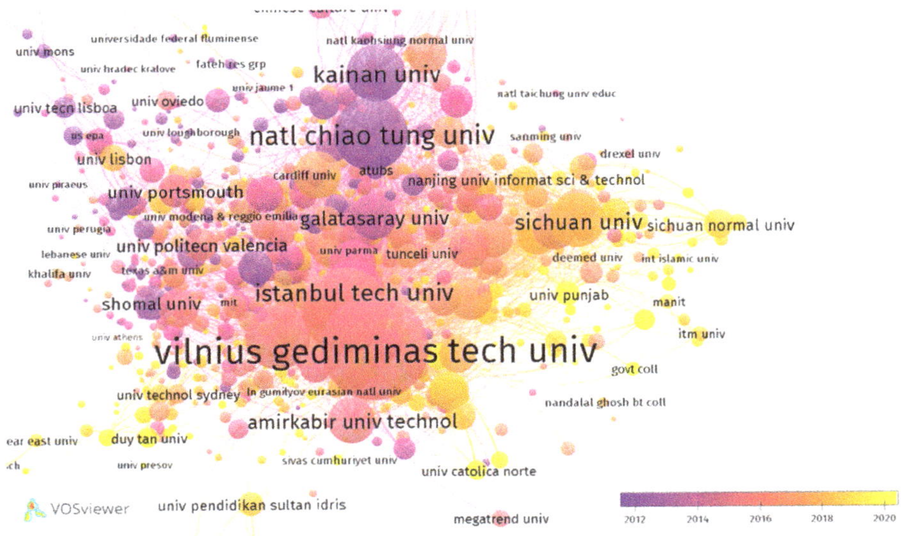

Figure 7. The network map of institutions involved in multi-criteria methods of decision-support research. Note: The colors of the circles are used to identify the clusters resulting from the analysis of the relations provided by the VOSviewer software.

Table 2. The top 10 countries/regions and institutions contributing to publications in multicriteria methods.

Rank	Country/ Region	Article Counts	Percentage (N/23,394), %	Total Citations	Percentage (TNC/373.732) %	Average Article Citations	Freq	SCP	MCP	MCP_Ratio	Institutions	Country	Article Counts	Percentage (N/23,394), %
1	China	4327	18.50	82,615	22.11	19.09	0.2035	3794	533	0.1232	Islamic Azad University	Iran	504	2.15
2	India	2485	10.62	23,643	6.33	9.51	0.1169	2338	147	0.0592	Vilnius Gediminas Technical University	Lithuania	456	1.95
3	Iran	1812	7.75	23,613	6.32	13.03	0.0852	1526	286	0.1578	National Institute of Technology	India	336	1.44
4	Turkey	1788	7.64	28,739	7.69	16.07	0.0841	1701	87	0.0487	University of Tehran	Iran	334	1.43
5	Taiwan	1192	5.10	32,535	8.71	27.29	0.0545	969	223	0.1126	Indian Institute of Technology System	India	265	1.13
6	USA	794	3.39	20,217	5.41	25.46	0.0380	633	161	0.2234	Istanbul Technical University	Turkey	243	1.04
7	Brazil	752	3.21	5584	1.49	7.43	0.0365	697	55	0.0861	University of Belgrade	Serbia	180	0.77
8	Spain	608	2.60	10,071	2.69	16.56	0.0294	496	112	0.2169	Yildiz Technical University	Turkey	176	0.75
9	Italy	555	2.37	8601	2.30	15.50	0.0272	463	92	0.1780	Sichuan University	China	157	0.67
10	Malaysia	493	2.11	6482	1.73	13.15	0.0244	389	104	0.2331	Central South University	China	150	0.64
	TOTAL	14,806	63.29	242,100	64.78								2801	11.97

Note: SCP: Single-country publications; MCP: Multiple-country publications.

This section provides a quick summary of the bibliometric findings. However, we chose to go beyond a typical bibliometric analysis by stratifying the investigation and providing the reader with specific information about the countries ranked in Figure 2. Table 3 lists the major research topics, universities, research funding organizations, notable authors, and the most relevant papers.

Table 3. Analytic picture of scientific production in the ten best-ranked countries.

Country	TOP 5				Studies
	Research Areas	**Universities**	**Research Sponsors (%)**	**Authors**	
China	Computer science, engineering, environmental sciences and ecology, operations research and management science, science technology, and other topics	Sichuan University, Central South University, North China Electric Power University, Hong Kong Polytechnic University, and Chinese Academy of Sciences	National Natural Science Foundation of China (48.75), Fundamental Research Funds For The Central Universities (7.77), China Postdoctoral Science Foundation (3.6), Ministry of Education China (2.68), and China Scholarship Council (1.9)	Jian-Qiang Wang, Zeshui Xu, Hu-chang Liao, Pei-De Liu, and Jing Wang	[67–76]
India	Engineering, computer science, environmental sciences and ecology, business economics, science technology, and other topics	National Institute of Technology, Indian Institute of Technology, Jadavpur University, Birla Institute of Technology Science Pilani, and National Institute of Technology Tiruchirappalli	Department of Science Technology India (2.097), University Grants Commission India (1.258), Council of Scientific Industrial Research India (0.779), National Natural Science Foundation of China (0.479), and Ministry of Human Resource Development Government of India (0.359).	Harish Garg, Ashwani Kumar, Sanjay Kumar, Shankar Chakraborty, and Samarjit Kar	[77–86]
Iran	Engineering, computer science, environmental sciences and ecology, business economics, science technology, and other topics	Islamic Azad University, University of Tehran, Amirkabir University of Technology, Tarbiat Modares University, and Iran University Science Technology	University of Tehran (0.925), National Natural Science Foundation of China (0.727), Austrian Science Fund (0.661), Islamic Azad University (0.528), and Iran National Science Foundation (0.462)	Seyed Meysam Mousavi, Maghsoud Amiri, Reza Tavakkoli-Moghaddam, Behnam Vahdani, and Abdolreza Yazdani-Chamzini	[87–96]
Turkey	Computer science; engineering, business economics, operations research and management science, and environmental sciences and ecology	Istanbul Technical University, Yildiz Technical University, Gazi University, Galatasaray University, and Karadeniz Technical University	Galatasaray University (3.628), Turkiye Bilimsel Ve Teknolojik Arastirma Kurumu Tubitak (2.243), Bagep Award of The Science Academy in Turkey (0.396), Erciyes University (0.396), and European Commission (0.396)	Cengiz Kahraman, Gulcin Buyukozkan, Basa Oztaysi, Ihsan Kaya, and Metin Dagdeviren	[97–106]
Taiwan	Computer science; engineering, operations research and management science, business economics, and environmental sciences and ecology	National Yang Ming Chiao Tung University, Nan Kai University Technology, National Taipei University, National Taipei University of Technology, and National Kaohsiung University of Science Technology	Ministry of Science and Technology Taiwan (18.635), Chang Gung Memorial Hospital (1.426), National Natural Science Foundation of China (1.426), Taiwan Ministry of Science and Technology (1.120), and Ministry Of Sciences And Technology In Taiwan (1.018)	Gwo-Hshiung Tzeng, James J. H. Liou, Chi-Yo Huang, Ming-Lang Tseng, and Ting-Yu Chen	[107–116]
United States	Engineering, computer science, operations research and management science, business economics, and environmental sciences and ecology	State University System of Florida, Pennsylvania Commonwealth System of Higher Education, University of California, University of Memphis, and La Salle University	National Natural Science Foundation of China (9.138), National Science Foundation (2.464), China Scholarship Council (1.437), Fundamental Research Funds for the Central Universities (1.335), and Portuguese Foundation for Science and Technology (1.027)	Madjid Tavana, Florentin Smarandache, Surendra M. Gupta, Joseph Sarkis, and Dursun Delen	[117–126]

Table 3. *Cont.*

Country	TOP 5				Studies
	Research Areas	Universities	Research Sponsors (%)	Authors	
Brazil	Engineering, computer science, business economics, operations research and management science, and environmental sciences and ecology	Universidade Federal de Pernambuco, Universidade Federal Fluminense, Universidade Federal do Rio De Janeiro, Universidade de São Paulo, and Universidade Tecnológica Federal do Paraná	National Council for Scientific and Technological Development (CNPQ) (22.18), Coordination for the Improvement of Higher Education Personnel (CAPES) (15.6), Foundation for Research Support of the State of São Paulo (FAPESP) (2.95), Foundation for the Support of Science and Technology of the State of Pernambuco (FACEPE) (1.39), and Foundation for Research Support of the State of Minas Gerais (FAPEMIG) (1.39)	Adiel Texeira de Almeida, Luiz Flavio Autran Monteiro Gomes, Danielle Costa Morais, Ana Paula Cabral Seixas Costa, and Helder Gomes Costa	[127–140]
Spain	Computer science, engineering, environmental sciences and ecology, operations research and management science, and business economics	The Polytechnic University of Valencia, Polytechnic University of Madrid, University of Granada, University of Oviedo, and Polytechnic University of Catalonia	European Commission (13.422), Spanish Government (8.555), National Natural Science Foundation of China (4.425), Spanish Ministry of Economy and Competitiveness (4.425), and Junta de Andalucia (2.507).	Morteza Yazdani, Juan Miguel Sanchez-Lozano, Monica Garcia-Melon, Maria Carmen Carnero, and Maria Teresa Lamata	[141–149]
Italy	Engineering, environmental sciences and ecology, computer science, science technology, other topics, and operations research and management science	University of Catania, University of Naples Federico II, University of Palermo, Polytechnic University of Turin, and University of Cassino	European Commission (3.303), Ministry of Education Universities and Research (2.385), National Natural Science Foundation of China (0.917), Ministry of Science and Higher Education Poland (0.734), and European Commission Joint Research Centre (0.550).	Salvatore Greco, Antonella Petrillo, Fabio De Felice, Fausto Cavallaro, and Silvia Carpitella	[150–159]
Malaysia	Engineering, computer science, science technology, other topics, environmental sciences and ecology, and operations research and management science	Universiti Teknologi Malaysia, Universiti Malaya, University Putra Malaysia, University Pendidikan Sultan Idris, and University Sains Malaysia	Ministry of Education Malaysia (4.48), University Teknologi Malaysia (2.83), University Sains Malaysia (2.12), University Kebangsaan Malaysia (1.18), and University Malaya (0.94).	Bilal Bahaa Zaidan, Aos Ala Zaidan, Lazim Abdullah, Osamah Shihab Albahri, and Mardini Abbas	[160–169]

3.2. *Overview of the Leading Journals and Papers That Disseminate Research on Multi-Criteria Methods*

Six thousand one hundred and five journals have published research on multi-criteria methods over the past forty-four years. As seen in Table 3, the top ten journals published 2180 of the total 20,861 studies on multi-criteria techniques (10.40%). *Expert Systems with Applications*, *Sustainability*, and *Journal of Cleaner Production* are the top three journals, accounting for over 4.67 percent of all indexed material. The journal with the highest impact factor (IF) is the *Journal of Cleaner Production* (7246), followed *by Applied Soft Computing* (5472), and *Expert Systems with Applications* (5041). (5.452). Five journals are classified as Q1 by the JCR 2019 standards, two as Q2, and three as Q3. In the eighth column of Table 4, the number of citations for each journal is displayed as an example.

Table 4. Top 10 most-active journals that published research articles on multicriteria methods (sorted by count).

Rank	Journal Title	Percentage (N/23,394), %	IF [2019]	Quartile in Category [2019]	H-Index	Article Counts	Total Number of Citations	Average Number of Citations	Percentage (TNC/373.732), %	Top 5 Countries by Source
1	*Expert Systems with Applications*	1.70	5.452	Q1	91	356	26,410	74.19	7.88	Taiwan, Turkey, China, USA, England
2	*Sustainability*	1.68	2.576	Q3	25	352	2978	8.46	0.89	China, Italy, Spain, Taiwan, Lithuania
3	*Journal of Cleaner Production*	1.29	7.246	Q1	43	270	7627	28.25	2.28	China, India, Iran, USA, Denmark
4	*European Journal of Operational Research*	1.26	4.213	Q1	76	264	22,144	83.88	6.61	France, England, USA, Belgium, Greece
5	*Journal of Intelligent & Fuzzy Systems*	1.07	1.851	Q3	26	225	2508	11.15	0.75	China, Turkey, Pakistan, Iran, India
6	*Applied Soft Computing*	0.79	5.472	Q1	48	166	6557	39.50	1.96	China, Iran, Turkey, Taiwan, India
7	*Computers & Industrial Engineering*	0.69	4.135	Q1	40	146	5165	35.38	1.54	China, Iran, Turkey, USA, Taiwan
8	*Soft Computing*	0.68	3.050	Q2	22	142	1402	9.87	0.42	China, Turkey, India, Iran, Taiwan
9	*Symmetry-Basel*	0.66	2.645	Q2	21	138	1407	10.20	0.42	China, Serbia, Lithuania, Pakistan, Taiwan
10	*International Journal of Information Technology & Decision Making*	0.58	1.894	Q3	24	121	2254	18.63	0.67	China, Taiwan, Turkey, USA, Iran
	Total	10.4				2180	78,452		23.42	

Figure 8 depicts the inter-relationship between the Journals, which was developed based on the researchers' preferences and referencing publications from sources with a high impact factor. The diameter of the circles is directly related to the number of citations, while the colors represent the identified clusters. In the eleventh column of Table 4, we can observe the five countries that published the most in each source. The maximum number of articles is from China, occupying the first position in eight out of the ten journals. The analysis of the highly cited papers shows that *Renewable and Sustainable Energy Reviews*, *Expert Systems with Applications*, and the *International Journal of Production Economics* have an incredible scientific impact on all scholars and have articles with more than 800 citations (Table 1).

3.3. Analysis of the Most Influential Authors Who Discuss the Topic of the Multi-Criteria Methods

Zavadskas E, Wang J, Tzeng G, Wang Y, and Kahraman C are among the top ten authors out of 29,050 who have published the most articles on this topic (Table 5). Edmundas Kazimieras Zavadskas is the first vice-rector of Vilnius Gediminas Technical University (VGTU). In addition, he is a member of the VGTU Senate, a professor, and the head of the Department of Construction Technology and Management. He has co-written over fifty novels in Lithuanian, Russian, German, and English. Corporations and academic institutions commissioned over forty research papers. The professor's primary research interests include building life cycles, decision-support systems, and multi-criteria optimization methods in construction technology and management.

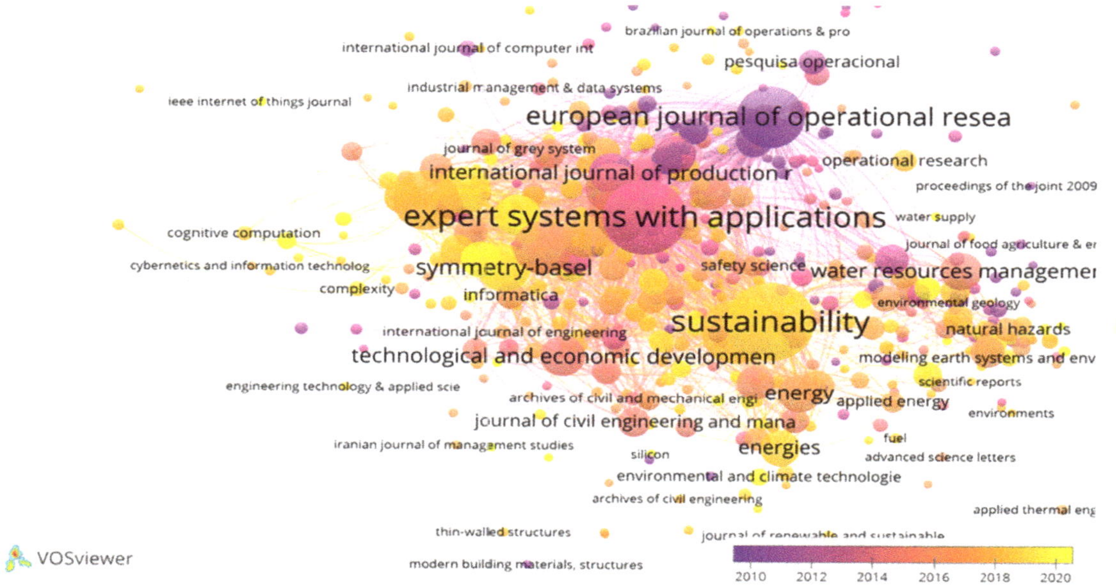

Figure 8. The network map of co-cited journals. Note: The colors of the circles are used to identify the clusters resulting from the analysis of the relations produced by the VOSviewer software.

Table 5. Ranking of authors with the highest scientific production on multicriteria methods.

Rank	Authors	Country	University	H_Index	G_Index	Article Counts	Total Number of Citations	Average Number of Citations	First Author Counts	First Author Citations Counts	Average First Author Citations Counts
1	ZAVADSKAS E	Lithuania	Vilnius Gediminas Technical University	57	87	240	9955	41.48	50	1806	36.12
2	WANG J	China	Central South University	46	68	211	5785	27.42	65	1946	29.93
3	TZENG G	Taiwan	National Taipei University	44	97	191	9814	51.38	5	1621	324.2
4	WANG Y	China	Qinghai Normal University	28	57	161	3419	21.24	75	2222	29.62
5	KAHRAMAN C	Turkey	Istanbul Technical University	34	68	145	4980	34.34	39	1939	49.71
6	CHEN Y	China	Chongqing University	29	53	124	3036	24.48	42	1173	27.92
7	ZHANG H	China	Central South University	37	59	104	3620	34.81	27	552	20.44
8	XU Z	China	Sichuan University	31	64	95	4178	43.98	12	832	69.33
9	WANG X	China	Central South University	20	33	94	1321	14.05	28	526	18.78

Table 5. *Cont.*

Rank	Authors	Country	University	H_Index	G_Index	Article Counts	Total Number of Citations	Average Number of Citations	First Author Counts	First Author Citations Counts	Average First Author Citations Counts
10	TURSKIS Z	Lithuania	Vilnius Gediminas Technical University	34	63	93	4264	45.85	10	273	27.3
Total						1458	50,372	34.54	353	12,890	36.51

Figure 9 depicts a group of 160 authors grouped into six clusters based on two essential criteria about the authors' academic output: the minimum number of citations (n ≥ 500) and the minimum number of documents (n ≥ 10). Each cluster, identified by a distinct color, indicates the authors' and co-authors' iterations. The number of links and the total links strength (TLS) are employed to determine the strength of the relationships. Each cluster's featured author is the author with the most links and the highest TLS. In this way, each cluster's information is presented: Cluster 1 (red) contains 37.5% of the sample, with an emphasis on authors Wang Y (Links = 112, TLS = 540) and Cheng Y (Links = 103, TLS = 394); Cluster 2 (green) contains 26.9% of the sample, with an emphasis on authors Wang J (Links = 140, TLS = 315), Xu Z (Links = 141, TLS = 2048), Zhang H (Links = 144, TLS = 1935), and Wang X (Links = 121, TLS = 658); Cluster 3 (blue) contains 10.6% of the sample, with an emphasis on author Kahraman C (Links = 143, TLS = 2548); Cluster 4 (yellow) contains 10% of the sample, with an emphasis on authors Zavadskas E (Links = 153, TLS = 9165) and Turskis Z (Links = 138, TLS = 4074); Cluster 5 (purple) contains 7.5% of the sample, and author Liu H stands out (Links = 122, TLS = 1395); Cluster 6 (light blue) has 7.5% of the sample, highlighting the author Tzeng G (Links = 139, TLS = 2167).

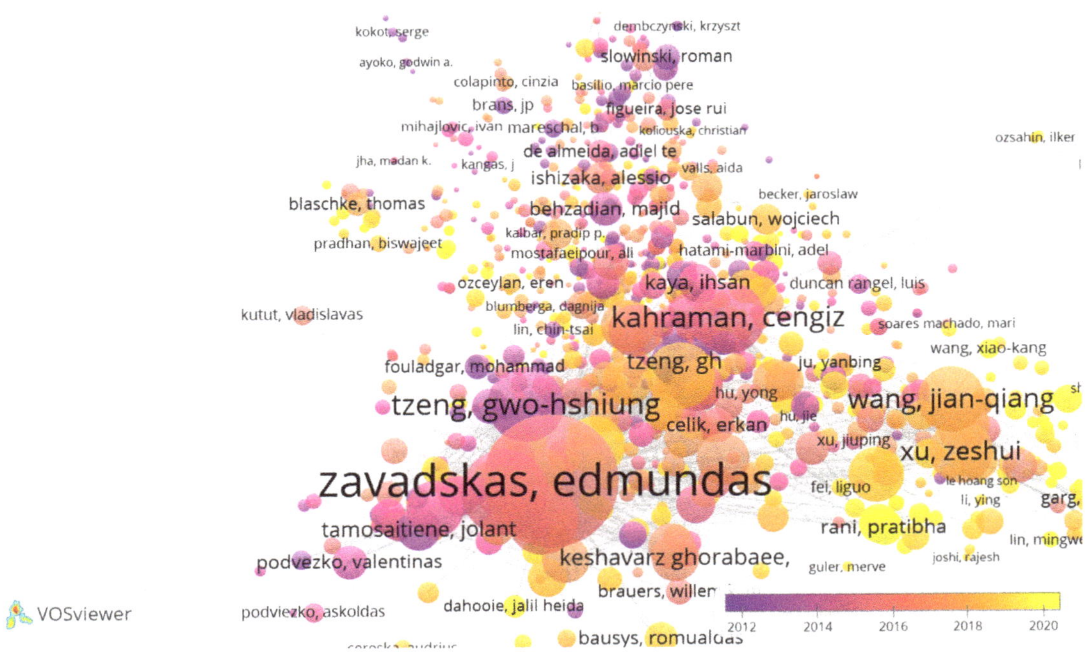

Figure 9. The network map of productive authors. Note: The colors of the circles are used to identify the clusters resulting from the analysis of the relations produced by the VOSviewer software.

3.4. Main Research Areas for the Application of Multi-Criteria Methods

The distribution of scientific production by research areas is depicted in Table 6. It is observed that there has been a shift in the preferences of academics in research fields over the past four decades. Table 7 displays the top five study areas by period. There was no change in the first five areas observed in the first two periods. From 1982 to 2002, research and applications of multi-criteria methods focused mainly on the following areas: operations research (1st), business economics (2nd), computer Science (3rd), engineering (4th), and mathematics (5th). With the increase in the volume of works published in the third decade under study, as shown in Figure 2, there was also a change in the research areas. From 2003 to 2012, the mathematics field was surpassed by environmental sciences ecology, which ranked fifth with 288 papers. Operations research, which held the number-one spot for two decades, was ranked third. The field of business economics lost its second place to computer science and fell to fourth place, followed by the ascent of engineering from fourth to first place. The most recent period analyzed was marked by a substantial increase in the number of published works. However, regarding the areas of interest of researchers, there has been a clear preference for engineering (1st) and computer science (2nd), followed by a change in preference as the traditional area of operations research has given way to environmental sciences ecology (3rd). In the fourth position, we find science technology, which has emerged with a greater level of interest from researchers due to the advancement of recent changes. The fifth place was occupied by business economics, a field in which scholars' interest has diminished over the past four decades.

Table 6. Distribution of scientific production by research areas.

Research Areas	Recorded Count	% of 26,376
Engineering	5101	19.34
Computer science	4706	17.84
Environmental sciences ecology	2133	8.09
Business economics	2122	8.05
Operations research	2010	7.62
Science technology	1635	6.20
Energy fuels	915	3.47
Mathematics	869	3.30
Water resources	579	2.20
Materials science	511	1.94
Total	20,581	78.02

Note: It is necessary to clarify the value indicated in the third column, "26,376" this is the total number of articles in the sample associated with the research areas. Each article can be related to more than one search area.

Table 7. Evolution of scientific production according to research areas in the analyzed periods.

Research Areas	Periods			
	1982 to 1992	1993 to 2002	2003 to 2012	2013 to 2022 (April 29)
	Ranking	Ranking	Ranking	Ranking
Engineering	4th	4th	1st	1st
Computer science	3rd	3rd	2nd	2nd
Environmental sciences ecology	-	-	5th	3rd
Science technology	-	-	-	4th
Business economics	2nd	2nd	4th	5th
Operations research	1st	1st	3rd	-
Mathematics	5th	5th	-	-

Note: Only data corresponding to the fifth position in each period were recorded.

In Section 3.1, a global overview of the scientific output on multi-criteria methods is provided, highlighting the significant countries and classifying each production. However, as seen in the case of research domains, the hegemony of the scientific output has also evolved differently between nations. The shift in emphasis in specific scientific fields and the consolidation of others directly impact the hegemony of nations. If we analyze Table 2, we can see the consolidation of engineering and computer science as prominent areas in the production of the ten countries explored and the emergence of interest in science and technology.

3.5. Most-Used Methods

Table 8 lists the 26 methods examined throughout the sample period. The publishing period in WoS/Scopus concerning the investigated method is recorded in column 3. The chronology was produced based on the evolution of multi-criteria approaches, as shown in Figure 10, using information from the starting period of each method's scientific output. The chronology depicts techniques that have been embedded in the literature and that continue to evolve, such as AHP, TOPSIS, PROMETHEE, ELECTRE, and others, such as SWARA, WASPAS, and FITRADEOFF, that have been published for up to ten years but are not yet well-known in academia. The publications of each studied technique are then noted in column 4. The AHP, TOPSIS, and VIKOR approaches have the most publications in the four decades studied. They are also the most commonly employed methods by professionals in solving multi-criteria related issues. Column 5 indicates the research areas wherein the specialists used the method the most. Computer science stands out among others because 47% of the researched methods address issues related to these areas, with the TOPSIS method being used the most. Engineering follows, with 35% of the methods, with the AHP method being the second most-used method. Business economics takes 11%, and operations research 8% respectively. In column 7, we build on the study to show a trend toward developing solutions that include one or more methodologies and the creation of hybrid models based on the data acquired. This section concludes by emphasizing that, despite the small number of applications, the scenario depicts the integration of multi-criteria methods with some machine learning techniques, which could be the beginning of a new trend in the coming years (see column 8).

Table 8. Characteristics of the methods most used by researchers.

N	Method	Publication Time	Recorded Count	Research Areas	Publication Time (Integrated/Hybrid Model)	Hybrid Model	New Technologies (Machine Learning)
1	AHP	1990–2021	6.835	Engineering (2.329)	1995–2021	1.388	38
2	TOPSIS	1991–2021	4.907	Computer science (1.797)	2003–2021	1.024	47
3	VIKOR	2002–2021	1.475	Computer science (519)	2009–2021	416	5
4	PROMETHEE	1989–2021	1.382	Engineering (445)	2001–2021	202	16
5	ANP	2000–2021	1.262	Engineering (428)	2006–2021	488	10
6	ELECTRE	1991–2021	1.005	Computer science (331)	2003–2021	120	6
7	DEMATEL	2007–2021	888	Computer science (289)	2007–2021	476	5
8	GOAL PRO-GRAMMING	1983–2021	553	Operations research (202)	1993–2021	147	3
9	SAW	1997–2021	403	Engineering (137)	2007–2021	67	5
10	TODIM	1999–2021	306	Computer science (171)	2013–2021	56	2
11	COPRAS	2006–2021	294	Business economics (83)	2011–2021	100	2
12	WASPAS	2012–2021	214	Engineering (68)	2013–2020	67	0
13	MULTIMOORA	2011–2021	198	Computer science (75)	2011–2021	43	0
14	SWARA	2011–2021	181	Business economics (46)	2011–2021	90	1
15	MAUT	1984–2021	164	Engineering (56)	2007–2021	19	0
16	MACBETH	1999–2021	162	Computer science (47)	1999–2021	27	0
17	WSM	1994–2021	87	Engineering (29)	2014–2021	17	2
18	DRSA	2002–2021	85	Computer science (51)	2012–2021	20	4
19	WPM	1997–2021	57	Computer science (23)	2014–2021	7	0
20	CBR	1996–2021	40	Computer science (25)	2006–2020	10	1

Table 8. *Cont.*

N	Method	Publication Time	Recorded Count	Research Areas	Publication Time (Integrated/Hybrid Model)	Hybrid Model	New Technologies (Machine Learning)
21	CONDORCET	1999–2021	35	Business economics (9)	-	0	1
22	FITRADEOFF	2016–2021	29	Computer science (14)	-	0	0
23	UTADIS	1998–2020	27	Operations research (14)	2005–2016	2	0
24	SMART	1996–2021	22	Engineering (9)	2021	2	0
25	PAPRIKA	2014–2021	12	Computer science (4)	2020	1	0
26	THOR	2008–2021	5	Engineering (2)	-	0	0

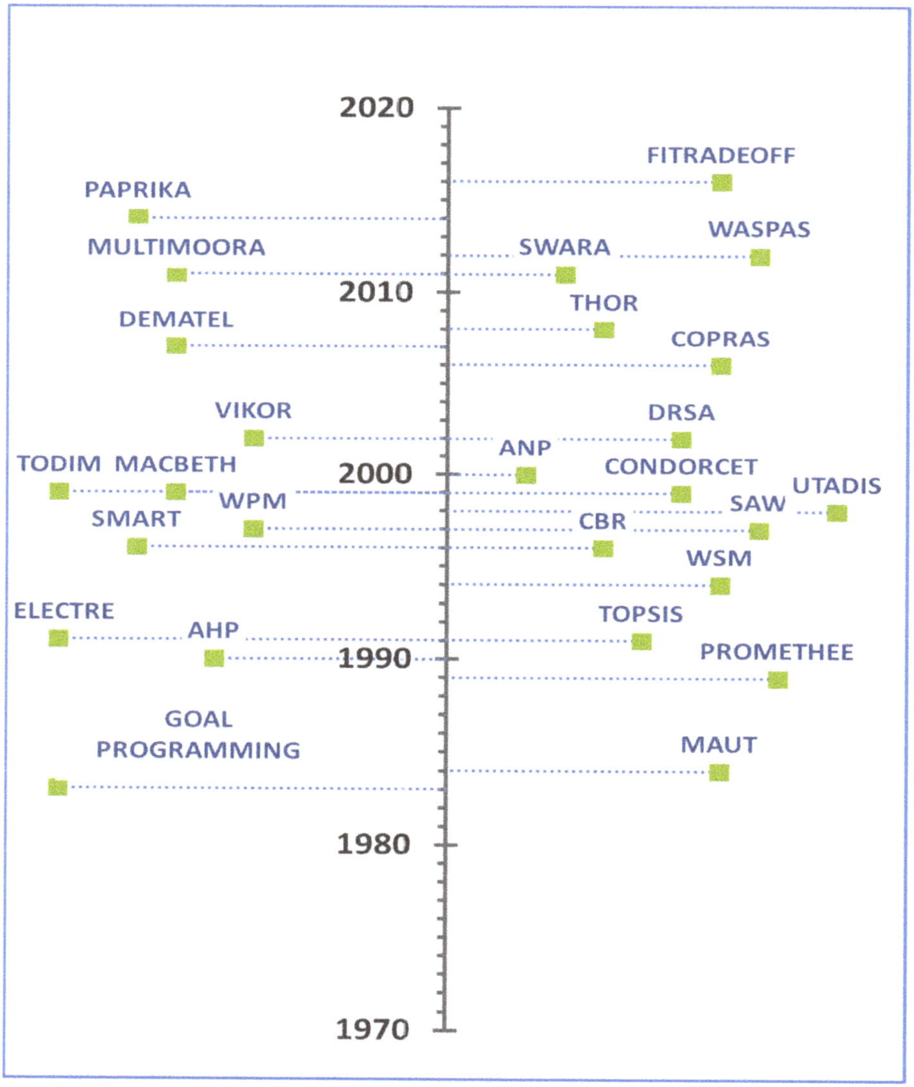

Figure 10. Evolution of scientific production classified by method over the period analyzed.

3.6. Mapping the Evolution of Themes

Cobo et al. [170] assert the set of identified themes of the subperiod t, with U ∈ T^t representing each detected theme in the subperiod t. Let V ∈ T^(t + 1) represent each theme found in the subsequent subperiod t + 1. It is argued that there is a thematic progression from topic U to theme V if both related thematic networks contain the same keywords. Thus, V can be considered a development of U. Additionally, the keyword cluster k ∈ U ∩ V is regarded as a "thematic nexus" or "conceptual nexus".

Figure 11 was created using the "thematicEvolution" function of the Bibliometrix R package. The evolution of themes associated with multi-criteria methods is depicted in Figure 11 across the five time periods. In the first period, i.e., between 1977 to 1986, three themes are recorded. As the rectangles represented the same region during this period, it may be deduced that there was a balance in disseminating topics. In the second phase (1987–1995), there are twelve topics, of which eight had no foundation in the first period, such as "AHP", "TOPSIS", and "fuzzy set theory". These methods have their earliest publication record in 1990/1991 (Table 8). Still, researchers favor them, as in the case of TOPSIS, which has the same rectangular area as "GOAL PROGRAMMING", one of the three primary subjects of the program. During the third era (1996–2004), we recorded fourteen themes that originated in or branched from the preceding period. In this third period, the focus is on the AHP method, which is the most influential subject, as indicated by a distinct set of four keywords ("ahp", "analytic hierarchy process", and "analytical hierarchy process (ahp)"). It is important to note that the "GOAL PROGRAMMING" theme has become less popular and that the PROMETHEE and ELECTRE methods have become more popular. Despite being published for the first time in 1989/1991, they did not emerge as a topic until the third period. The themes decreased from fourteen to nine for the fourth phase (2005–2013). Two AHP-related concepts continue to hold the apex of importance. In addition to the PROMETHEE method, the TOPSIS methods, which did not emerge in the third era, reappeared distinctly. The final period evaluated between 2014–2022 continues with a reduction from nine to six themes presented in a balanced way, reflecting the preference for topics associated with the AHP and TOPSIS methods. The use of the theme-evolution map allowed us to graphically confirm the choice of specialists in solving multi-criteria problems using original tools in the AHP and TOPSIS methods during the study period.

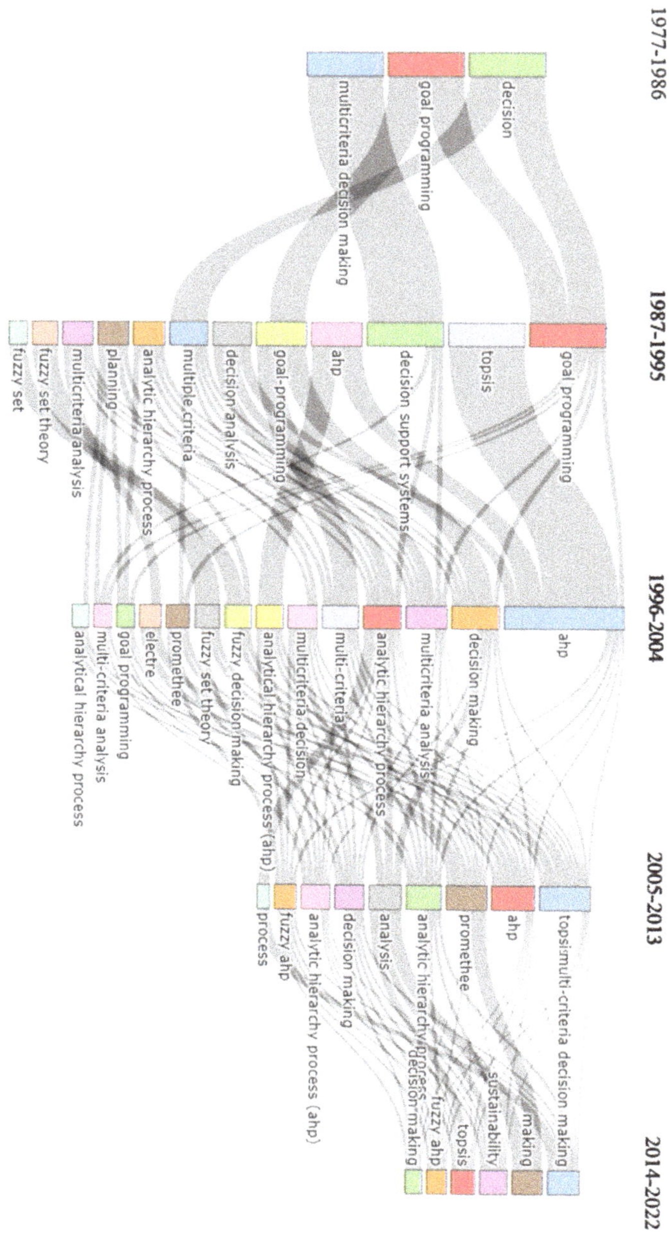

Figure 11. The evolution of themes built with the authors' keywords.

4. Discussion

This research article presents a bibliometric analysis of the multi-criteria methods from 1977 to 29 April 2022. The bibliographic data was obtained from the Scopus and Web of Science (WoS) databases. The bibliometric analysis was conducted using the Bibliometrix R tool and the VOSviewer software to investigate the essential characteristics of the studies done so far, including publications; citations, citation structure; influential authors; co-citation contributors and burst detection analysis; author-keywords; co-occurrence analyses; and timeline-view analysis. The ability to make judgments is a distinguishing characteristic of a person. Man makes spontaneous and intuitive decisions based on his brain's information-processing skills. We judge the color of our ties for a business meeting as to whether or not to invest millions of dollars in a specific project. We realize that we face two distinct types of decisions: simple and complex. We can make straightforward decisions with few variables and little trouble. However, when the problem involves a matrix ($n \times m$) variable, we require methodologies and computer capabilities to systematize, arrange, and rank the best options to aid decision-making. Accordingly, the objective of this study was to comprehend the global evolution of research on the creation and use of multi-criteria decision methods.

With a scientific production growth rate of 14.18% each year, it is clear that the academic community is interested in researching and publishing publications on multi-criteria decision-making approaches. Moreover, 60.93% of all publications were concentrated in only ten nations, with China leading the way with 18.50%, India coming in second with 10.62%, and Iran coming in third with 7.75%. In addition, the remaining 39% of publications have an average production rate of less than 1%, suggesting that the dissemination of multi-criteria approach research in such nations could enhance academic output. The top 10 countries in terms of citations follow a consistent pattern, accounting for 62.48% of all citations made during the research period. Among the top 10 countries in terms of multi-country collaboration (MCP) in publications, Turkey has the lowest MCP ratio with 0.0487, indicating a limited partnership with researchers from other nations, followed by India (0.0592) and Brazil (0.0861). Malaysia leads multi-country collaboration, with an MCP ratio of 0.2331, followed by the United States (0.2234) and Spain (0.2169).

Regarding sites that publish articles on multi-criteria techniques, the study reveals the top ten journals that have published approximately 10.4% of the subject's total publications. China, India, Iran, and Turkey, the four nations with the most publications on multi-criteria techniques, account for around 80% of the university-based publications on multi-criteria methods. These universities account for 11.79% of academic output, with the Islamic Azad University of Iran contributing 2.14% and Vilnius Gediminas Technical University of Lithuania accounting for 2.18%. Surprisingly, Lithuania is not among the top ten nations regarding scientific output. However, among the other authors in this survey, Prof. Edmundas Kazimieras Zavadskas of Lithuania ranks first with 240 articles on multi-criteria approaches.

The journal *Expert Systems with Applications* has published 1.70% of all articles to date, followed by *Sustainability* with 1.68 percent and the *Journal of Cleaner Production* with 1.30%. The leading journals in terms of citations are *Expert Systems with Applications*, with an average of 7.88 citations per paper, followed by the *European Journal of Operational Research*, with 6.61 citations per article. Regarding the origin of publications, eight of the top ten countries publish most of their articles in the ten highest-ranked journals. In contrast, the *European Journal of Operational Research* ratio is 2 out of 10.

Regarding the most influential authors in this field, approximately 0.034% of 33,201 authors are responsible for 6.98% of publications over the past forty-four years, with ZAVADSKAS E having the most publications, with 240, followed by WANG J with 211 articles and TZENG G with 191 articles. This bibliometric analysis reveals that six of the top ten authors are Chinese, with the Central South University author affiliation standing out.

In addition to identifying writers with higher academic production, this study includes a comprehensive summary of the countries, funding sources, and the five multi-criteria approaches, i.e., AHP, TOPSIS, VIKOR PROMETHEE, and ANP, most frequently utilized by the authors in their respective studies. Engineering and computer science are the most prominent subjects in terms of research fields. One trend identified was the expansion of multi-criteria technique integration and the formation of hybrid models.

This paper gives a complete overview of multi-criteria methods through a bibliometric study, enabling scholars to comprehend the current state and future development patterns of multi-criteria decision-making methods research. As an indication for prospective research, we can emphasize the need to understand the emergence and regionalization of specific techniques and their variations, expand research within the identified countries to gain a deeper understanding of their scientific production on the issue investigated, apply topic modeling to find latent themes in the researched database, and systematize method variants and their interfaces with other research areas, such as machine learning.

Author Contributions: Conceptualization, M.P.B.; data curation, V.P.; formal analysis, M.S.; investigation, V.P.; methodology, M.P.B.; project administration, M.S.; supervision, H.G.C.; validation, V.P. writing—original draft, M.P.B.; writing—review and editing, M.P.B. and A.G. All authors have read and agreed to the published version of the manuscript.

Funding: This research received no external funding.

Data Availability Statement: https://github.com/marciobasilio/bibliometric_multicriteria (accessed on 24 April 2022).

Conflicts of Interest: The authors declare no conflict of interest.

Abbreviations

The following abbreviations are used in this manuscript:

AHP	Analytic Hierarchy Process
ANP	Analytical Network Process
COMET	Characteristic Objects Method
COPRAS	Complex Proportional Assessment
DRSA	Dominance-based Rough Set Approach
ELECTRE	ÉLimination et Choix Traduisant la REalité (French)
MACBETH	Measuring Attractiveness by a Categorical Based Evaluation Technique
MCDA	Multi-Criteria Decision Analysis
MCDM	Multi-Criteria Decision Making
MODM	MultiObjective Decision Making
MOORA	Multi-Objective Optimization by Ratio Analysis
MULTIMOORA	MOORA plus the full Multiplicative Form
NAIADE	Novel Approach to Imprecise Assessment and Decision Environment
PCCA	Pairwise Criterion Comparison Approach
PROMETHEE	Preference Ranking Organization Method for Enrichment of Evaluation
WASPAS	Weighted Aggregated Sum Product Assessment
TODIM	Tomada de Decisão Interativa Multicritério (Portuguese)
TOPSIS	Technique for Order of Preference by Similarity to Ideal Solution
VIKOR	VlseKriterijumska Optimizacija I Kompromisno Resenje (Serbian)

References

1. Basilio, M.P.; Pereira, V.; de Oliveira, M.W.C.M.; da Costa Neto, A.F.; de Moraes, O.C.R.; Siqueira, S.C.B. Knowledge discovery in research on domestic violence: An overview of the last fifty years. *Data Technol. Appl.* **2021**, *55*, 480–510. [CrossRef]
2. Simon, H.A. A Behavioral Model of Rational Choice. *Q. J. Econ.* **1955**, *69*, 99–118. [CrossRef]
3. Sałabun, W.; Wątróbski, J.; Shekhovtsov, A. Are MCDA Methods Benchmarkable? A Comparative Study of TOPSIS, VIKOR, COPRAS, and PROMETHEE II Methods. *Symmetry* **2020**, *12*, 1549. [CrossRef]
4. Behzadian, M.; Otaghsara, S.K.; Yazdani, M.; Ignatius, J. A state-of-the-art survey of TOPSIS applications. *Expert Syst. Appl.* **2012**, *39*, 13051–13069. [CrossRef]
5. Kahraman, C.; Onar, S.C.; Oztaysi, B. Fuzzy Multicriteria Decision-Making: A Literature Review. *Int. J. Comput. Intell. Syst.* **2015**, *8*, 637–666. [CrossRef]
6. Govindan, K.; Jepsen, M.B. ELECTRE: A comprehensive literature review on methodologies and applications. *Eur. J. Oper. Res.* **2016**, *250*, 1–29. [CrossRef]
7. Wang, J.-J.; Jing, Y.-Y.; Zhang, C.-F.; Shi, G.-H.; Zhang, X.-T. A fuzzy multi-criteria decision-making model for trigeneration system. *Energy Policy* **2008**, *36*, 3823–3832. [CrossRef]
8. Greco, S.; Figueira, J.; Ehrgott, M. *Multiple Criteria Decision Analysis*; Springer: Berlin/Heidelberg, Germany, 2016.

9. Basilio, M.P.; Pereira, V.; Costa, H.G. Classifying the integrated public safety areas (IPSAs): A multi-criteria based approach. *J. Model. Manag.* **2019**, *14*, 106–133. [CrossRef]
10. Basilio, M.; Pereira, V. Operational research applied in the field of public security: The ordering of policing strategies such as the ELECTRE IV. *J. Model. Manag.* **2020**, *15*, 1227–1276. [CrossRef]
11. Basilio, M.P.; Pereira, V.; de Oliveira, M.W.C.; da Costa Neto, A.F. Ranking policing strategies as a function of criminal complaints: Application of the PROMETHEE II method in the Brazilian context. *J. Model. Manag.* **2020**, *16*, 1185–1207. [CrossRef]
12. Moreira, M.L.; Gomes, C.F.S.; dos Santos, M.; Basilio, M.P.; Costa, I.P.D.A.; Junior, C.d.S.R.; Jardim, R.R.-A.J. Evaluation of drones for public security: A multicriteria approach by the PROMETHEE-SAPEVO-M1 systematic. *Procedia Comput. Sci.* **2022**, *199*, 125–133. [CrossRef]
13. Roy, B. Decision-aid and decision-making. *Eur. J. Oper. Res.* **1990**, *45*, 324–331. [CrossRef]
14. Youd, S.; Fuchs-Hanusch, D. A bibliometric-based survey on AHP and TOPSIS techniques. *Expert Syst. Appl.* **2017**, *78*, 158–181. [CrossRef]
15. de Almeida, I.D.P.; Corriça, J.V.P.; Costa, A.P.A.; Costa, I.P.A.; Maêda, S.M.N.; Gomes, C.F.S.; Santos, M. Study of the Location of a Second Fleet for the Brazilian Navy: Structuring and Mathematical Modeling Using SAPEVO-M and VIKOR Methods. In *Production Research. ICPR-Americas 2020. Communications in Computer and Information Science*; Rossit, D.A., Tohmé, F., Mejía Delgadillo, G., Eds.; Springer: Cham, Switzerland, 2021; Volume 1408. [CrossRef]
16. Basilio, M.P.; Pereira, V.; Costa, H.G. Review of the Literature on Multicriteria Methods Applied in the Field of Public Security. *Univers. J. Manag.* **2017**, *5*, 549–562. [CrossRef]
17. Guitouni, A.; Martel, J.-M. Tentative guidelines to help choosing an appropriate MCDA method. *Eur. J. Oper. Res.* **1998**, *109*, 501–521. [CrossRef]
18. Zanakis, S.H.; Solomon, A.; Wishart, N.; Dublish, S. Multi-attribute decision making: A simulation comparison of select methods. *Eur. J. Oper. Res.* **1998**, *107*, 507–529. [CrossRef]
19. Gershon, M. The role of weights and scales in the application of multiobjective decision making. *Eur. J. Oper. Res.* **1984**, *15*, 244–250. [CrossRef]
20. Wątróbski, J.; Jankowski, J.; Ziemba, P.; Karczmarczyk, A.; Zioło, M. Generalised framework for multi-criteria method selection. *Omega* **2019**, *86*, 107–124. [CrossRef]
21. Cinelli, M.; Kadziński, M.; Gonzalez, M.; Słowiński, R. How to support the application of multiple criteria decision analysis? Let us start with a comprehensive taxonomy. *Omega* **2020**, *96*, 102261. [CrossRef]
22. Fossile, D.K.; Frej, E.A.; da Costa, S.E.G.; de Lima, E.P.; de Almeida, A.T. Selecting the most viable renewable energy source for Brazilian ports using the FITradeoff method. *J. Clean. Prod.* **2020**, *260*, 121107. [CrossRef]
23. Siksnelyte-Butkiene, I.; Zavadskas, E.K.; Streimikiene, D. Multi-Criteria Decision-Making (MCDM) for the Assessment of Renewable Energy Technologies in a Household: A Review. *Energies* **2020**, *13*, 1164. [CrossRef]
24. Akhtar, N.; Ishak, M.; Ahmad, M.; Umar, K.; Yusuff, M.M.; Anees, M.; Qadir, A.; Almanasir, Y.A. Modification of the Water Quality Index (WQI) Process for Simple Calculation Using the Multi-Criteria Decision-Making (MCDM) Method: A Review. *Water* **2021**, *13*, 905. [CrossRef]
25. Syan, C.S.; Ramsoobag, G. Maintenance applications of multi-criteria optimization: A review. *Reliab. Eng. Syst. Saf.* **2019**, *190*, 106520. [CrossRef]
26. Costa, I.P.A.; Basilio, M.P.; Maeda, S.M.N.; Rodrigues, M.V.G.; Moreira, M.A.L.; Gomes, C.F.S.; Santos, M. Bibliometric Studies on Multi-Criteria Decision Analysis (MCDA) Applied in Personnel Selection. In *Modern Management Based on Big Data II and Machine Learning and Intelligent Systems III—Proceedings of MMBD 2021 and MLIS 2021*; Tallón-Ballesteros, A.J., Ed.; IOS Press: Amsterdam, The Netherlands, 2021; Volume 341, pp. 119–125. [CrossRef]
27. Salih, M.; Zaidan, B.; Zaidan, A.; Ahmed, M. Survey on fuzzy TOPSIS state-of-the-art between 2007 and 2017. *Comput. Oper. Res.* **2019**, *104*, 207–227. [CrossRef]
28. Pelissari, R.; Oliveira, M.C.; Abackerli, A.J.; Ben-Amor, S.; Assumpção, M.R.P. Techniques to model uncertain input data of multi-criteria decision-making problems: A literature review. *Int. Trans. Oper. Res.* **2021**, *28*, 523–559. [CrossRef]
29. Moreno-Calderón, A.; Tong, T.S.; Thokala, P. Multi-criteria Decision Analysis Software in Healthcare Priority Setting: A Systematic Review. *Pharmacoeconomics* **2020**, *38*, 269–283. [CrossRef]
30. Heidari, M.D.; Gandasasmita, S.; Li, E.; Pelletier, N. Proposing a framework for sustainable feed formulation for laying hens: A systematic review of recent developments and future directions. *J. Clean. Prod.* **2021**, *288*, 125585. [CrossRef]
31. Cunha, V.H.C.; Caiado, R.G.G.; Corseuil, E.T.; Neves, H.F.; Bacoccoli, L. Automated compliance checking in the context of Industry 4.0: From a systematic review to an empirical fuzzy multi-criteria approach. *Soft Comput.* **2021**, *25*, 6055–6074. [CrossRef]
32. Serugga, J.; Kagioglou, M.; Tzortzopoulos, P. A Utilitarian Decision—Making Approach for Front End Design—A Systematic Literature Review. *Buildings* **2020**, *10*, 34. [CrossRef]
33. Aria, M.; Cuccurullo, C. bibliometrix: An R-tool for comprehensive science mapping analysis. *J. Informetr.* **2017**, *11*, 959–975. [CrossRef]
34. Rousseau, D.M. *The Oxford Handbook of Evidence-Based Management*; Oxford University Press: Oxford, UK; New York, NY, USA, 2012.
35. Derviş, H. Bibliometric Analysis using Bibliometrix an R Package. *J. Scientometr. Res.* **2019**, *8*, 156–160. [CrossRef]

36. Wang, C.; Ghadimi, P.; Lim, M.K.; Tseng, M.-L. A literature review of sustainable consumption and production: A comparative analysis in developed and developing economies. *J. Clean. Prod.* **2019**, *206*, 741–754. [CrossRef]
37. Wang, C.; Lim, M.K.; Zhao, L.; Tseng, M.-L.; Chien, C.-F.; Lev, B. The evolution of Omega-The International Journal of Management Science over the past 40 years: A bibliometric overview. *Omega* **2020**, *93*, 102098. [CrossRef]
38. Ghadimi, P.; Wang, C.; Lim, M.K. Sustainable supply chain modeling and analysis: Past debate, present problems, and future challenges. *Resour. Conserv. Recycl.* **2019**, *140*, 72–84. [CrossRef]
39. Inamdar, Z.; Raut, R.; Narwane, V.S.; Gardas, B.; Narkhede, B.; Sagnak, M. A systematic literature review with bibliometric analysis of big data analytics adoption from period 2014 to 2018. *J. Enterp. Inf. Manag.* **2020**, *34*, 101–139. [CrossRef]
40. Merigó, J.M.; Yang, J.-B. A bibliometric analysis of operations research and management science. *Omega* **2017**, *73*, 37–48. [CrossRef]
41. Ratten, V.; Manesh, M.F.; Pellegrini, M.M.; Dabic, M. The Journal of Family Business Management: A bibliometric analysis. *J. Fam. Bus. Manag.* **2020**, *11*, 137–160. [CrossRef]
42. Borgatti, S.P.; Mehra, A.; Brass, D.J.; Labianca, G. Network analysis in the social sciences. *Science* **2009**, *323*, 892–895. [CrossRef]
43. Barabási, A.L.; Jeong, H.; Néda, Z.; Ravasz, E.; Schubert, A.; Vicsek, T. Evolution of the social net-work of scientific collaborations. *Phys. A Stat. Mech. Its Appl.* **2002**, *311*, 590–614. [CrossRef]
44. González-Alcaide, G.; Pinargote, H.; Ramos, J.M. From cut-points to key players in coauthorship networks: A case study in ventilator-associated pneumonia research. *Scientometrics* **2020**, *123*, 707–733. [CrossRef]
45. Yan, E.; Ding, Y. Scholarly network similarities: How bibliographic coupling networks, citation networks, co-citation networks, topical networks, coauthorship networks, and co-word networks relate to each other. *J. Am. Soc. Inf. Sci. Technol.* **2012**, *63*, 1313–1326. [CrossRef]
46. Hernández, J.M.; Dorta-González, P. Interdisciplinarity Metric Based on the Co-Citation Network. *Mathematics* **2020**, *8*, 544. [CrossRef]
47. Perianes-Rodriguez, A.; Waltman, L.; van Eck, N.J. Constructing bibliometric networks: A comparison between full and fractional counting. *J. Informetr.* **2016**, *10*, 1178–1195. [CrossRef]
48. Callon, M.; Courtial, J.-P.; Turner, W.A.; Bauin, S. From translations to problematic networks: An introduction to co-word analysis. *Soc. Sci. Inf.* **1983**, *22*, 191–235. [CrossRef]
49. Dai, S.; Duan, X.; Zhang, W. Knowledge map of environmental crisis management based on key-words network and co-word analysis, 2005–2018. *J. Clean. Prod.* **2020**, *262*, 121168. [CrossRef]
50. Cobo, M.J.; López-Herrera, A.G.; Herrera-Viedma, E.; Herrera, F. SciMAT: A new science mapping analysis software tool. *J. Am. Soc. Inf. Sci. Technol.* **2012**, *63*, 1609–1630. [CrossRef]
51. Cheng, B.; Wang, M.; Mørch, A.I.; Chen, N.S.; Spector, J.M. Research on e-learning in the workplace 2000–2012: A bibliometric analysis of the literature. *Educ. Res. Rev.* **2014**, *11*, 56–72. [CrossRef]
52. Leung, X.Y.; Sun, J.; Bai, B. Bibliometrics of social media research: A co-citation and co-word analysis. *Int. J. Hosp. Manag.* **2017**, *66*, 35–45. [CrossRef]
53. Ravikumar, S.; Agrahari, A.; Singh, S.N. Mapping the intellectual structure of scientometrics: A co-word analysis of the journal Scientometrics (2005–2010). *Scientometrics* **2015**, *102*, 929–955. [CrossRef]
54. De la Hoz-Correa, A.; Muñoz-Leiva, F.; Bakucz, M. Past themes and future trends in medical tour-ism research: A co-word analysis. *Tour. Manag.* **2018**, *65*, 200–211. [CrossRef]
55. Van Eck, N.J.; Waltman, L. CitNetExplorer: A new software tool for analyzing and visualizing citation networks. *J. Informetr.* **2014**, *8*, 802–823. [CrossRef]
56. Van Eck, N.J.; Waltman, L. Software survey: VOSviewer, a computer program for bibliometric mapping. *Scientometrics* **2010**, *84*, 523–538. [CrossRef] [PubMed]
57. Persson, O.; Danell, R.; Schneider, J.W. How to use Bibexcel for various types of bibliometric analysis. In *Celebrating Scholarly Communication Studies: A Festschrift for Olle Persson at His 60th Birthday*; Astrom, F., Danell, R., Eds.; 2009; pp. 9–24. Available online: https://www.researchgate.net/publication/285473885_How_to_use_Bibexcel_for_various_types_of_bibliometric_analysis (accessed on 29 April 2022).
58. Sci2 Team. Science of Science (Sci2) Tool. Indiana University and SciTech Strategies. 2009. Available online: https://sci2.cns.iu.edu (accessed on 24 April 2022).
59. Chen, C. CiteSpace II: Detecting and visualizing emerging trends and transient patterns in scientific literature. *J. Assoc. Inf. Sci. Technol.* **2006**, *57*, 359–377. [CrossRef]
60. Basilio, M.P.; Pereira, V.; de Oliveira, M.W.C.M. Knowledge discovery in research on policing strategies: An overview of the past fifty years. *J. Model. Manag.* **2021**. [CrossRef]
61. Ghosh, A.; Prasad, V.K.S. Off-grid Solar energy systems adoption or usage—A Bibliometric Study using the Bibliometrix R tool. *Libr. Philos. Pract.* **2021**, 5673. Available online: https://digitalcommons.unl.edu/libphilprac/5673 (accessed on 24 April 2022).
62. van Laarhoven, P.; Pedrycz, W. A fuzzy extension of Saaty's priority theory. *Fuzzy Sets Syst.* **1983**, *11*, 229–241. [CrossRef]
63. Brans, J.P.; Vincke, P.; Mareschal, B. How to select and how to rank projects: The Promethee method. *Eur. J. Oper. Res.* **1986**, *24*, 228–238. [CrossRef]
64. Chen, S.-M.; Tan, J.-M. Handling multicriteria fuzzy decision-making problems based on vague set theory. *Fuzzy Sets Syst.* **1994**, *67*, 163–172. [CrossRef]

55. Opricovic, S.; Tzeng, G.-H. Compromise solution by MCDM methods: A comparative analysis of VIKOR and TOPSIS. *Eur. J. Oper. Res.* **2004**, *156*, 445–455. [CrossRef]
56. Pohekar, S.D.; Ramachandran, M. Application of multi-criteria decision making to sustainable energy planning—A review. *Renew. Sustain. Energy Rev.* **2004**, *8*, 365–381. [CrossRef]
57. Wu, X.; Liao, H.; Xu, Z.; Hafezalkotob, A.; Herrera, F. Probabilistic Linguistic MULTIMOORA: A Multicriteria Decision Making Method Based on the Probabilistic Linguistic Expectation Function and the Improved Borda Rule. *IEEE Trans. Fuzzy Syst.* **2018**, *26*, 3688–3702. [CrossRef]
58. Yuan, Y.; Xu, Z.; Zhang, Y. The DEMATEL–COPRAS hybrid method under probabilistic linguistic environment and its application in Third Party Logistics provider selection. *Fuzzy Optim. Decis. Mak.* **2021**, *21*, 137–156. [CrossRef]
59. Wang, J.-Q.; Wu, J.-T.; Wang, J.; Zhang, H.-Y.; Chen, X.-H. Multi-criteria decision-making methods based on the Hausdorff distance of hesitant fuzzy linguistic numbers. *Soft Comput.* **2016**, *20*, 1621–1633. [CrossRef]
60. Liao, H.; Xu, Z.; Zeng, X.-J. Hesitant Fuzzy Linguistic VIKOR Method and Its Application in Qualitative Multiple Criteria Decision Making. *IEEE Trans. Fuzzy Syst.* **2014**, *23*, 1343–1355. [CrossRef]
71. Liu, X.; Ma, Y. A method to analyze the rank reversal problem in the ELECTRE II method. *Omega* **2021**, *102*, 102317. [CrossRef]
72. Nie, R.-X.; Tian, Z.-P.; Wang, J.-Q.; Zhang, H.-Y.; Wang, T.-L. Water security sustainability evaluation: Applying a multistage decision support framework in industrial region. *J. Clean. Prod.* **2018**, *196*, 1681–1704. [CrossRef]
73. Liang, X.; Chen, T.; Ye, M.; Lin, H.; Li, Z. A hybrid fuzzy BWM-VIKOR MCDM to evaluate the service level of bike-sharing companies: A case study from Chengdu, China. *J. Clean. Prod.* **2021**, *298*, 126759. [CrossRef]
74. Peng, J.-J.; Wang, J.-Q.; Zhang, H.-Y.; Chen, X.-H. An outranking approach for multi-criteria decision-making problems with simplified neutrosophic sets. *Appl. Soft Comput.* **2014**, *25*, 336–346. [CrossRef]
75. Wang, P.; Zhu, Z.; Wang, Y. A novel hybrid MCDM model combining the SAW, TOPSIS and GRA methods based on experimental design. *Inf. Sci.* **2016**, *345*, 27–45. [CrossRef]
76. Xu, Z.; Zhang, X. Hesitant fuzzy multi-attribute decision making based on TOPSIS with incomplete weight information. *Knowl.-Based Syst.* **2013**, *52*, 53–64. [CrossRef]
77. Luthra, S.; Govindan, K.; Kannan, D.; Mangla, S.K.; Garg, C.P. An integrated framework for sustainable supplier selection and evaluation in supply chains. *J. Clean. Prod.* **2017**, *140*, 1686–1698. [CrossRef]
78. Pati, R.K.; Vrat, P.; Kumar, P. A goal programming model for paper recycling system. *Omega* **2008**, *36*, 405–417. [CrossRef]
79. Jaiswal, P.; Singh, A.; Misra, S.C.; Kumar, A. Barriers in implementing lean manufacturing in Indian SMEs: A multi-criteria decision-making approach. *J. Model. Manag.* **2021**, *16*, 339–356. [CrossRef]
80. Kumar, A.; Sah, B.; Singh, A.R.; Deng, Y.; He, X.; Kumar, P.; Bansal, R.C. A review of multi criteria decision making (MCDM) towards sustainable renewable energy development. *Renew. Sustain. Energy Rev.* **2017**, *69*, 596–609. [CrossRef]
81. Choudhary, D.; Shankar, R. An STEEP-fuzzy AHP-TOPSIS framework for evaluation and selection of thermal power plant location: A case study from India. *Energy* **2012**, *42*, 510–521. [CrossRef]
82. Chatterjee, P.; Chakraborty, S. Material selection using preferential ranking methods. *Mater. Des.* **2012**, *35*, 384–393. [CrossRef]
83. Chakraborty, S.; Zavadskas, E.K. Applications of WASPAS Method in Manufacturing Decision Making. *Informatica* **2014**, *25*, 1–20. [CrossRef]
84. Vinodh, S.; Prasanna, M.; Prakash, N.H. Integrated Fuzzy AHP–TOPSIS for selecting the best plastic recycling method: A case study. *Appl. Math. Model.* **2014**, *38*, 4662–4672. [CrossRef]
85. Garg, H. Novel intuitionistic fuzzy decision making method based on an improved operation laws and its application. *Eng. Appl. Artif. Intell.* **2017**, *60*, 164–174. [CrossRef]
86. Chatterjee, K.; Kar, S. A multi-criteria decision making for renewable energy selection using Z-numbers in uncertain environment. *Technol. Econ. Dev. Econ.* **2018**, *24*, 739–764. [CrossRef]
87. Hatefi, M.A. BRAW: Block-wise Rating the Attribute Weights in MADM. *Comput. Ind. Eng.* **2021**, *156*, 107274. [CrossRef]
88. Shemshadi, A.; Shirazi, H.; Toreihi, M.; Tarokh, M. A fuzzy VIKOR method for supplier selection based on entropy measure for objective weighting. *Expert Syst. Appl.* **2011**, *38*, 12160–12167. [CrossRef]
89. Kadhim, M.H.; Mardukhi, F. A Novel IoT Application Recommendation System Using Metaheuristic Multi-Criteria Analysis. *Comput. Syst. Sci. Eng.* **2021**, *37*, 149–158. [CrossRef]
90. Govindan, K.; Khodaverdi, R.; Jafarian, A. A fuzzy multi criteria approach for measuring sustainability performance of a supplier based on triple bottom line approach. *J. Clean. Prod.* **2013**, *47*, 345–354. [CrossRef]
91. Rezaeisaray, M.; Ebrahimnejad, S.; Khalili-Damghani, K. A novel hybrid MCDM approach for outsourcing supplier selection. *J. Model. Manag.* **2016**, *11*, 536–559. [CrossRef]
92. Ghasemi, P.; Mehdiabadi, A.; Spulbar, C.; Birau, R. Ranking of Sustainable Medical Tourism Destinations in Iran: An Integrated Approach Using Fuzzy SWARA-PROMETHEE. *Sustainability* **2021**, *13*, 683. [CrossRef]
93. Jahanshahloo, G.; Lotfi, F.H.; Izadikhah, M. An algorithmic method to extend TOPSIS for decision-making problems with interval data. *Appl. Math. Comput.* **2006**, *175*, 1375–1384. [CrossRef]
94. Hashemi, S.H.; Karimi, A.; Tavana, M. An integrated green supplier selection approach with analytic network process and improved Grey relational analysis. *Int. J. Prod. Econ.* **2015**, *159*, 178–191. [CrossRef]
95. Behzadian, M.; Kazemzadeh, R.; Albadvi, A.; Aghdasi, M. PROMETHEE: A comprehensive literature review on methodologies and applications. *Eur. J. Oper. Res.* **2010**, *200*, 198–215. [CrossRef]

96. Hashemi, H.; Mousavi, S.M.; Zavadskas, E.K.; Chalekaee, A.; Turskis, Z. A New Group Decision Model Based on Grey-Intuitionistic Fuzzy-ELECTRE and VIKOR for Contractor Assessment Problem. *Sustainability* **2018**, *10*, 1635. [CrossRef]
97. Boran, F.E.; Genç, S.; Kurt, M.; Akay, D. A multi-criteria intuitionistic fuzzy group decision making for supplier selection with TOPSIS method. *Expert Syst. Appl.* **2009**, *36*, 11363–11368. [CrossRef]
98. Özceylan, E.; Erbaş, M.; Çetinkaya, C.; Kabak, M. Analysis of Potential High-Speed Rail Routes: A Case of GIS-Based Multicriteria Evaluation in Turkey. *J. Urban Plan. Dev.* **2021**, *147*, 04021012. [CrossRef]
99. Durak, I.; Arslan, H.M.; Özdemir, Y. Application of AHP–TOPSIS methods in technopark selection of technology companies: Turkish case. *Technol. Anal. Strat. Manag.* **2021**, 1–15. [CrossRef]
100. Önüt, S.; Soner, S. Transshipment site selection using the AHP and TOPSIS approaches under fuzzy environment. *Waste Manag.* **2008**, *28*, 1552–1559. [CrossRef]
101. Kahraman, C.; Ruan, D.; Doğan, I. Fuzzy group decision-making for facility location selection. *Inf. Sci.* **2003**, *157*, 135–153. [CrossRef]
102. Kaya, T.; Kahraman, C. Multicriteria renewable energy planning using an integrated fuzzy VIKOR & AHP methodology: The case of Istanbul. *Energy* **2010**, *35*, 2517–2527. [CrossRef]
103. Gencer, C.; Gürpinar, D. Analytic network process in supplier selection: A case study in an electronic firm. *Appl. Math. Model.* **2007**, *31*, 2475–2486. [CrossRef]
104. Dağdeviren, M.; Yavuz, S.; Kılınç, N. Weapon selection using the AHP and TOPSIS methods under fuzzy environment. *Expert Syst. Appl.* **2009**, *36*, 8143–8151. [CrossRef]
105. Büyüközkan, G.; Güleryüz, S. An integrated DEMATEL-ANP approach for renewable energy resources selection in Turkey. *Int. J. Prod. Econ.* **2016**, *182*, 435–448. [CrossRef]
106. Colak, E.H.; Memisoglu, T.; Gercek, Y. Optimal site selection for solar photovoltaic (PV) power plants using GIS and AHP: A case study of Malatya Province, Turkey. *Renew. Energy* **2020**, *149*, 565–576. [CrossRef]
107. Chen, C.-T. Extensions of the TOPSIS for group decision-making under fuzzy environment. *Fuzzy Sets Syst.* **2000**, *114*, 1–9. [CrossRef]
108. Lin, Y.-F. Construction of Consistent Comparison Matrix by Macharis' Method Revisit. *Math. Probl. Eng.* **2021**, *2021*, 5585662. [CrossRef]
109. Chiu, W.-Y.; Manoharan, S.H.; Huang, T.-Y. Weight Induced Norm Approach to Group Decision Making for Multiobjective Optimization Problems in Systems Engineering. *IEEE Syst. J.* **2020**, *14*, 1580–1591. [CrossRef]
110. Opricovic, S.; Tzeng, G.-H. Extended VIKOR method in comparison with outranking methods. *Eur. J. Oper. Res.* **2007**, *178*, 514–529. [CrossRef]
111. Chen, T. Enhancing the efficiency and accuracy of existing FAHP decision-making methods. *EURO J. Decis. Process.* **2020**, *8*, 177–204. [CrossRef]
112. Yang, C.-C.; Shen, C.-C.; Lin, Y.-S.; Lo, H.-W.; Wu, J.-Z. Sustainable Sports Tourism Performance Assessment Using Grey-Based Hybrid Model. *Sustainability* **2021**, *13*, 4214. [CrossRef]
113. Tzeng, G.-H.; Chiang, C.-H.; Li, C.-W. Evaluating intertwined effects in e-learning programs: A novel hybrid MCDM model based on factor analysis and DEMATEL. *Expert Syst. Appl.* **2007**, *32*, 1028–1044. [CrossRef]
114. Chen, F.-H.; Hsu, T.-S.; Tzeng, G.-H. A balanced scorecard approach to establish a performance evaluation and relationship model for hot spring hotels based on a hybrid MCDM model combining DEMATEL and ANP. *Int. J. Hosp. Manag.* **2011**, *30*, 908–932. [CrossRef]
115. Liou, J.J.; Tamošaitienė, J.; Zavadskas, E.K.; Tzeng, G.-H. New hybrid COPRAS-G MADM Model for improving and selecting suppliers in green supply chain management. *Int. J. Prod. Res.* **2015**, *54*, 114–134. [CrossRef]
116. Chen, T.-Y. Comparative analysis of SAW and TOPSIS based on interval-valued fuzzy sets: Discussions on score functions and weight constraints. *Expert Syst. Appl.* **2012**, *39*, 1848–1861. [CrossRef]
117. Hong, D.H.; Choi, C.-H. Multicriteria fuzzy decision-making problems based on vague set theory. *Fuzzy Sets Syst.* **2000**, *114*, 103–113. [CrossRef]
118. Dymova, L.; Kaczmarek, K.; Sevastjanov, P.; Sułkowski, Ł.; Przybyszewski, K. An Approach to Generalization of the Intuitionistic Fuzzy Topsis Method in the Framework of Evidence Theory. *J. Artif. Intell. Soft Comput. Res.* **2021**, *11*, 157–175. [CrossRef]
119. Mousavi, M.M.; Lin, J. The application of PROMETHEE multi-criteria decision aid in financial decision making: Case of distress prediction models evaluation. *Expert Syst. Appl.* **2020**, *159*, 113438. [CrossRef]
120. Tam, M.C.; Tummala, V. An application of the AHP in vendor selection of a telecommunications system. *Omega* **2001**, *29*, 171–182. [CrossRef]
121. Pires, A.; Chang, N.-B.; Martinho, G. An AHP-based fuzzy interval TOPSIS assessment for sustainable expansion of the solid waste management system in Setúbal Peninsula, Portugal. *Resour. Conserv. Recycl.* **2011**, *56*, 7–21. [CrossRef]
122. Rani, P.; Mishra, A.R.; Pardasani, K.R.; Mardani, A.; Liao, H.; Streimikiene, D. A novel VIKOR approach based on entropy and divergence measures of Pythagorean fuzzy sets to evaluate renewable energy technologies in India. *J. Clean. Prod.* **2019**, *238*, 117936. [CrossRef]
123. Saaty, T.L. The Modern Science of Multicriteria Decision Making and Its Practical Applications: The AHP/ANP Approach. *Oper. Res.* **2013**, *61*, 1101–1118. [CrossRef]

124. Saaty, T.L.; Ergu, D. When is a Decision-Making Method Trustworthy? Criteria for Evaluating Multi-Criteria Decision-Making Methods. *Int. J. Inf. Technol. Decis. Mak.* **2015**, *14*, 1171–1187. [CrossRef]
125. Abdel-Basset, M.; Saleh, M.; Gamal, A.; Smarandache, F. An approach of TOPSIS technique for developing supplier selection with group decision making under type-2 neutrosophic number. *Appl. Soft Comput.* **2019**, *77*, 438–452. [CrossRef]
126. Tavana, M.; Li, Z.; Mobin, M.; Komaki, M.; Teymourian, E. Multi-objective control chart design optimization using NSGA-III and MOPSO enhanced with DEA and TOPSIS. *Expert Syst. Appl.* **2016**, *50*, 17–39. [CrossRef]
127. Gaviao, L.O.; Sant'Anna, A.P.; Lima, G.B.A.; Garcia, P.A.D.A.; Kostin, S.; Asrilhant, B. Selecting a Cargo Aircraft for Humanitarian and Disaster Relief Operations by Multicriteria Decision Aid Methods. *IEEE Trans. Eng. Manag.* **2020**, *67*, 631–640. [CrossRef]
128. Maêda, S.M.D.N.; Basílio, M.P.; Costa, I.P.D.A.; Moreira, M.L.; dos Santos, M.; Gomes, C.F.S.; de Almeida, I.D.P.; Costa, A.P.D.A. Investments in Times of Pandemics: An Approach by the SAPEVO-M-NC Method. In *Modern Management Based on Big Data II and Machine Learning and Intelligent Systems III—Proceedings of MMBD 2021 and MLIS 2021*; Tallón-Ballesteros, A.J., Ed.; IOS Press: Amsterdam, The Netherlands, 2021; Volume 341, pp. 162–168. [CrossRef]
129. Drumond, P.; Basílio, M.P.; Costa, I.P.D.A.; Pereira, D.A.D.M.; Gomes, C.F.S.; dos Santos, M. Multicriteria Analysis in Additive Manufacturing: An ELECTRE-MOr Based Approach. In *Modern Management Based on Big Data II and Machine Learning and Intelligent Systems III—Proceedings of MMBD 2021 and MLIS 2021*; Tallón-Ballesteros, A.J., Ed.; IOS Press: Amsterdam, The Netherlands, 2021; Volume 341, pp. 126–132. [CrossRef]
130. Costa, I.P.D.A.; Basílio, M.P.; Maêda, S.M.D.N.; Rodrigues, M.V.G.; Moreira, M.L.; Gomes, C.F.S.; dos Santos, M. Algorithm Selection for Machine Learning Classification: An Application of the MELCHIOR Multicriteria Method. In *Modern Management Based on Big Data II and Machine Learning and Intelligent Systems III—Proceedings of MMBD 2021 and MLIS 2021*; Tallón-Ballesteros, A.J., Ed.; IOS Press: Amsterdam, The Netherlands, 2021; Volume 341, pp. 154–161. [CrossRef]
131. Basilio, M.; Brum, G.S.; Pereira, V. A model of policing strategy choice: The integration of the Latent Dirichlet Allocation (LDA) method with ELECTRE I. *J. Model. Manag.* **2020**, *15*, 849–891. [CrossRef]
132. Maêda, S.M.D.N.; Basílio, M.P.; Costa, I.P.D.A.; Moreira, M.L.; dos Santos, M.; Gomes, C.F.S. The SAPEVO-M-NC Method. In *Modern Management Based on Big Data II and Machine Learning and Intelligent Systems III—Proceedings of MMBD 2021 and MLIS 2021*; Tallón-Ballesteros, A.J., Ed.; IOS Press: Amsterdam, The Netherlands, 2021; Volume 341, pp. 89–95. [CrossRef]
133. Krohling, R.A.; de Souza, T.T. Combining prospect theory and fuzzy numbers to multi-criteria decision making. *Expert Syst. Appl.* **2012**, *39*, 11487–11493. [CrossRef]
134. Silva, M.D.C.; Gavião, L.O.; Gomes, C.F.S.; Lima, G.B.A. Global Innovation Indicators analysed by multicriteria decision. *Braz. J. Oper. Prod. Manag.* **2020**, *17*, 1–17. [CrossRef]
135. Soares, L.d.M.B.; Moreira, M.L.; Basilio, M.P.; Gomes, C.F.S.; dos Santos, M.; Costa, I.P.D.A. Strategic Analysis for the Installation of Field Hospitals for COVID-19 Control: An Approach Based on P-Median Model. In *Modern Management Based on Big Data II and Machine Learning and Intelligent Systems III—Proceedings of MMBD 2021 and MLIS 2021*; Tallón-Ballesteros, A.J., Ed.; IOS Press: Amsterdam, The Netherlands, 2021; Volume 341, pp. 112–118. [CrossRef]
136. de Almeida, A.T. Multicriteria decision model for outsourcing contracts selection based on utility function and ELECTRE method. *Comput. Oper. Res.* **2007**, *34*, 3569–3574. [CrossRef]
137. Morais, D.C.; de Almeida, A.T. Group decision making on water resources based on analysis of individual rankings. *Omega* **2012**, *40*, 42–52. [CrossRef]
138. Barata, J.F.F.; Quelhas, O.L.G.; Costa, H.G.; Gutierrez, R.H.; Lameira, V.D.J.; Meiriño, M.J. Multi-Criteria Indicator for Sustainability Rating in Suppliers of the Oil and Gas Industries in Brazil. *Sustainability* **2014**, *6*, 1107–1128. [CrossRef]
139. Pereira, V.; Costa, H.G. Nonlinear programming applied to the reduction of inconsistency in the AHP method. *Ann. Oper. Res.* **2015**, *229*, 635–655. [CrossRef]
140. Basilio, M.; De Freitas, J.G.; Kämpffe, M.G.F.; Rego, R. Investment portfolio formation via multicriteria decision aid: A Brazilian stock market study. *J. Model. Manag.* **2018**, *13*, 394–417. [CrossRef]
141. Liu, H.; Rodríguez, R.M. A fuzzy envelope for hesitant fuzzy linguistic term set and its application to multicriteria decision making. *Inf. Sci.* **2014**, *258*, 220–238. [CrossRef]
142. Jato-Espino, D.; Castillo-Lopez, E.; Rodriguez-Hernandez, J.; Canteras-Jordana, J.C. A review of application of multi-criteria decision making methods in construction. *Autom. Constr.* **2014**, *45*, 151–162. [CrossRef]
143. Cárdenas-Gómez, J.; Gonzales, M.B.; Lazo, C.D. Evaluation of Reinforced Adobe Techniques for Sustainable Reconstruction in Andean Seismic Zones. *Sustainability* **2021**, *13*, 4955. [CrossRef]
144. Casas-Rosal, J.C.; Segura, M.; Maroto, C. Food market segmentation based on consumer preferences using outranking multicriteria approaches. *Int. Trans. Oper. Res.* **2021**. [CrossRef]
145. Luna, M.; Llorente, I.; Cobo, A. A fuzzy approach to decision-making in sea-cage aquaculture production. *Int. Trans. Oper. Res.* **2020**, *29*, 1025–1047. [CrossRef]
146. Romero, C. Extended lexicographic goal programming: A unifying approach. *Omega* **2001**, *29*, 63–71. [CrossRef]
147. Sánchez-Lozano, J.; García-Cascales, M.; Lamata, M. GIS-based onshore wind farm site selection using Fuzzy Multi-Criteria Decision Making methods. Evaluating the case of Southeastern Spain. *Appl. Energy* **2016**, *171*, 86–102. [CrossRef]
148. Bilbao-Terol, A.; Arenas-Parra, M.; Cañal-Fernández, V.; Antomil-Ibias, J. Using TOPSIS for assessing the sustainability of government bond funds. *Omega* **2014**, *49*, 1–17. [CrossRef]

149. Bana e Costa, C.A.; Carnero, M.C.; Oliveira, M.D. A multi-criteria model for auditing a Predictive Maintenance Programme. *Eur J. Oper. Res.* **2012**, *217*, 381–393. [CrossRef]
150. Braglia, M.; Frosolini, M.; Montanari, R. Fuzzy TOPSIS approach for failure mode, effects and criticality analysis. *Qual. Reliab Eng. Int.* **2003**, *19*, 425–443. [CrossRef]
151. La Fata, C.; Giallanza, A.; Micale, R.; La Scalia, G. Ranking of occupational health and safety risks by a multi-criteria perspective Inclusion of human factors and application of VIKOR. *Saf. Sci.* **2021**, *138*, 105234. [CrossRef]
152. Bottero, M.; Comino, E.; Riggio, V. Application of the Analytic Hierarchy Process and the Analytic Network Process for the assessment of different wastewater treatment systems. *Environ. Model. Softw.* **2011**, *26*, 1211–1224. [CrossRef]
153. Zoghi, M.; Rostami, G.; Khoshand, A.; Motalleb, F. Material selection in design for deconstruction using Kano model, fuzzy-AHP and TOPSIS methodology. *Waste Manag. Res. J. Sustain. Circ. Econ.* **2021**, *40*, 410–419. [CrossRef] [PubMed]
154. Corrente, S.; Greco, S.; Leonardi, F.; Słowiński, R. The hierarchical SMAA-PROMETHEE method applied to assess the sustainability of European cities. *Appl. Intell.* **2021**, *51*, 6430–6448. [CrossRef]
155. Beccali, M.; Cellura, M.; Mistretta, M. Decision-making in energy planning. Application of the Electre method at regional level for the diffusion of renewable energy technology. *Renew. Energy* **2003**, *28*, 2063–2087. [CrossRef]
156. Formisano, A.; Mazzolani, F.M. On the selection by MCDM methods of the optimal system for seismic retrofitting and vertical addition of existing buildings. *Comput. Struct.* **2015**, *159*, 1–13. [CrossRef]
157. Norese, M.F. ELECTRE III as a support for participatory decision-making on the localisation of waste-treatment plants. *Land Use Policy* **2006**, *23*, 76–85. [CrossRef]
158. Barrios, M.A.O.; De Felice, F.; Negrete, K.P.; Romero, B.A.; Arenas, A.Y.; Petrillo, A. An AHP-Topsis Integrated Model for Selecting the Most Appropriate Tomography Equipment. *Int. J. Inf. Technol. Decis. Mak.* **2016**, *15*, 861–885. [CrossRef]
159. Cavallaro, F. Fuzzy TOPSIS approach for assessing thermal-energy storage in concentrated solar power (CSP) systems. *Appl. Energy* **2010**, *87*, 496–503. [CrossRef]
160. Azadnia, A.H.; Saman, M.Z.M.; Wong, K.Y. Sustainable supplier selection and order lot-sizing: An integrated multi-objective decision-making process. *Int. J. Prod. Res.* **2015**, *53*, 383–408. [CrossRef]
161. Umer, R.; Touqeer, M.; Omar, A.H.; Ahmadian, A.; Salahshour, S.; Ferrara, M. Selection of solar tracking system using extended TOPSIS technique with interval type-2 pythagorean fuzzy numbers. *Optim. Eng.* **2021**, *22*, 2205–2231. [CrossRef]
162. Mardani, A.; Jusoh, A.; MD Nor, K.; Khalifah, Z.; Zakwan, N.; Valipour, A. Multiple criteria decision-making techniques and their applications—A review of the literature from 2000 to 2014. *Econ. Res.-Ekon. Istraž.* **2015**, *28*, 516–571. [CrossRef]
163. Khoso, A.R.; Yusof, A.M.; Khahro, S.H.; Abidin, N.I.A.B.; Memon, N.A. Automated two-stage continuous decision support model using exploratory factor analysis-MACBETH-SMART: An application of contractor selection in public sector construction. *J. Ambient Intell. Humaniz. Comput.* **2021**, 1–31. [CrossRef]
164. Rostamzadeh, R.; Govindan, K.; Esmaeili, A.; Sabaghi, M. Application of fuzzy VIKOR for evaluation of green supply chain management practices. *Ecol. Indic.* **2015**, *49*, 188–203. [CrossRef]
165. Abdullah, L.; Najib, L. A new type-2 fuzzy set of linguistic variables for the fuzzy analytic hierarchy process. *Expert Syst. Appl.* **2014**, *41*, 3297–3305. [CrossRef]
166. Mardani, A.; Jusoh, A.; Zavadskas, E.K. Fuzzy multiple criteria decision-making techniques and applications—Two decades review from 1994 to 2014. *Expert Syst. Appl.* **2015**, *42*, 4126–4148. [CrossRef]
167. Zaidan, A.; Zaidan, B.; Al-Haiqi, A.; Kiah, M.; Hussain, M.; Abdulnabi, M. Evaluation and selection of open-source EMR software packages based on integrated AHP and TOPSIS. *J. Biomed. Inform.* **2015**, *53*, 390–404. [CrossRef] [PubMed]
168. Mir, M.A.; Ghazvinei, P.T.; Sulaiman, N.M.N.; Basri, N.E.A.; Saheri, S.; Mahmood, N.Z.; Jahan, A.; Begum, R.A.; Aghamohammadi, N. Application of TOPSIS and VIKOR improved versions in a multi-criteria decision analysis to develop an optimized municipal solid waste management model. *J. Environ. Manag.* **2016**, *166*, 109–115. [CrossRef]
169. Adiat, K.A.N.; Nawawi, M.N.M.; Abdullah, K. Assessing the accuracy of GIS-based elementary multi-criteria decision analysis as a spatial prediction tool—A case of predicting potential zones of sustainable groundwater resources. *J. Hydrol.* **2012**, *440–441*, 75–89. [CrossRef]
170. Cobo, M.J.; López-Herrera, A.G.; Herrera-Viedma, E.; Herrera, F. An approach for detecting, quantifying, and visualizing the evolution of a research field: A practical application to the Fuzzy Sets Theory field. *J. Informetr.* **2011**, *5*, 146–166. [CrossRef]

MDPI
St. Alban-Anlage 66
4052 Basel
Switzerland
Tel. +41 61 683 77 34
Fax +41 61 302 89 18
www.mdpi.com

Electronics Editorial Office
E-mail: electronics@mdpi.com
www.mdpi.com/journal/electronics

www.ingramcontent.com/pod-product-compliance
Lightning Source LLC
Chambersburg PA
CBHW061806210926
51850CB00006B/628
* 9 7 8 3 0 3 6 5 6 7 8 8 4 *